D1243747

TROPICAL RAIN FORESTS

# Tropical Rain Forests

## An Ecological and Biogeographical Comparison

Richard Primack and
Richard Corlett

Blackwell
Publishing

BLACKWELL PUBLISHING
350 Main Street, Malden, MA 02148-5020, USA
108 Cowley Road, Oxford OX4 1JF, UK
550 Swanston Street, Carlton, Victoria 3053, Australia



First published 2005 by Blackwell Science Ltd

*Library of Congress Cataloging-in-Publication Data*

Primack, Richard B., 1950–
Tropical rain forests: an ecological and biogeographical comparison / Richard Primack & Richard Corlett.
p. cm.
Includes bibliographical references and index.
ISBN 0-632-04513-2 (pbk. : alk. paper)
1. Rain forests. 2. Rain forest ecology. I. Corlett, Richard. II. Title.
QH86.P75 2005
577.34—dc22 2004009785

A catalogue record for this title is available from the British Library.

Set in 9/11½ pt Meridien
by Graphicraft Limited, Hong Kong
Printed and bound in the United Kingdom
by TJ International, Padstow, Cornwall

The publisher's policy is to use permanent paper from mills that operate a sustainable forestry policy, and which has been manufactured from pulp processed using acid-free and elementary chlorine-free practices. Furthermore, the publisher ensures that the text paper and cover board used have met acceptable environmental accreditation standards.

For further information on
Blackwell Publishing, visit our website:
www.blackwellpublishing.com

# Contents

## Primate Communities: a Key to Understanding Biogeography and Ecology, 75

## Carnivores and Plant-eaters, 98

## Birds: Linkages in the Rain Forest Community, 133

## Fruit Bats and Gliding Animals in the Tree Canopy, 178

## 7 Insects: Diverse, Abundant, and Ecologically Important, 198

## 8 The Future of Rain Forests, 233

# Preface

In the popular imagination, the tropical rain forest consists of giant trees towering above a tangle of vines and beautiful orchids below, with colorful birds, tree frogs, and monkeys everywhere abundant. Scientists and visitors quickly realize that this image is not accurate: animal life, while highly diverse, is not necessarily strikingly abundant, and flowers are often very small and hard to find. But beyond the difference between perception and reality, there are tremendous differences among regions. Biologists working in one area rapidly recognize the special features of the biological community in their area, yet they would find themselves in highly unfamiliar terrain should they move, for example, from their accustomed study site in Borneo to a seemingly similar location in New Guinea. Indeed, even in locations within the same overall zone, such as the Amazon and Central American forests, there can be differences both dramatic and subtle from place to place. There are unique plants and animals in every community, and even those organisms common to other regions are part of a distinctive mixture of species that interact in ways readily distinguishable from other forests. Thus, the tendency of both popular media and science to make sweeping statements about "the rain forest" is highly misleading.

On a larger scale, rain forests on different continents have fundamentally different characteristics that make each of them unique. In many earlier books on rain forests, authors such as Paul Richards in *The Tropical Rainforest* and Tim Whitmore in *An Introduction to Tropical Rain Forests* tried to describe the unifying properties of rain forests on each continent—features such as the high diversity of trees species and the low nutrient status of the soils. They took comparable examples from each region and emphasized certain principles of tropical ecology that are true on all continents. However, this emphasis on commonalities meant that readers could—and often did—easily overlook the fact that each of these rain forests has its own unique features of plants, animals, climate, topography, and past history. Our goal in writing this book is therefore to redress this oversight by emphasizing the ways in which the major rain forest areas are special.

We believe this approach can suggest new research questions that can be investigated in comparative studies of rain forests in different regions. At the end of each chapter, we suggest specific new approaches, sometimes involving experimental methods, that could be used to develop new research questions. Finally, in the last chapter we consider the unique threats faced by rain forests in each area of the world and suggest strategies for conservation. Such topics may have relevance to policy initiatives aimed at protecting rain forest habitats —initiatives that currently are based upon a misunderstanding that the various communities would respond in the same manner to the same methods of management.

We hope that readers will come away with an appreciation that our planet is host to not one monolithic tract of rain forest, but many unique tropical rain forest habitats, all worthy of study and protection.

Richard Primack and Richard Corlett

# Acknowledgments

Anyone attempting to write a book that makes simple clear statements about the whole range of tropical ecology, biogeography, and conservation, quickly reaches the limits of their own knowledge. Books can be studied and articles examined. But in the end, it is the tropical ecologists themselves who must be consulted for the accuracy of impressions and ideas, and to explain facts which seem confusing. Therefore, we have many people to thank for their expertise in making this book as accurate as possible, for providing colorful examples, and also for saying when we really do not have the data to make definitive statements.

The book is greatly improved by the numerous images supplied to us from photographers throughout the world. These people are identified by name next to their photographs. We would especially like to thank Tim Laman in making his collection of outstanding slides available to us. The photographs were organized by Dan Primack.

An extended overview of this topic was read and commented on by Tim Laman, Ted Fleming, Robin Chazdon, Rob Colwell, Linus Chen, Mike Sorenson, Cam Webb, and Mark Leighton. Individual chapters were read by J.R. Flenley, Geoff Hope, Chris Schneider (Introduction), Ian Turner, John Kress, Ghillean Prance, Chris Dick (Chapter 2), John Fleagle, Cheryl Knott, Colin Chapman (Chapter 3), Tom Kunz, John Hart (Chapter 4), Mercedes Foster, David Pearson, Fred Wasserman (Chapter 5), Tom Kunz (Chapter 6), James Traniello, Phil DeVries, David Roubik (Chapter 7), and Kamal Bawa and Peter Feinsinger (Chapter 8). Tigga Kingston assisted in writing an early draft of parts of Chapters 4 and 6.

Vivi Tran and Elizabeth Platt were the principal research assistants for the project, and read and commented on all the chapters. Kerry Falvey provided advice on organization. Two reviewers engaged by Blackwell to read the manuscript gave us many useful suggestions. We greatly appreciate all their efforts. A Bullard Fellowship from Harvard University and a Visiting Lectureship from the University of Hong Kong to one of us (Primack) provided time for research and writing.

# Many Tropical Rain Forests

It is easy to make generalizations about tropical rain forests. Travel posters, magazine articles, and television programs give the casual observer the impression that tropical rain forests from any spot in the world are one interchangeable mass of tall, wet trees—the canopy filled with brightly colored birds, chirping tree frogs, and acrobatic monkeys, the ground level home to silent predators, tangled vines, and exotic flowers. This popular perception has been useful because it creates a compellingly attractive image of an untamed, beautiful place that is, on the one hand, a source of infinite mystery and adventure—and on the other, a fragile natural treasury that must be protected. Both conservation and ecotourism rely on this generalized image to promote the idea of tropical rain forests as a "good thing" to be preserved and enjoyed. Yet, beneficial as this image may be in encouraging travel and conservation, it obscures the fact that the world's tropical rain forests have major differences from one another in addition to their obvious similarities.

A major drawback of this generalized image is that it encourages a belief that saving "the rain forest" is a single problem with a single, universally applicable, set of answers. Nothing could be further from the truth. There are many different rain forests, all of which need action for protection, but this action must be targeted at the specific threats present in each region and adapted to the specific ecological characteristics of each rain forest. Policies, tactics, and techniques that work in one region may prove ineffective or even disastrous in another. The major differences between the tropical rain forests in different regions also mean that successes in one region will not compensate for losses in another. The task we are faced with is not "saving *the* rain forest", but "saving the *many* rain forests".

Scientists also have usually emphasized the common appearance of rain forests on different continents and highlighted examples of similar-looking species in separate regions of the world (Whitmore 1984, 1998; Richards 1996a). This emphasis on the common features of rain forests worldwide has had the unintended effect of discouraging research that makes comparisons between regions. The assumption that all rain forests are alike has also led to a tendency to fill gaps in the scientific understanding of one rain forest region by reference to studies in other regions. This tendency in turn gives the false impression that our understanding of rain forests is greater than it really is, so that scientific research that could fill the gaps is given a low priority. It also implies that differences between regions are minor, at least in comparison with the similarities.

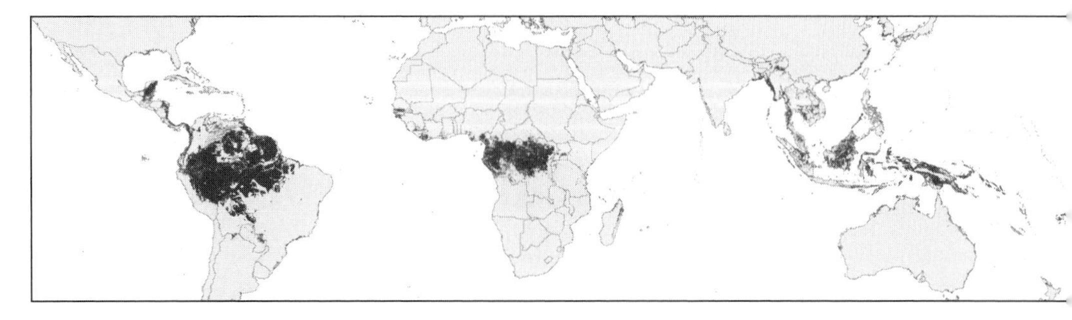

**Fig. 1.1** The currect global distribution of lowland tropical rain forests. (Courtesy of UNEP World Conservation Monitoring Centre, 2004.)

Our principal reason for writing this book is that we do not believe that the differences between rain forests are minor. It is our contention that the various rain forest regions are sufficiently distinct from one another that they merit individual consideration. In this book, therefore, we will compare the major rain forest regions of the world. The three largest of these rain forest regions are in the Amazon basin of South America, in the Congo River basin of Central Africa, and on the everwet peninsula and islands of Southeast Asia (Fig. 1.1). There are also two smaller and very distinctive rain forest regions on the giant islands of Madagascar and New Guinea. We will show that the rain forests of these five regions are unique biogeographical and ecological entities, each with many distinctive plants, animals, and ecological interactions that are not found in the other regions.

Rain forests occur outside of these core areas as well, but they are less extensive in area and usually less diverse in species. There are rain forests in Central America and coastal Brazil that are basically similar in species composition to those found in the Amazon, but have fewer species and occupy a much smaller area. Similarly, the rain forests of Sri Lanka and the Western Ghats of India resemble those of Southeast Asia, and Australian rain forests have many similarities to the more extensive and diverse forests of New Guinea. Each of these many areas has numerous noteworthy features and unique species, which we will mention in this book, but the focus will be on the differences among the five main regions. In addition, there are also small but distinctive areas of rain forest on many tropical oceanic islands.

Each rain forest region has different geographical, geological, and climatic features; each region supports plants and animals with separate evolutionary histories; and each region has experienced different past and present human impacts (Table 1.1) (Mittermeier et al. 1999). These differences have important implications for understanding how rain forests work and deciding how they should be exploited or conserved. Results from scientific research in one region may not apply in the others. At the end of each chapter, we suggest comparative investigations and experiments that could provide deeper insights into rain forest biology. Similarly, conservation measures or methods of sustainable exploitation that are successful in one area may not work as well in another. Each rain forest area, and even local areas within each of these regions, must be viewed

**Table 1.1** Some key characteristics of the main rain forest regions.

| | Neotropics | Africa | Madagascar | Southeast Asia | New Guinea |
|---|---|---|---|---|---|
| Main geographical feature | Amazon River basin and Andes Mountains | Congo River basin | Forests along eastern edge of island | Peninsula and islands on Sunda Shelf | Large, mountainous island |
| Distinctive biological features* | Bromeliad epiphytes, high bird diversity, small primates | Low plant richness, forest elephants, many forest browsers | Lemurs, low fruit abundance | Dipterocarp tree family, mast fruiting of trees, large primates | Marsupial mammals, birds of paradise |
| Annual rainfall (mm)† | 2000–3000 | 1500–2500 | 2000–3000 | 2000–3000, often >3000 | 2000–3000, often >3000 |
| Largest country | Brazil | Democratic Republic of Congo | Malagasy Republic | Indonesia | Papua New Guinea |

* Unfamiliar terms are explained in the text.
† Rainfall is highly variable within each region. These are the ranges over most of the core rainforest area (1000 mm equals 40 inches).
From Times Books (1994).

separately when scientific investigation, conservation efforts, and responsible development are undertaken. The issues of human impacts, conservation, and development will be considered in detail in the final chapter.

## What are tropical rain forests?

Tropical rain forests are the tall, dense, evergreen forests that form the natural vegetation cover of the wet tropics, where the climate is always hot and the dry season is short or absent. This broad definition allows for a considerable range of variation, as is necessary for any global comparison. One important variable is the proportion of deciduous trees in the forest canopy. We have excluded the predominantly deciduous tropical forests that occur in areas with a long dry season (Bullock et al. 1995), but many forests that we and others call rain forest have some deciduous trees in the canopy. Where this proportion is large, the forests can be called semievergreen (or semideciduous!) rain forest. Another important distinction is between lowland and montane rain forests. On high mountains in the wet tropics, forests extend from sea level to around 4000 m (13,000 feet), but the typical tall, lowland rain forest is confined below an altitude of 900–1200 m (Ashton 2003). Rain forests above this altitude are termed "montane" and have a distinctive ecology of their own (Hamilton et al. 1995; Whitmore 1998; Nadkarni & Wheelwright 2000).

Precise definitions are difficult in ecology, and there are large areas of forest in the tropics that some ecologists call rain forest while others do not. Definitions also differ between regions, with less strict, more inclusive, definitions in areas where rain forests are less extensive. Thus, on the very dry continent of Australia, almost any area of closed forest is called rain forest (Bowman 2000). Conversely, foresters familiar with the everwet rain forests of equatorial Southeast Asia might exclude much of what their African counterparts call rain forest. In this book, we focus on hot, wet, tall, and largely evergreen tropical rain forests, and we have made it clear when we are referring to unusual or marginal types.

## Where are the tropical rain forests?

On a simpler planet, tropical rain forests would form a broad belt around the equator, extending 5–10° to the north and south. On our untidy Earth, interactions between wind direction and mountain ranges, variations in sea surface temperature, and various other factors exclude rain forest from parts of this belt —notably most of East Africa—and, in other places, extend it for some distance outside. At least, that was the situation until very recently. During the last few hundred years, and particularly in the last few decades, between one-third to one-half of this rain forest has been converted into other land uses, ranging from productive farmland or tree plantations to urban areas or unproductive grasslands. This book is mostly about the rain forests that still survive, but the devastating effects of human impacts are considered in the last chapter.

### The Neotropics

Approximately half of the world's tropical rain forests are in tropical America (Gentry 1990; Newman 1990), the region that biologists call the Neotropics (literally "new tropics" or New World tropics). The Neotropical rain forests form three main blocks. The single largest block of tropical rain forest in the world covers the adjoining basins of the Amazon and Orinoco Rivers (Fig. 1.2). The Amazon River basin is centered on northern and central Brazil. It stretches more than 3000 km (2000 miles) from the foothills of the Andes Mountains in western South America, across the entire South American continent, until the Amazon empties into the Atlantic Ocean. The Orinoco River basin drains eastern Colombia and Venezuela and adjoins the Amazon basin along the Brazilian border. This rain forest block also continues to the northeast of Brazil into the countries of Guyana, Surinam, and French Guiana.

In addition to this giant block of rain forest centered on the Amazon basin, there are—or were, until recently—two other major blocks of tropical rain forest in the Americas. The Brazilian Atlantic Forest ran along the southeast coast of Brazil from Recife south to São Paulo (Dean 1995; Mittermeier et al. 1999; Galindo-Leal & Camara 2003). This almost unbroken band of forest was over 2000 km long and generally less than 160 km wide. It was separated from the Amazon forest block by hundreds of kilometers of dry scrub and savanna. This forest has now been reduced to less than 5% of its previous area. A third block extended from the Pacific coast of northwest South America through Central America to southernmost Mexico (Mast et al. 1999; Mittermeier et al. 1999). The rain forests of northwest South America became separated from the Amazon basin rain forest by the uplift of the Andes beginning around 25 million

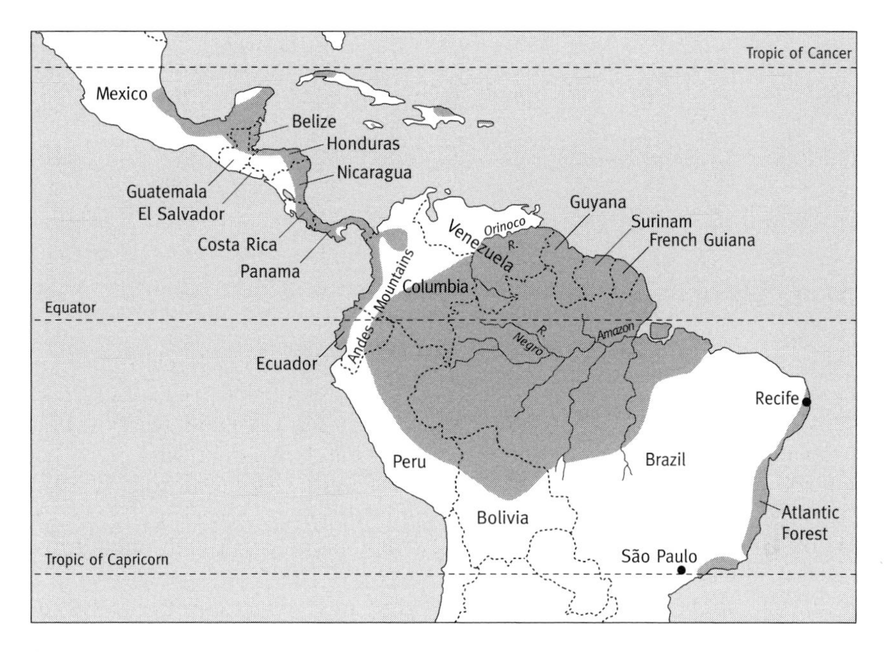


**Fig. 1.2** The estimated historical extent of tropical rain forest in South and Central America. (From multiple sources.)

years ago, and then became continuous with those of Central America when the final marine barrier between them disappeared 3 million years ago (see below). Only fragments of this rain forest block survive today. There were also smaller areas of rain forest on many of the Caribbean islands, where very little now remains (Hemphill et al. 1999).

## Africa

The second largest block of tropical rain forest is in Africa, centered on the Congo River basin (Fig. 1.3) (White 2001). About half of this rain forest is in the Democratic Republic of the Congo (Congo-Kinshasa, formerly Zaire), with most of the rest divided between the Republic of the Congo (Congo-Brazzaville), Gabon, and Cameroon. This Central African rain forest block formerly extended northwest into southern Nigeria, but little of this now remains. Rain forest also extended until recently as a belt, up to 350 km (200 miles) wide, along the coast of West Africa, from Ghana through Côte D'Ivoire (Ivory Coast) and Liberia to the eastern margin of Sierra Leone (Martin 1991; Bakarr et al. 1999). Most of this has now gone and the remaining fragments are threatened by logging, agriculture, and rapidly expanding human populations. The larger Central and smaller West African rain forest blocks were separated by 300 km of dry woodland and savanna at the Dahomey Gap, in Togo, Benin, and eastern Ghana. There are also outlying "islands" of rain forest in East Africa, mostly centered on mountains, and surrounded by a "sea" of dry woodland. Although the total area of these East African rain forest patches is small—approximately 10,000 km$^2$—some of them, on older mountains, are apparently of great age and have been

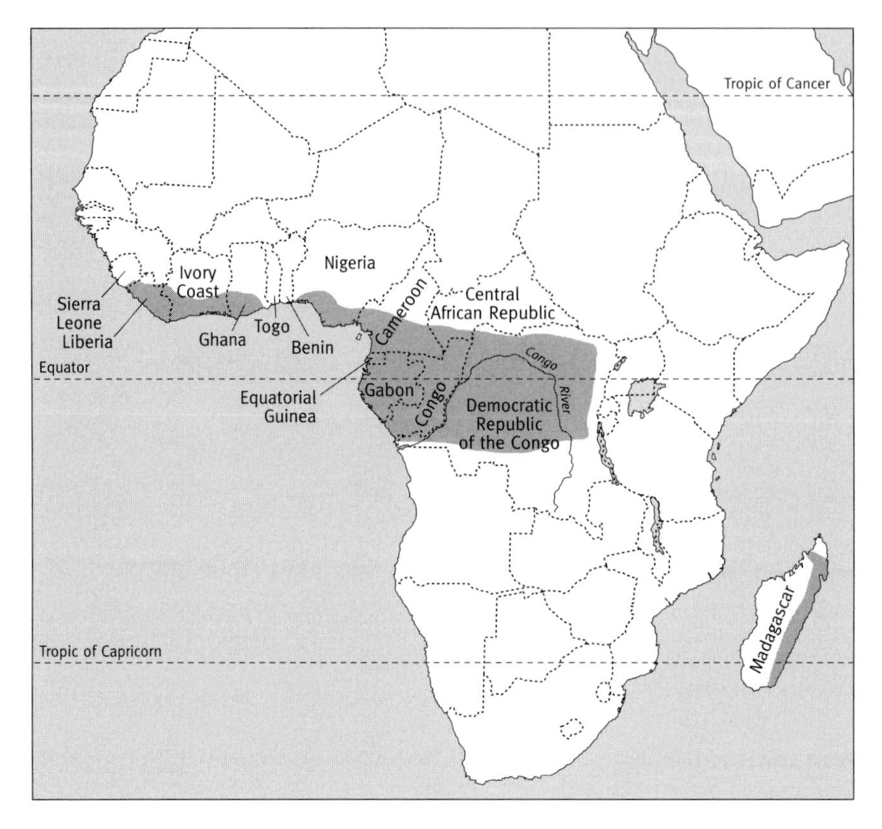


**Fig. 1.3** The estimated historical extent of tropical rain forest in Africa and Madagascar. (From multiple sources.)

isolated from the forests of West and Central Africa for millions of years (Lovett & Wasser 1993). As a result, many of their plant and animals species are endemic (i.e. occur nowhere else). On a longer timescale, all the rain forests of modern Africa can be seen as remnants of the much more extensive rain forest that spanned the entire continent until around 30 million years ago (Davis et al. 2002).

## Asia

The third largest rain forest area until recently occupied most of the Malay Peninsula and the large islands of Borneo, Sumatra, and Java (Fig. 1.4) (Whitmore 1984, 1987). Ecologists call this region "Sundaland", after the surrounding Sunda continental shelf, and we will follow the convention in this book. Despite the large expanses of sea between the major islands, the Sundaland rain forests are surprisingly uniform and can be considered as a single block. The island of Java has been almost entirely cleared of its forests due to the fertile volcanic soils and consequent dense rural population. Most of the forests that remain there are centered on volcanic mountains. The Malay Peninsula, Borneo, and Sumatra still retain considerable forest areas, although the combination of a growing population, high levels of logging, and increased clearance for plantation crops, has resulted in the degradation and elimination of lowland rain forests over

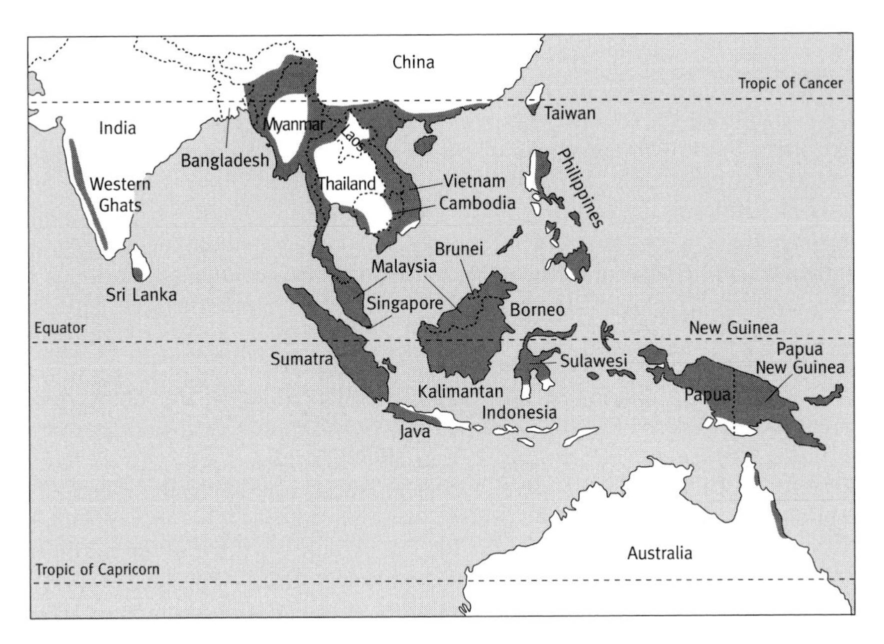


**Fig. 1.4** The estimated historical extent of tropical rain forest in Asia, New Guinea, and Australia. (From multiple sources.)

a wide area. Sulawesi and many of the smaller Indonesian islands between Borneo and New Guinea also have considerable areas of rain forest, but these are also being affected by human activity (Whitten et al. 1999). The rain forest that covered much of the Philippines was nearly completely destroyed by massive, uncontrolled logging and agricultural clearance from the early 1950s until the mid 1970s.

Rain forest also extended north from Sundaland into the more seasonal climates of mainland Southeast Asia, including most of Cambodia, Laos, and Vietnam, and much of Thailand and Myanmar (formerly Burma). However, large areas in the interior of Myanmar and Thailand did not support rain forest as a result of rainshadows caused by several long north–south mountain chains. Rain forest also once covered much of tropical southern China, in a mosaic with drier forest types, east to the southern tip of Taiwan. To the north, this tropical rain forest merged gradually into the subtropical and warm temperate forests. Most of this northern rain forest has now been cleared, and what remains is under severe threat.

The rain forests of Southeast Asia extend westward through Myanmar into northeastern India. India also had a completely separate rain forest area as a long narrow strip, 50–100 km wide, running parallel to the west coast for 1500 km along the crest of the Western Ghats (Kumar et al. 1999). This now fragmented rain forest band occurs on the tops and sides of these hills, where sea mists can drench the plants even during the dry season. Apart from this isolated ridge, the remainder of India south of the Himalayas is too dry to support rain forest. Just across the Palk Strait lies the island of Sri Lanka. Formerly much of the southwest of the island supported rain forest, but now only small fragments remain, the most significant being Sinharaja Forest in the very southwest of the country (Kumar et al. 1999).

### New Guinea and Australia

The fourth largest, and now most intact, block of rain forest covers most of the large island of New Guinea, except for the dry southern and eastern margins, and the highest mountain peaks (Fig. 1.4) (Paijmans 1976). Although biologically and culturally uniform, the island is divided politically into two halves: the western half forms the Indonesian province of Papua (previously Irian Jaya) and the eastern half is the independent country of Papua New Guinea. The neighboring and largely dry continent of Australia also supports a small area of rain forest in the northeast. The largest block of rain forest occurs along the coast between Cooktown and Townsville, with the best-known region being the Atherton Tablelands, but there are also numerous smaller patches.

Australia's tiny rain forest area has many similarities to the much more extensive rain forests of New Guinea, but also many differences, including several distinctive endemic plant genera. These differences reflect the very different histories of the two regions, despite their proximity and intermittent contacts. Rain forest covered much of northern Australia during the early to middle Miocene (23 to 15 million years ago), but has become restricted to the northeast because of the subsequent drying of the continent (Long et al. 2002). Most of the rain forest in New Guinea, by contrast, occupies land that was uplifted above sea level only in the last 10–15 million years, making this the youngest of the major rain forest blocks.

New Guinea remains largely covered by rain forest today because of the rugged terrain, late contact with the outside world, and strong clan ownership rights to the land. However, expanding human populations, logging, and mining now pose an increasing threat, particularly in the Indonesian half of the island. Since the arrival of Europeans, the area of rain forest in Australia has become even smaller because of logging and clearance for agriculture, but the remnants are now among the best-protected rain forests in the world.

### Madagascar

The final major area of rain forest is on the island of Madagascar (now the Malagasy Republic) (see Fig. 1.3). Although Madagascar has about two-thirds the land area of New Guinea, most of the island is very dry, and rain forest was confined to a 120 km (75 miles) wide band along the eastern edge (Mittermeier et al. 1999). Humans first arrived in Madagascar around 2000 years ago, resulting in mass extinctions among the larger and more vulnerable vertebrates (Burney et al. 2003). Rain forests seem to have been the last habitat to be settled by humans, but most of the rain forest band has now been cleared, and what remains is fragmented and in many places highly degraded.

## Rain forest environments

### Rainfall

The very name "tropical rain forest" suggests a steamy jungle that is unfailingly hot and wet, every day of the year. In reality, there is no place on Earth where it rains every day, and rain forests are found in a surprisingly wide range of climates. In the tropical lowlands, these forests grow on almost all soil types where the annual rainfall is well distributed and greater than about 1800 mm (70 inches). Mean annual rainfall can be as low as 1500 mm on sites with soils that can hold water well or where dry-season water stress is moderated by cloud or low temperatures. At the other extreme, the mean annual rainfall is greater than 10,000 mm (10 m!) at Cherrapunji in northeast India, Ureka in Equatorial Guinea, and in parts of the Chocó region of western Colombia.

To a first approximation, the amount and timing of rainfall in the tropics is controlled by the seasonal movements of the Intertropical Convergence Zone (ITCZ)—a band of low pressure, cloudiness, and rainfall that migrates north and south a month or two behind the overhead sun (McGregor & Nieuwolt 1998). The ITCZ results from rising warm air masses over regions where the sun is most directly overhead at midday. In this simple model the equatorial region is continuously influenced by the proximity of the ITCZ, so it is wet all year, with two rainfall peaks, a month or so after the equinoxes (Fig. 1.5). Away from the

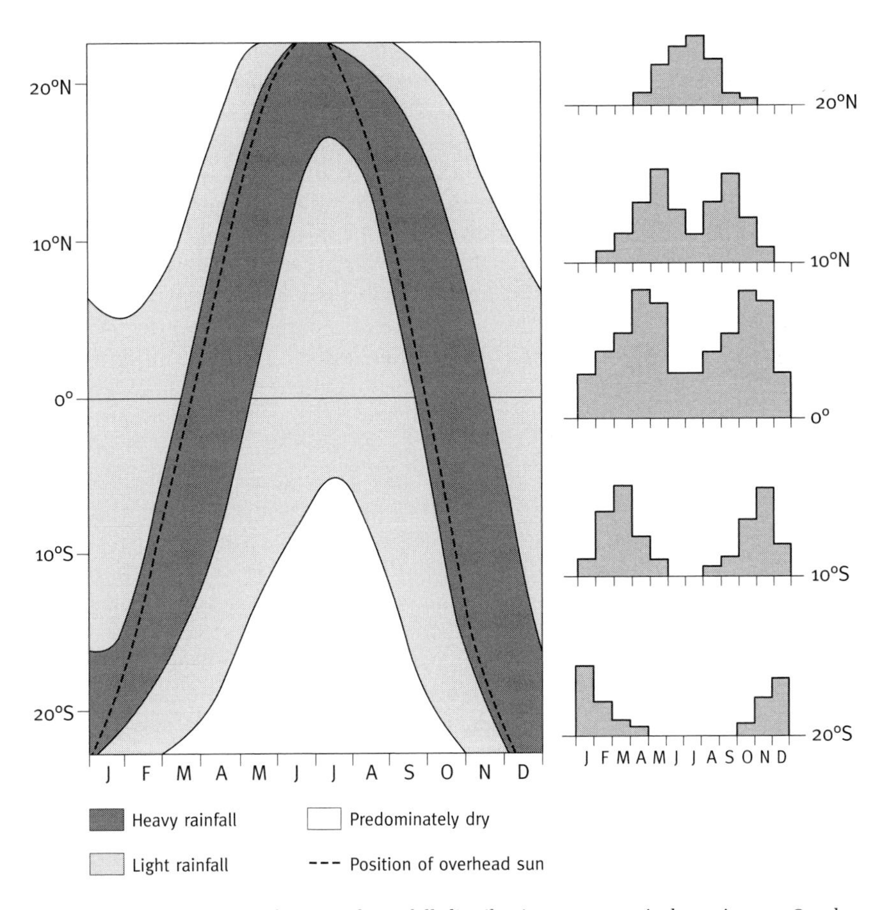


**Fig. 1.5** A simple model of seasonal rainfall distribution over tropical continents. On the right are idealized seasonal patterns of rainfall at five different latitudes. Real rainfall patterns differ from these ideals for reasons mentioned in the text. (From McGregor & Nieuwolt 1998.)

equator, rainfall is concentrated in the periods when the ITCZ is present and there are dry periods when it moves away. With increasing distance from the equator, the two rainfall peaks move together and the "winter" period of low rainfall becomes longer and drier. Near the margins of the tropics, there is only one, relatively short, wet season, and a long, rainless dry season.

This simple model works fairly well over the oceans, but many factors introduce complications over land. While all rainfall results from upward movements of moist air, surface heating by the overhead sun is by no means the only mechanism that can cause this. The most important disruptions to the general pattern described above occur when moist air is forced to rise over a mountain range, producing rainfall on the windward slopes. Where mountains face a sustained flow of moist air throughout the year, this effect can produce an everwet climate well way from the equator. This explains the presence of rain forest in eastern Madagascar, at latitudes where we might expect a seasonally dry climate. Other examples of high rainfall resulting from such "orographic" uplift include parts of the Caribbean coast of Central America, eastern Brazil, West Africa, the west coast of India, coastal Queensland, and many tropical islands. There are also anomalies in the opposite direction: dry climates at latitudes we expect to be wet. The most striking interruption to the equatorial belt of rain forest climates is in East Africa, where a combination of relatively dry monsoon air flows and large latitudinal movements of the ITCZ makes most of the region too dry for rain forest. Cold ocean currents produce anomalously dry climates in western Ecuador, while other dry areas are in the lee of mountain ranges. As a result of these and other factors, the overall patterns of rainfall and rainfall seasonality in the tropics can be very complex. Interested readers should refer to more detailed accounts by Walsh (1996) and McGregor and Nieuwolt (1998).

All the major rain forest regions have relatively dry and relatively wet areas, but the amount of rainfall in the most extensive forest type varies between regions (Fig. 1.6). In general, the wettest rain forests are those of equatorial Southeast Asia, centered on the core Sundaland region of western Indonesia and Malaysia, and those on the island of New Guinea (Times Books 1994). In both these regions, the mean annual rainfall exceeds 3000 mm over large areas. In contrast, most Madagascan and American rain forest receives 2000–3000 mm, although there are wetter areas receiving 3000 mm or more in the upper Amazon basin, western Colombia, and the eastern slopes of Central America. The Atlantic Coastal Forest of Brazil is mostly somewhat drier, with less than 2000 mm of rain. Most African rain forests are distinctly drier than rain forests elsewhere, with an annual rainfall of only 1500–2000 mm, except in narrow fringes along the coast where the rainfall can exceed 4000 mm. Within particular regions, there is often considerable variation in rainfall, determined by distance from the coast, elevation, land use patterns, and other climatic factors. For example, in the Amazon River basin, rainfall varies from less than 1200 mm per year to over 6400 mm, with higher rainfall along the coast and in the northwestern edge of the basin (Sombroek 2001).

Even in regions with very high total rainfall, most rain forests experience some dry months, when the water lost by evaporation and transpiration—around 100 mm per month—is greater than the amount of rain that falls. In tropical Asia outside the everwet Sundaland core, in almost all African rain forests, and in most of those in tropical America, there is an annual dry season lasting 1–4 months. This dry season is harsher in tropical America and continental Asia,

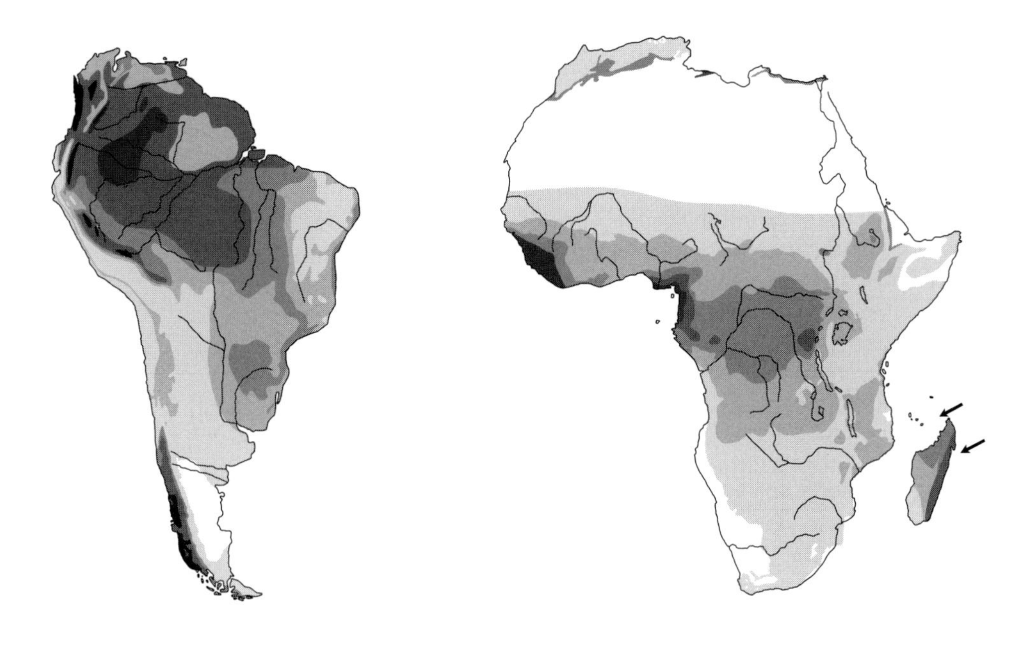

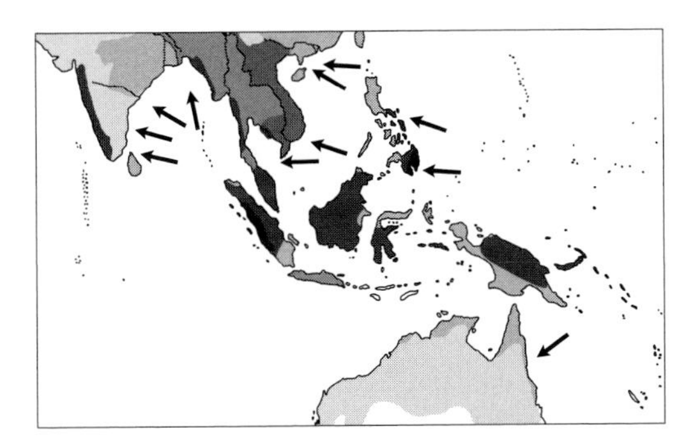

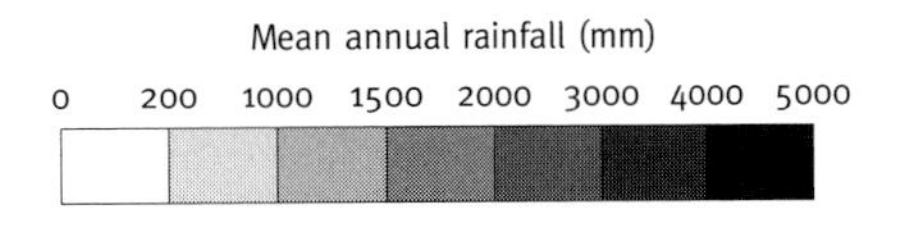


**Fig. 1.6** Worldwide distribution of rainfall over land, emphasizing the variability in rainfall within regions, and the greater rainfall of Southeast Asia than Africa. Also shown are the major tropical storm tracks, heading toward Madagascar, northeastern Australia, the Philippines, eastern India, and Southeast Asia. Note that South America has no major storm tracks; storms pass further north toward the Caribbean and Central America. (From Times Books 1994.)

where it is usually accompanied by cloudless skies, than in much of the African rain forest, where the rainless months are often misty and overcast (McGregor & Nieuwolt 1998). The predictable rainfall seasonality in these forests is reflected in all aspects of their biology, with more or less regular annual peaks in leafing, flowering, fruiting, and animal reproduction.

In striking contrast to the typical African or American rain forest, there are large areas of Sundaland and on the island of New Guinea where there is no regular dry season, and the mean rainfall for every month is over 100 mm (Fig. 1.6). In some places, dry months are very rare, and longer dry periods are unknown. In most Sundaland rain forests, however, dry periods occur every few years, with important consequences for their biology. These forests show what can only be described as "multiyear seasonality". Just as the fall in temperate zones brings on specific biological changes, such as leaf-dropping by deciduous trees and hibernation in some mammals, these less frequent events in Sundaland rain forests trigger dramatic biological changes, including mass flowering of many tree species, increased reproduction in many animal species, and large-scale migrations in others (Curran & Leighton 2000; Sakai 2002).

In Borneo and the eastern side of the Malay Peninsula, these dry periods are associated with the El Niño–Southern Oscillation cycle. The term El Niño means "Christ child" and was originally used by Peruvian fishermen to describe the warm current appearing off the western coast of Peru around Christmas time. Today, El Niño refers to the warm phase of a naturally occurring sea surface temperature oscillation in the tropical Pacific Ocean. The El Niño cycle is associated with the Southern Oscillation—a seesaw shift in surface air pressure between Darwin, Australia, and the Pacific island of Tahiti. Hence, the name El Niño–Southern Oscillation or ENSO cycle for this complex coupled cycle in the ocean–atmospheric system (Nash 2002). There are three phases in the cycle: (i) a normal or neutral phase; (ii) the El Niño phase with unusually warm sea surface temperatures in the tropical Pacific; and (iii) the opposite La Niña ("little girl") phase with unusually cold sea surface temperatures. La Niña events occur after some, but not all, El Niños.

El Niño events generally occur at intervals of 2–8 years, although their intensity varies greatly. The strongest recent events—and probably the strongest for the last century—were in 1982–83 and 1997–98. A strong El Niño brings low rainfall to large areas of Indonesia and Malaysia, and coincides with mass flowering, then synchronized fruiting, by most individuals of the dominant tree family, the dipterocarps (Dipterocarpaceae), across huge areas (Sakai 2002) (see Chapter 2 for more details). The precise nature of the trigger for mass flowering is still debated, but one possibility is a brief period of low night-time temperatures in the preceding weeks (Yasuda et al. 1999). Fruiting then occurs 7–9 months later, at the peak of the El Niño phase. Many nondipterocarps flower and fruit at the same time so, for a few weeks only, the usually green and monotonous rain forest experiences a burst of colorful reproductive activity. The superabundance of flowers and fruits, in turn, leads to an inflow of nomadic animals, including giant honeybees, parakeets, and bearded pigs, and a burst of reproduction in the resident animal species.

The influence of the ENSO cycle weakens towards the west of the Sundaland region. The rain forests of northern Sumatra and the western side of the Malay Peninsula show the same multiyear seasonality of dry periods and associated biological events as further east, but this has no clear relationship with the ENSO cycle and the effects are usually more localized (Wich & van Schaik 2000). The absence of the large-scale synchrony of flowering and fruiting shown in the ENSO-dominated east may make it easier for West Sundaland animals to track peaks in their food supply by moving shorter distances in the rain forest to places where trees are flowering and fruiting.

Strong El Niño episodes lead to dry periods of increased severity in most other rain forests as well, including those of New Guinea, Africa, Central America, and Amazonia. However, the impact of these extreme events on the forest seems to be much greater in the Sundaland rain forests, which lack a regular dry season, than in forests that are adapted to an annual shortage of water. At Lambir in

Sarawak, the long-term average rainfall of the driest month is 168 mm, which is not low enough to cause any water stress. During the exceptional El Niño-associated drought of 1998, in contrast, the total for the 3 months from January to March was only 139 mm, resulting in extensive leaf loss and a large increase in tree mortality (Potts 2003). The same El Niño episode in central Amazonia merely intensified the annual dry season, causing wilting and leaf fall but only a modest increase in tree mortality (Williamson et al. 2000).

El Niño episodes in the Neotropics also increase fruit production, but there is no equivalent of the mass, synchronized fruiting at multiyear intervals seen in the Sundaland rain forests (Williamson & Ickes 2002). High fruit production during El Niño episodes on Barro Colorado Island, Panama, is followed by low fruit production in the wet, cloudy years that often follow, leading to increased mortality among primates and other fruit-eating animals (Wright et al. 1999). This Neotropical pattern of a regular fruit supply, with famines at multiyear intervals, is in striking contrast to the situation in the rain forests of Sundaland, where fruit famines are the normal situation and feasts occur at multiyear intervals. We will see in later chapters how these contrasting patterns of fruit (and flower) supply have apparently resulted in contrasting adaptations in the animals of these forests.

Before all differences between rain forests are attributed to the varying effects of the ENSO cycle, a note of caution is necessary. Reliable historical records of El Niño episodes go back only a century or so, but a variety of indirect sources of information suggest that both the frequency and intensity of El Niño events has varied considerably during the Holocene (the last 10,000 years) (Moy et al. 2002). This raises the interesting possibility that the ENSO-associated patterns of plant and animal reproduction that can be observed today, particularly in the Sundaland rain forests, are not necessarily typical of even the last few thousand years. Variability on longer timescales is likely to be even greater, so it is unlikely that plant and animal responses to the ENSO cycle are as finely tuned as they sometimes appear to be.

### Temperature

Wetness is only half of the tropical rain forest equation: the other half is warmth. Typical equatorial lowland rain forests have a mean annual temperature of 25–26°C (77°F) and very little seasonal variation. At Danum in Sabah, Malaysia, for instance, the difference between the highest daytime temperature and the lowest nighttime temperature is 8–9°C (47°F), but the difference between the average temperature of the hottest and coolest months is less than 2°C (36°F) (Walsh & Newbery 1999). Even near the equator, however, brief incursions of cold air cause surprisingly low minimum temperatures in some rain forest areas. In the upper Amazon basin, cold air from temperate South America moves northwards along the Andes and can bring temperatures as low as 11°C to Iquitos, just north of the equator (Walsh 1996). Even lower minimum temperatures (8°C) have been recorded in the lowland rain forest at Cocha Cashu, in Manu National Park, Peru, 12° south of the equator. Annual temperature ranges increase with distance from the equator because of reduced solar radiation in winter, as well as an increased impact of these "cold waves", known as "friagems" in Brazil. Near the southern margins of the tropical rain forest in South America and Australia, and the northern margins in Asia, frosts (i.e. subzero temperatures) can occur down to sea level, resulting in selective defoliation and shoot dieback in sensitive tropical plant species.

Nevertheless, the latitudinal limits of tropical rain forest are, in most places, set by drought rather than cold. Only in East Asia is there a continuous belt of lowland forest climates from the equator to the Arctic, without an intervening belt of climates too dry to support forest. Forests that closely resemble the typical tropical rain forests of Southeast Asia in terms of structure, floristics, and diversity, extend north of the Tropic of Cancer in southwestern China, northern Myanmar, and northeast India (Proctor et al. 1998). The climate in southwest China is extreme for tropical rain forest, with a mean annual rainfall as low as 1500 mm and a long and very dry winter, with minimum temperatures regularly falling below 10°C (Zhu 1997). The low winter temperatures and the frequent thick fog reduce water stress, so the forest is still largely evergreen (Liu et al. 2004). An unforgettable experience for a tropical rain forest ecologist in China is to watch the fog clear on a wintry morning in Xishuangbanna, to reveal a rain forest with emergent dipterocarp trees rising to 60 m (200 feet). The dipterocarps all belong to one species—*Shorea chinensis*—in contrast to the dozens of coexisting dipterocarp species in Bornean rain forests, but the whole appearance of the forest is distinctly tropical (Cao et al. 1996).

Temperatures also decline with increasing altitude above sea level, but in this case there is no associated increase in seasonality. On equatorial mountains, such as Mt Wilhelm in Papua New Guinea, the mean annual temperature at the altitudinal tree limit, at 4000 m (13,000 feet) above sea level, is only around 5°C (41°F), but there is very little seasonal variation (Hnatiuk et al. 1976). The temperature falls to near zero every night and rises above 10°C during the day: a climate that has been aptly termed "summer every day, winter every night". With increasing altitude above the lowlands, the rain forest becomes shorter, tree heights more even, the crowns and leaves smaller, rooting more shallow, and cold-intolerant plant families, such as dipterocarps and figs, progressively drop out (Whitmore 1998). The direct effects of declining temperature may, however, be less important than changes in other factors, such as soil conditions, and a marked increase in soil organic matter is the most consistent environmental change at the upper limits of the lowland rain forest (Ashton 2003). The most dramatic vegetation changes often coincide with the zone of persistent cloud cover, where trunks and branches become gnarled and bryophytes—mosses and liverworts—cover all surfaces. This vegetation is often referred to as "cloud forest" or "mossy forest", although the bryophytes are mostly liverworts rather than mosses. In this book we focus our attention on lowland rain forests, but the ecology of tropical cloud forests has been covered by several recent publications (e.g. Hamilton et al. 1995; Nadkarni & Wheelwright 2000).

## Wind

Another climatic factor with a major influence on the structure, if not the distribution, of tropical rain forest is wind. The combination of very tall trees and shallow root systems makes rain forests particularly vulnerable to strong winds. All rain forests are subject to occasional squalls of strong wind that may blow down single trees or, more rarely, fell large swathes of forest. Indeed, the ecological importance of these rare but widespread blowdown events may have been underestimated (Nelson et al. 1994; Proctor et al. 2001). The most dramatic effects of wind, however, are in the rain forest areas subject to tropical cyclones. Such cyclones are absent from the region approximately 10° either side of the equator that contains most tropical rain forest, but these storms affect with varying frequency the rain forests of the Caribbean, much of Central America, Madagascar, northern Southeast Asia (particularly the northern Philippines), northeastern Australia, and many oceanic islands (see Fig. 1.6). Sustained wind speeds during a major cyclone can exceed 70 m per second (150 mph), with brief gusts of much higher speeds.

The short-term impact of a single, severe, hurricane-strength cyclone is dramatic, with a large proportion of canopy trees uprooted or snapped off in the worst affected areas and almost complete defoliation in less damaged areas (Whitmore 1984; Brokaw & Walker 1991). Most of these damaged and defoliated trees will soon put out a new crop of leaves. However, "super cyclones" of extreme intensity occur at longer intervals in some regions, and may kill trees over a large area (Nott & Hayne 2001). In the longer term, repeated cyclone damage may allow an increased proportion of light-demanding tree species to persist in the forest. In areas with a very high frequency of cyclones, such as the island of Mauritius and parts of Queensland, a distinct "cyclone forest" may develop. Such a forest is dominated by short-lived, rapidly reproducing, light-demanding species that are either less easily damaged by strong winds or able to complete their life cycle between successive hurricanes. Thus, the impact of a single cyclone will depend not only on its severity but also on the time since the previous one. Even rare cyclones eliminate the advantages for a tree of being taller than its neighbors are, so rain forests in the cyclone belt tend to be relatively short (de Gouvenain & Silander 2003). Outside the cyclone zone, trees can grow taller and the canopy is more continuous, so the seedlings of many dominant canopy species are adapted to grow in more shady conditions.

## Sunlight

In the equatorial region the sun is high in the sky throughout the year, but cloudiness and the high water vapor content of the air greatly reduce the amount of solar radiation reaching the forest canopy (Walsh 1996). A perhaps surprising consequence of this is that the availability of light—rather than water, temperature, or soil nutrients—can limit plant growth at certain times of the year. This was neatly demonstrated by installing high-intensity lamps above a canopy tree species, *Luehea seemannii*, in semievergreen rain forest in Panama (Graham et al. 2003). Trees given extra lighting during the cloudiest periods of the wet season grew more than those receiving only natural light. Light availability in the rain forest canopy is known to vary between sites, between seasons, and between phases of the ENSO cycle, but the potential consequences of this variation have not yet been investigated. It has been suggested, for instance, that above-average light intensities during El Niño events, as a result of reduced cloud cover, may be at least partly responsible for the enhanced fruit production observed in many rain forests (Wright et al. 1999).

## Soils

There is an increasing amount of evidence that soil factors control plant distributions in tropical rain forests on both local and regional scales (Sollins 1998; Ashton 2003; Tuomisto et al. 2003; Palmiotto et al. 2004). Soil characteristics also strongly influence plant biomass and, indirectly, animal biomass (Peres 2000). Exactly which soil factors are most important is still not certain, however, since soil texture, drainage, nutrients, and surface topography are all usually correlated and few studies in lowland rain forests have looked at the full range of possible factors.

Soil properties depend, in part, on the nature of the geological substrate from which they are formed. Over time, however, soil depth increases and the weatherable minerals in the soil are lost, leaving only quartz and clays. Those soil nutrients that were derived from weathering of the parent rock are either leached out of the soil (calcium, magnesium, potassium) or, in the case of phosphorous, form insoluble compounds that are unavailable to plants (Baillie 1996). High temperatures and rainfall in the humid tropics speed up these processes, but even then it can take several million years before the final stages are reached (Chadwick et al. 1999; Hedin et al. 2003). Soils this old are found only in geologically stable areas, such as the Amazon basin, Central Africa, and those parts of tropical Asia that are furthest from the margins of tectonic plates. Elsewhere, as in much of Central America, Southeast Asia, and New Guinea, tectonic movements or volcanic eruptions reset the clock at intervals, keeping the soils relatively young. The annual influx of river-borne sediments has the same effect in the floodplains of major rivers.

Deep, old, highly leached and weathered soils are acid and infertile, with very low levels of plant-available phosphorous, calcium, potassium, and magnesium, and high levels of potentially toxic aluminum (Baillie 1996). Such soils are unsuitable for most forms of permanent agriculture, yet can support tall, dense, hyperdiverse rain forests. This apparent paradox reflects the ability of undisturbed rain forests on poor soils to recycle nutrients with very little loss (Cuevas 2001). Most nutrients are withdrawn before leaves are dropped and the nutrients released in the litter layer are rapidly taken up by a dense mat of roots and their associated mycorrhizal fungi. If there is no unweathered parent material left within the root zone, the inevitable small losses of nutrients from the forest ecosystem must be replenished from the atmosphere, in dust and rain, and by nitrogen fixation. A variety of evidence suggest that phosphorous is typically the most limiting nutrient for plant growth on such soils.

Tropical rain forests occur on a wide range of soil types, by no means all of which are unsuitable for permanent agriculture. Relatively fertile soils occur in a variety of situations, such as in the volcanic areas of Java and on the floodplains of whitewater rivers in the Amazon region. Unfortunately, rain forests on these more fertile soils are particularly prone to clearance, while long-term protection is mostly likely for forests on the least fertile sites. Deforestation is thus concentrated in the areas that support the highest plant and animal biomass, so the impact on biodiversity and carbon storage is even greater than crude estimates of percentage area loss imply (see Chapter 8).

Variations in soil texture, drainage, and chemistry affect the botanical composition of the rain forest, but only the most extreme soil types support distinctly different vegetation types (Whitmore 1998; Turner 2001b). Most distinctive are the heath forests, which are also known by a variety of different local names, such as caatinga in Amazonia and kerangas in Southeast Asia. Heath forests develop on infertile, drought-prone, sandy soils derived from coastal deposits or the weathering of sandstone. Compared with typical tropical lowland rain forests, they are lower in stature and the trees have smaller, harder leaves. The streams that drain these forests are blackish or dark brown as a result of the presence of particulate and colloidal organic matter. Heath forests are found in all the major rain forest regions but they are most extensive in the upper reaches of the appropriately named Rio Negro (Black River) in South America (Prance 2001). Other distinctive, but more variable, forest types occur on soils derived from limestone, as well as those on ultramafic (iron and nickel-rich) rocks (Proctor 2003). Forests on these soils are also typically low in stature, with many distinctive plant species.

Different forest types also develop on sites where peat, consisting largely of partly decomposed plant material, has built up to such a depth that the forest is isolated from the ground water (Whitmore 1998). These peat swamp forests are totally dependent on nutrient input from the rainfall, which also saturates and preserves the peat, and both the height and species diversity of the vegetation decrease with increasing peat depth. Raised, deep peat beds are found only in areas with high rainfall and without a long dry season, and are particularly extensive on the islands of Borneo, Sumatra, and New Guinea.

### Flooding

Flooding by river water produces an array of different forest types depending on whether the floods are permanent or periodic, and whether the periodicity is daily, monthly, or annual. Freshwater swamp forests, known locally as várzea, are most extensive along the Amazon River, which has annual floods and is also influenced by tides up to 900 km (600 miles) from its mouth (Goulding 1989; Prance 2001). Extensive, but little studied, swamp forests also occur in the Congo River basin, and there are smaller areas in New Guinea (Paijmans 1976) and Southeast Asia (Whitmore 1998). Freshwater swamp forests generally support a lower diversity of plant species than dryland forests, presumably because of the problems of dispersal, germination, establishment, and growth in an environment that experiences such seasonal extremes (Lopez & Kursar 2003). However, várzea swamp forests on the fertile floodplains of Amazonian whitewater rivers support a significantly higher biomass of vertebrates than adjacent dryland, or terra firme, forests (Peres 2000) and the same is likely to be true in other rain forest regions. Near the mouths of major rivers, freshwater swamp forests are replaced by brackish-water swamp forests and then mangrove forest, which has a much simpler structure and lower diversity than other tropical forest types.

## Rain forest histories

### Plate tectonics and continental drift

To understand the similarities and differences between modern rain forests, it is necessary to learn about their pasts. Most of the land masses that currently support tropical rain forest have a common origin in the ancient southern supercontinent of Gondwana (Fig. 1.7) (Morley 2000). Gondwana means "land of the Gonds", and is named after a tribe from southern India whose land provided the first evidence that India had been part of the supercontinent. The core of modern Southeast Asia is made up of continental blocks that broke away from Gondwana between 400 and 160 million years ago, which is too early for them to have carried modern groups of plants and animals (Metcalfe 1998). Australia, India, Madagascar, Africa, and South America, in contrast, separated from each other and drifted north during the late Jurassic, Cretaceous, and early Tertiary (160 to 50 million years ago). This is the period during which many modern groups of plants and animals originated, so the sequence and timing of the break-up has had a significant influence on modern biogeographical patterns (McLoughlin 2001). New Caledonia, an island east of Australia, and the Seychelles, an archipelago east of Africa in the Indian Ocean, are also fragments of Gondwana.

India broke away early (130 million years ago) and had fused with Asia by the Eocene (40 million years ago), so its modern flora and fauna are Asian. Africa's longer period of isolation in the late Cretaceous and early Tertiary produced a spectacular radiation of forms in an endemic clade of mammals, the Afrotheria, represented today by fewer than a hundred species in six orders: Proboscidea (elephants), Hyracoidea (hyraxes), Macroscelidea (elephant shrews), Tubulidentata (aardvark), Afrosoricida (golden moles and Madagascan tenrecs), and Sirenia (dugong and manatees) (Madsen et al. 2001). Although Africa has now been physically connected to Asia for at least 20 million years, climatic barriers such as large arid zones (represented today by the Sahara Desert) have allowed only limited exchange of rain forest taxa between the two continents for most of this period. Madagascar has remained isolated for 90 million years, and its relatively small rain forest area has developed on its own unique evolutionary path.

South America remained isolated from other continents for over 70 million years. This presented the opportunity for peculiar forms to evolve that are found nowhere else in the world, including the sloths, armadillos, and anteaters, in the endemic order Xenarthra, the New World monkeys (see Chapter 3), and endemic radiations of marsupials and ungulates (see Chapter 4) (Whitmore & Prance 1987; Terborgh 1992; Nowak 1999). South America's long isolation finally ended when the Isthmus of Panama rose 3 million years ago, connecting the two American continents and allowing the intermingling of North American and South American biotas (Fig. 1.8). This dramatic event has become known as the Great American Interchange (Marshall 1988; Webb 1997). Some movement of animals between continents via rafting, chance dispersal, and island hopping appears to have begun as early as 8–10 million years ago. The North American invaders, which included such familiar groups as the cats, tapirs, deer, and squirrels, underwent explosive diversification in South America and, today, their descendents make up more than half the mammal fauna.

Central America is a composite of geological units of different ages and origins (Coates 1997) whose biota, before the Interchange, was dominated by North American organisms. The rain forests of Central America today, however, are overwhelmingly dominated by plants and animal groups that are shared with South America, and the effects of the deep-sea barrier that separated Central and South America until 3 million years ago have been almost totally erased (Webb 1997).

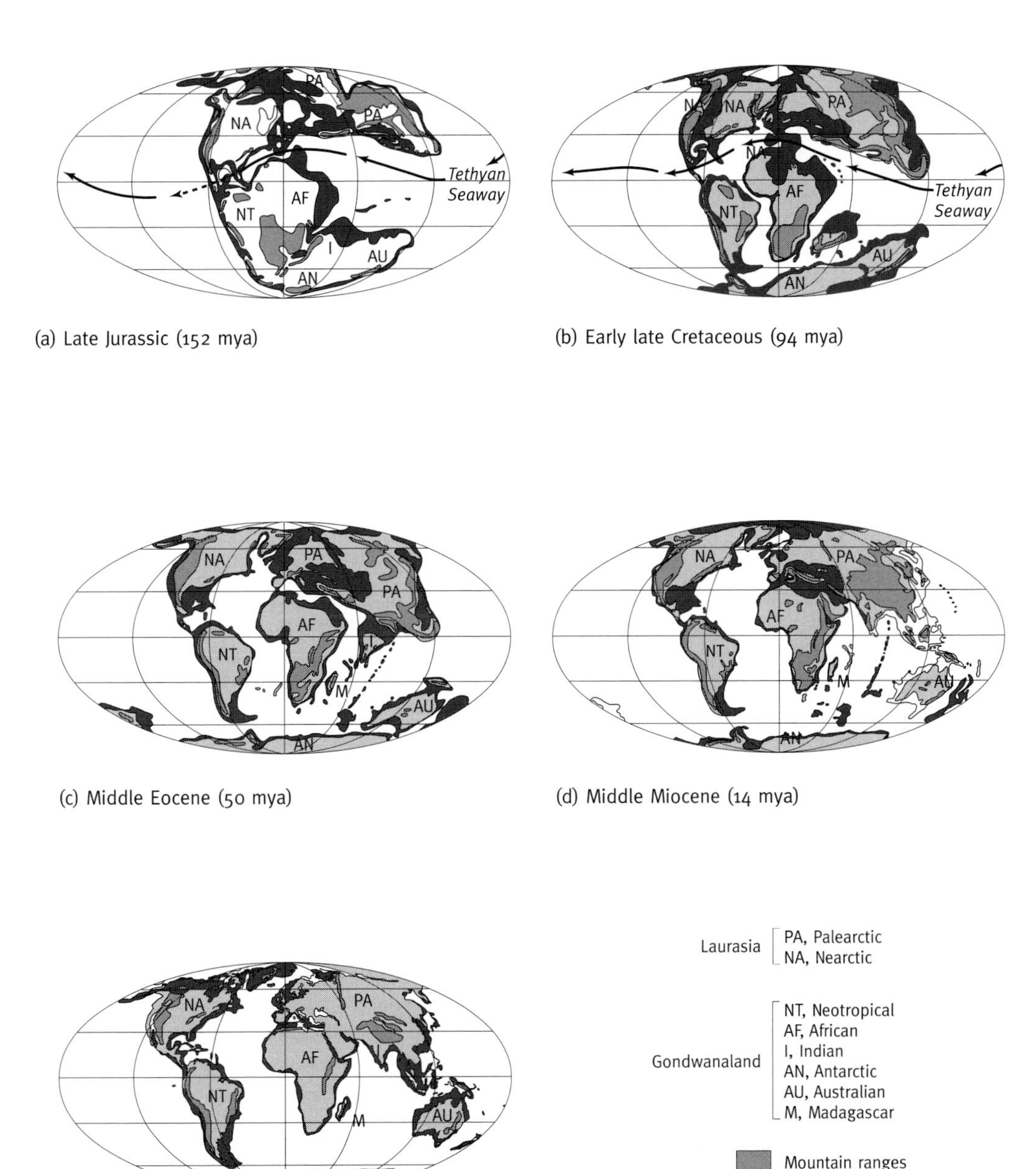


**Fig. 1.7** Continents have changed their position over past geological time periods. (a) In the later Jurassic (152 million years ago), the southern continents were connected into the supercontinent of Gondwana, and species could migrate among these areas. (b) By the early part of the late Cretacous (94 million years ago), South America, Africa, India, and Australia had separated from one another, and connections with northern continents were severed. (c) In the middle Eocene (50 million years ago), Madagascar had separated from Africa, India was moving toward Asia, and New Guinea had emerged. (d) By the middle Miocene (14 million years ago), the Central American land bridge connecting North and South America was beginning to form, with a water barrier still present; India had merged with Asia; New Guinea, Borneo, Sumatra, and neighboring islands had emerged; and Africa was touching Asia. (e) The modern world, showing the establishment of Central America and the firm contact of Africa and Asia. (From Brown & Lomolino 1998; after Chris Scotese, PALEOMAP project.)

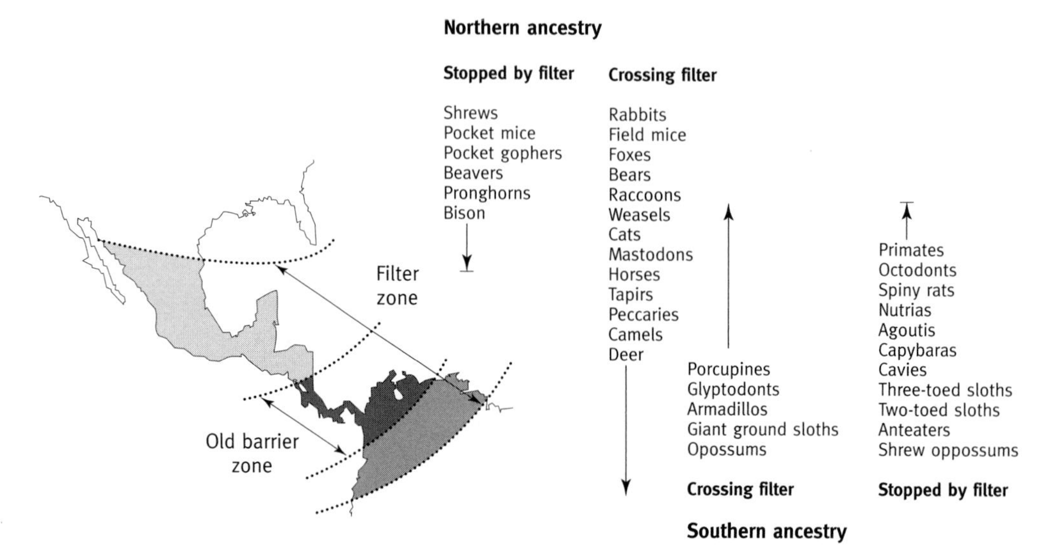


**Fig. 1.8** The Central American land bridge, which was formed 3 million years ago, acted as a filter, preventing some North American animal families from crossing the barrier and allowing many others to pass on to South America. In contrast, many South American families did not disperse to North America. (From Brown & Lomolino 1998.)

Australia and New Guinea are still isolated from Southeast Asia by marine barriers, but the northward movement of Australia in the Miocene—continuing to this day—gave rise to the Indonesian archipelago, which has permitted an increasing interchange with Southeast Asia for organisms that can disperse from island to island, such as birds, bats, insects, and many plants. The distinctive differences in animal communities between Southeast Asia and New Guinea were first noted in the 19th century by the great naturalist Alfred Russel Wallace (Wallace 1859). In recognition of his discovery, the line of separation between these two biotic regions is now called "Wallace's line" and the region between Borneo and New Guinea, with its numerous islands, is known as "Wallacea" (Fig. 1.9). Geologists estimate that Australia and New Guinea will become connected to Asia in about 40 million years time, allowing their biotas to mix more completely in a "Great Australasian Interchange".

## Changes in climate and sea level

Over the last few million years, rain forests expanded and contracted in area depending on the climate of the times (Flenley 1998; Morley 2000). Fossil evidence suggests that the tropical lowlands were both substantially cooler and, in large areas of the tropics, a lot drier during the glacial periods (ice ages) that occupied most of the last 2 million years. Lower atmospheric carbon dioxide levels during the glacial periods may also have been an important factor and may complicate the interpretation of the climatic records. These changes altered the composition of tropical rain forests and reduced their ranges. Pollen records from many sites show that montane or savanna plants were more widespread in glacial times and that rain forests disappeared from marginal areas (Mayle et al.

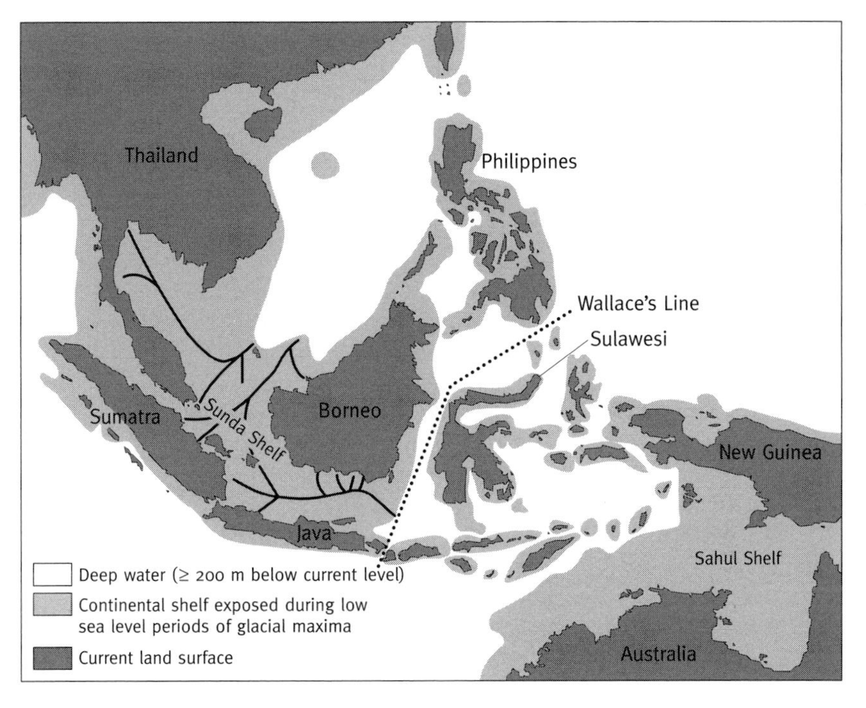


**Fig. 1.9** The Sunda Shelf (shaded) was exposed during the peaks of glaciation, allowing the movement of animals and plants along the river valleys connecting Borneo, Sumatra, Java, the Malay Peninsula, and smaller islands. These river valleys (shown as solid black lines) are submerged today beneath the South China Sea. Note that the Philippines is still separated from the Sunda Shelf. The Sahul Shelf, which surrounds Australia and New Guinea, is also shown. The area between the Sunda Shelf and the Sahul Shelf is known as Wallacea; the western boundary of Wallacea was described by Wallace as the eastern limit of distribution for many species of Asian animals, and is known today as Wallace's line. (From Brown & Lomolino 1998.)

2004). Rainfall remained high in many upland areas, but lower temperatures probably made these areas unsuitable as "refuges" for lowland rain forest organisms (Colinvaux et al. 2001).

Australia and Africa show the strongest evidence for glacial drying and rain forest contraction. In Africa, rain forest was reduced to perhaps 10% of its area during the glacial maximum, persisting only in a few areas with high rainfall and as gallery forests along river margins (Morley 2000). This desiccation of African rain forests is a major determinant of the far lower current diversity of palms, orchids, epiphytic species, and amphibians in African rain forests in comparison with rain forests in the Amazon and Asia (Richards 1973; Gentry 1993; Goldblatt 1993). The impact of glacial changes was less severe in the Asian and American tropics, and on the islands of Madagascar and New Guinea, due to higher levels of rainfall associated with mountain ranges. Such upland areas are notably lacking in the Congo basin, and are found only as isolated mountains in East Africa. The high diversity of South American plants and animals, in particular, is probably due in part to the persistence of large areas of rain forest during glacial periods in those areas of the Amazon basin that remained wet or were near rivers and other wetland areas (Haffer 1997). Some present patterns of species distribution may still reflect this past pattern of persistence.

Low sea levels (up to 130 m (430 feet) below the present level) during glacial maxima linked the major land masses of insular Southeast Asia—Sumatra, Java, and Borneo—to the Malay Peninsula and Asian mainland, and greatly increased the exposed land area (Fig. 1.9) (Whitmore 1987). Although much of Southeast Asia seems to have been too dry to support rain forest during glacial times, there is evidence that large areas of lowland rain forest persisted to the north of Kalimantan and in West Sumatra (Morley 2000). Past rain forest connections between the major modern land areas of Southeast Asia are reflected in the similarities of their modern forest biotas, although recent studies suggest that migration across these glacial land bridges was more limited than previously assumed (Gorog et al. 2004). Connections among the islands could have occurred via migration along river valleys that extended from the coastal plains of northwestern Borneo, Sumatra, western Java, and the east coast of the Malay Peninsula onto the exposed Sunda Shelf. The modern, relatively short, rapidly flowing rivers of northwestern Borneo, Sumatra, Java, and the Malay Peninsula are just the upper reaches of an extensive river system that merged on the exposed Sunda Shelf and flowed northward into the South China Sea (see Fig. 1.9). In contrast, most of the Philippines, Sulawesi, and the smaller islands between Sulawesi and New Guinea remained as islands, albeit often larger and connected among themselves, even at the lowest sea levels.

New Guinea had land connections to Australia during the glacial maxima, with the most recent connection being interrupted only 8000 years ago. However, the cooler, drier climate of those periods restricted interchange to those species that could migrate along the forested margins of waterways. Although New Guinea and Australia share rain forest species of tree kangaroo, possums, birds, snakes, frogs, and even fishes, New Guinea, with its larger rain forest area, has largely developed its own unique biota, which is far richer than that now found in Australia (Adam 1992).

A key point here is that the drier, cooler glacial episodes contracted African forests and divided them into smaller, isolated blocks, and probably had a similar effect on the Amazon basin as well. In contrast, glacial periods created land connections among the islands of Southeast Asia and the Asian mainland, and between New Guinea and Australia. As a result, the rain forests of Southeast Asia are more uniform, and those of Africa and South America less uniform, than might be expected from their present-day geography.

## Human occupation

One of the most important, but least understood, differences between rain forest regions is in their histories of human occupation. The popular idea that tropical rain forests were untouched, virgin ecosystems until the 20th century is a myth that has proved very hard to dispel, even among scientists (Mercader 2003). All rain forests have been modified by people and they cannot be understood if this fact is ignored.

The broad picture of the evolution and spread of humans across the Earth's surface is now quite well documented (Fig. 1.10). Modern humans originated in Africa and spread early to the warmer regions of Asia. New Guinea, Australia,

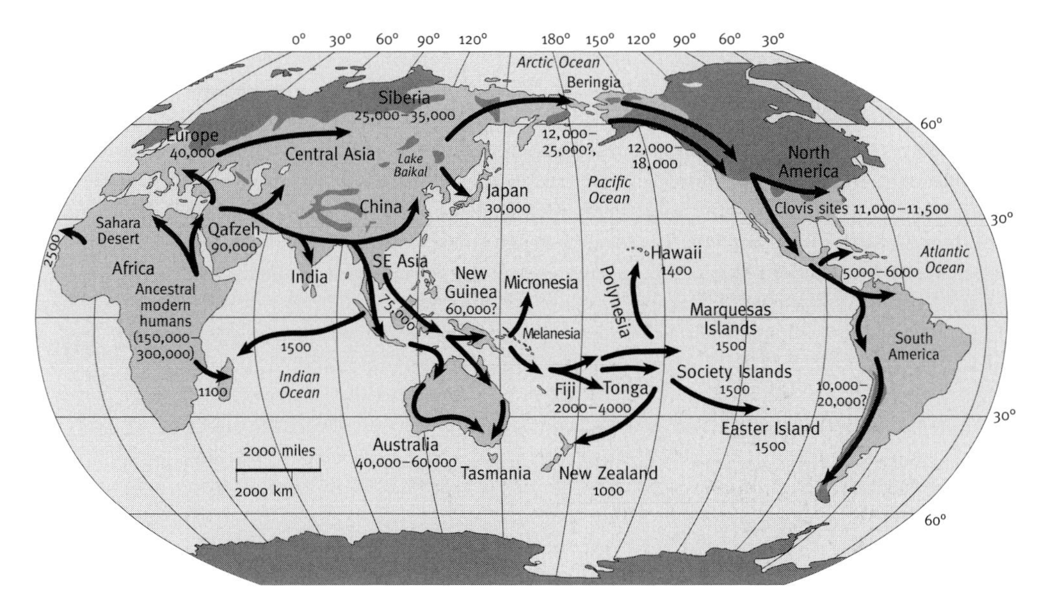


**Fig. 1.10** The spread of humans over the world's surface. Values are the number of years that humans have been in a place; for example, humans arrived in Australia 40,000–60,000 years ago. (After Gamble 1994, from Brown & Lomolino 1998.)

and the Americas were colonized during the last glacial period, when low sea levels eliminated or reduced the water gaps that had set a limit to earlier spread. Larger water gaps remained a barrier until the development of improved boating technology within the last few thousand years. Madagascar was the last major tropical land mass to be reached by humans, a mere 1500–2000 years ago. Some smaller, more isolated tropical islands remained uninhabited until the last few hundred years.

The presence of humans on a continent does not necessarily mean that they occupied the rain forest. Dense tropical forests are one of the least attractive environments for human occupation, since only a small proportion of the edible plant and animal material is accessible from ground level. Heavy rainfall washes away soil and mineral nutrients, requiring specialized techniques for agriculture, such as shifting cultivation and tree farming. Tropical rain forests may, however, have appeared more attractive before the first human hunters eliminated the most vulnerable ground animals, and recent archeological work has tended to push back the dates of first occupation (Mercader 2003). Indeed, it is possible that an earlier emigrant from Africa, *Homo erectus*, occupied rain forests in Southeast Asia a million or more years before the arrival of modern humans, although there is no direct archaeological evidence for this (Semah et al. 2003).

Surprisingly, perhaps, human impact on rain forest faunas is greater the more recently the first humans arrived (Martin & Steadman 1999). African and Asian rain forests still retain their original megafaunas: elephants are still found in both African and Asian forests, and rhinoceroses are still present in the rain forests of Asia, although they are restricted to nonforest environments in Africa. The largest animals, however, were eliminated from areas that were colonized later, most recently and dramatically from Madagascar. This counterintuitive result probably reflects the ability of African and Asian animals to adapt over millennia to the gradual improvements in human hunting techniques and technology, whereas the humans who invaded the other rain forest regions were armed and dangerous upon arrival.

Although hunters were the first humans to enter rain forests, evidence for cultivation, in the form of charcoal layers, broken pottery, crop remains, and modified soils, shows that most rain forest regions have also supported agricultural populations for millennia. In Amazonia, pre-Columbian agriculture left scattered patches of rich, black soil, known as *terra preta do Indo* (Indian dark earth), which are still prized by farmers for their fertility (Mann 2002). The denser, settled populations that agriculture permits, in turn, increase the hunting pressure on the local wildlife. It is no longer just the large animals that are in danger. In recent decades, the availability of modern weapons and a ready market for meat among expanding rural and urban populations has led to massive overhunting in many rain forest regions, threatening numerous species that had survived millennia of subsistence exploitation (see Chapter 8) (Milner-Gulland & Bennett 2003).

## Origins of the similarities and differences in rain forests

The similarities and differences among the tropical rain forests in different regions can be explained in a variety of ways. These explanations can be grouped into two major types: "ecological" explanations relate the similarities and differences between rain forests to similarities and differences between their present-day environments, while "historical" explanations relate them to events that happened in the past (Crisci et al. 2003). Each of these major types of explanation can, in turn, involve a huge range of possible factors; for example, soil nutrients and rainfall seasonality are ecological factors, while the movement of tectonic plates, changes in climate and sea level, and past human impacts are all historical factors. Further complications arise when ecological and historical factors interact as, for instance, when past human impacts have been concentrated on the most fertile soils.

To distinguish between the many possible explanations for the distribution of a particular group of organisms, we need to know several things. First, we need to know the pattern of branching of lineages during evolution of the group—its phylogeny. The phylogeny for the primates, for instance, tells us that New World primates arose as a branch of the Old World primates (see Chapter 3), while that for army ants shows that the New and Old World species evolved from a common ancestor (see Chapter 7) (Brady 2003). Second, we need to know the timing of these branching events in relation to the availability of dispersal routes between rain forest regions. If, as current evidence suggests, the New World primates branched from their Old World ancestors around 30 million years ago, then they must have crossed the sea to reach the Americas since no land route was available during this period. In contrast, the split between the two main army ant lineages appears to be old enough for it to have occurred while there were still dryland connections between Africa and South America. Finally, we need to know the pattern of extinctions. Today, only New World rain forests have scavenging vultures, because the unrelated Old World vultures are confined to open habitats (see Chapter 5). It would be tempting to attribute this difference to the inability of the forest-adapted New World vultures to cross the oceans to Africa and Asia, were it not for fossil evidence that both groups of vultures were once present in the Old World (Houston 1994).

In the past, most discussions of these issues were largely speculation. In the last few decades, however, molecular techniques have greatly improved our understanding of the phylogenies of major plant and animal groups (e.g. APG 2003; Barker et al. 2004; Springer et al. 2004). These techniques can also provide approximate dates for branching events, by counting the number of mutations that have accumulated since, although this "molecular clock" must be calibrated from the very incomplete fossil record (Magallón 2004). The DNA evidence so far has provided a surprising result: most lowland rain forest species examined appear to have evolved prior to the Pleistocene glaciations of the last 2–3 million years, indicating that these biological communities are very old and speciation processes are slower than expected (Moritz et al. 2000). This new understanding of phylogeny has coincided with the equally striking advances in our understanding of the Earth's history that have come from developments in plate tectonics and paleoclimatology (the study of past climates) (Crisci et al. 2003). Together, these developments have given us new insights into rain forest communities.

### Similarities

Some similarities between rain forests in different parts of the world, such as the many shared families and genera of plants, may be inherited from ancient Gondwana. For a group of organisms to have reached all the major parts of Gondwana before it broke up, it would need to have originated at least 130 million years ago, but land connections persisted longer between some fragments than others, so such regions would be expected to have more similarities than areas that broke away completely at an earlier time (McLoughlin 2001). Moreover, dispersal between the fragments of Gondwana would have been relatively easy while they were still close together in the late Cretaceous and early Tertiary. A Gondwanan origin has therefore been suggested at one time or another for many groups of rain forest organisms, with a ride north on India providing a plausible route into tropical Asia. However, very few of these suggestions have been backed up by evidence—from fossils or molecular clocks—that the organisms involved existed early enough to take advantage of these opportunities. One recent exception, mentioned above, is the molecular evidence that the army ants of the Old and New Worlds had a common origin in Gondwana around 105 million years ago (see Chapter 7) (Brady 2003).

In the early Tertiary, the extension of frost-free climates to much higher latitudes than at present provided an alternative, northern, route between some of the rain forest regions. Around 50 million years ago, during the early Eocene, tropical forests grew at latitudes that today would be considered temperate, and a land bridge across the North Atlantic via southern Greenland provided a frost-free link between western Eurasia and North America (see Fig. 1.7c) (Morley 2003). Many rain forest plant genera are old enough to have used this route. Cooling at the end of the Eocene broke this link forever by making the high latitudes too cold for tropical organisms.

Other similarities undoubtedly reflect later dispersal events as the Gondwanan fragments approached, and were eventually joined with, the northern continents:

first India (by 40 million years ago), then Africa (20 million years ago), and finally South America (2–3 million years ago). Although there is still no dry-land connection between Australia/New Guinea and Southeast Asia, the largest water gap that an organism would have needed to cross during the last period of low sea level was less than 70 km (45 miles), making dispersal relatively easy for some types of organisms. Dispersal across much larger water gaps is rare, but not impossible, particularly for plants, and its significance in explaining current distribution patterns may have been underestimated (Givnish & Renner 2004).

Many other similarities between rain forest regions are not a result of shared ancestry at all, but of convergent evolution—the development of similar adaptations by unrelated organisms because they inhabit similar environments. Similarities resulting from convergent evolution have received a lot of attention in the past, but they are often superficial, as we shall see later in the book. Common ancestry and convergence are not mutually exclusive explanations for similarities, since the more closely related two organisms are, the more likely they are to evolve similar adaptations to similar environments.

## Differences

The main theme of this book is not the similarities, but the many and important differences between rain forest regions. Such differences could have arisen for a variety of reasons, which are often impossible to disentangle. The simplest explanation for many of the biological differences is the ecological one: that they are a response to the differences in the physical environments of the various regions that have been outlined already. Even closely related organisms may evolve divergently in each region because the rain forest environments differ. For the same reason, some types of organisms may be able to invade and diversify in the rain forest in one region but not in another. The major differences among regions in the amount and seasonal distribution of rainfall have been called upon to explain many of the differences among rain forest regions. Rain forest organisms themselves also form an important part of the environment, and late arrivals may be unable to establish or diversify if their potential niches are already occupied. Thus the absence of specialist leaf-eaters among the New World primates may reflect their late arrival in a continent that already had leaf-eating sloths (see Chapter 3).

The different plate tectonic histories of the rain forest regions can also explain many of the differences in which groups of organisms are present or absent. Particular groups of plants and animals are shared between regions only if these regions were once connected or dispersal between them was once possible. Without the land bridge provided by the Isthmus of Panama, for instance, the rain forests of South America would be even more distinctive in comparison to rain forests elsewhere than they are now. Chance may play a major role here, particularly when dispersal on or over the open sea is involved. In many cases, the presence of a particular group of organisms in a region appears to be the result of a single, very unlikely event. The presence of primates in South American and Madagascan rain forests, and their absence in New Guinea, can perhaps be explained in this way.

Extinction is another cause of differences, and one that is often difficult to detect. Changes in climate, as a result of plate movements, mountain uplift, or global climate change, may eliminate sensitive organisms from one region while they survive in another. Past human impacts are another, often overlooked, source of differences between regions. In the Neotropics, Madagascar and New Guinea, the "megafaunas"—the very large animals—were largely eliminated by the first human arrivals, long before their ecological roles had been documented (Martin & Steadman 1999). In most cases, it is not even known if they inhabited rain forests.

Given time, evolution may fill the niches left vacant by the initial absence or subsequent extinction of a particular group of organisms. But evolution is a gradual process and can only act on the organisms that are present in the region. Particular niches may remain unfilled for a long time, or be filled in very different ways. When this happens, rain forests may differ not only in their biotas but also in the ways in which they function.

### Functional consequences

The major question about tropical rain forests that we attempt to answer in this book is: why is what where? This is a very traditional approach to ecology, but one that has benefitted greatly from the recent scientific advances outlined above. An alternative or complimentary approach would be to look at the functioning of the whole rain forest ecosystem: such attributes as the production of biomass, the cycling of nutrients, and the rate at which these processes recover after natural and human disturbance. Combining the two approaches, we can ask what, if any, are the functional consequences of the observed differences between the biotas of the different rain forest regions? Does the presence of fungus-growing termites in Old World rain forests affect nutrient cycling? Does the absence of primates from New Guinea affect seed dispersal?

The answer to the great majority of such questions has to be that we do not know. We can and do speculate, but identifying functional differences requires comparisons between sites that have been carefully matched for the major environmental factors. These comparisons have not yet been made. Identifying which of the numerous biological differences between regions are responsible for particular differences in function will require the experimental removal or addition of the organisms in question. In some cases, such experiments could be done quite easily (e.g. the exclusion of browsing herbivores from an area of rain forest), in others they have already been done by accident (e.g. the introduction of honeybees to tropical America), while many more would be too dangerous to carry out in practice and should remain forever as "thought experiments" (e.g. the introduction of leaf-cutter ants to the Old World).

## Many rain forests

For the reasons outlined above, the tropical rain forests of each region have distinctive characteristics and elements that give each a quality all its own. The Neotropical rain forest is the most extensive, most diverse, and in many ways the most distinctive. The richest Neotropical rain forest sites have more tree species (see Chapter 2), more bird species (see Chapter 5), more bat species (see Chapter 6), and more butterfly species (see Chapter 7) living together than rain forests elsewhere, and the same pattern is found in many, but not all, other groups of organisms. The effects of South America's long isolation have not been erased by the influx from the north after the formation of the Panama land bridge, and many characteristic groups of plants and mammals are found in no other rain forest region. The epiphytic plant family Bromeliaceae gives an unmistakable appearance to the forest and their water tanks provide a unique canopy resource that is exploited by numerous species (see Chapter 2). Hummingbirds (see Chapter 5) and the flowers that they pollinate (see Chapter 2) show a degree of evolutionary diversification that is unparalleled in other rain forests, while other New World endemic groups of birds dominate the insectivore, frugivore, and scavenging niches. Both primates (see Chapter 3) and caviomorph rodents (see Chapter 4) diversified along very different lines in the Neotropics from their ancestors in the Old World, with giant, long-legged rodents partly filling niches occupied by ungulates in Africa and Asia. The fruit bats (see Chapter 6) in Neotropical rain forests are an entirely separate evolutionary radiation from the fruit bats in all other rain forests, with different flight, sensory, and fruit-processing capabilities. Long columns of leaf-cutter ants (see Chapter 7) bringing cut wedges of leaf back to their underground nests are another distinctive feature of Neotropical rain forests that has no equivalent elsewhere.

African rain forests could hardly be more different. They are mostly drier, lower, and more open than rain forests elsewhere and have a relatively less diverse flora (see Chapter 2), apparently as a result of both present and past climates. Diversity is high in some other groups, however, including the primates (see Chapter 3) and termites (see Chapter 7). Perhaps the most distinctive feature of the African rain forests is the abundance and diversity of large, ground-living mammals, including many species of primates (see Chapter 3) and terrestrial herbivores (see Chapter 4). The African elephant is the largest of all rain forest mammals and the gorilla by far the largest primate. The bird fauna shares most major groups with Asian rain forests, but there is an endemic group of frugivores, the turacoes (see Chapter 5).

Most Asian rain forests can be characterized as "dipterocarp forests", because they are dominated by large trees in the family Dipterocarpaceae (see Chapter 2). Many dipterocarps are among the tallest trees in any rain forest. Probably because of this dominance by a single family, Southeast Asian dipterocarp forests show a unique pattern of mass flowering and fruiting at 2–7-year intervals, described in more detail in Chapter 2. The short periods of "feast" separated by long periods of "famine" for any animal that eats flowers, fruits, or seeds, appears to shape the whole ecology of the forest. Another peculiar feature of these forests—the abundance and diversity of gliding animals (see Chapter 6)—may be connected to this phenomenon.

The rain forests of New Guinea are a paradox, with a basically Asian flora (see Chapter 2), but a very un-Asian fauna (see Chapters 3 and 4). This is the only rain forest region without primates or placental carnivores. Indeed, bats and rodents are the only placental mammals, and marsupials occupy most other mammalian niches. New Guinea and Australia are also notable for several unique radiations of birds, including families such as the cassowaries, birds of paradise, and bowerbirds that could probably not have survived predation by placental carnivores.

The assembly of Madagascar's unique rain forests appears to owe a great deal to chance. The entire nonflying mammal fauna can be explained by four colonization events, involving, respectively, a single ancestral species of lemur (see Chapter 3), a mongoose-like carnivore (see Chapter 4), a rodent, and an insectivore. Most of the birds also represent endemic radiations from a very small number of initial colonizers (see Chapter 5) and the same pattern will probably be shown when modern molecular techniques are brought to bear on other groups of organisms.

## Conclusions

In this first chapter, the major rain forest regions of the world have been introduced. The rain forests differ in their biogeographical histories and in both their past and present environments. Most notably, the presence or absence of particular groups of animals and plants gives each region a distinctive character. However, the purpose of this book is not merely to list differences in biogeography, environment, and species, but to show how the differences impact on evolutionary and ecological relationships. And, in the end, the goal is to bring together the accumulated insights to examine how conservation strategies might use this information.

In the next chapter, plant communities are examined first, as plants represent the building blocks of the biological community. Subsequent chapters will consider animal communities.

## Further reading

Coates A.G. (1997) *Central America: a Natural and Cultural History.* Yale University Press, New Haven, CT.

McGregor G.R. & Nieuwolt S. (1998) *Tropical Climatology: an Introduction to the Climates of Low Latitudes,* 2nd edn. John Wiley & Sons, Chichester, UK.

Morley R.J. (2000) *Origin and Evolution of Tropical Rain Forests.* Wiley, Chichester, UK.

Richards P.W. (1996) *The Tropical Rainforest: an Ecological Study,* 2nd edn. Cambridge University Press, Cambridge, UK.

Terborgh J. (1992) *Diversity and the Tropical Rain Forest.* Scientific American Library, New York.

Whitmore T.C. (1998) *An Introduction to Tropical Rain Forests,* 2nd edn. Oxford University Press, Oxford.

Chapter 2

# Plants: Building Blocks of the Rain Forest

In a book about rain forest communities, it makes sense to cover plants before animals because plants are the foundation of any rain forest community. They provide the structure to the forest through the growth of trees, shrubs, vines, and herbs, as well as epiphytes—nonparasitic plants that grow perched on the branches of trees. The diversity of plants also provides a huge variety of food sources for the animal community that feeds on the flowers, fruits, seeds, leaves, twigs, bark, roots, and other plant parts. With a wide variety of plants, it is possible for many animal species to coexist by specializing on individual plant species or groups of related species. One beetle species, for example, could feed exclusively on the roots of ginger plants, while another species could live on the young leaves of a passionflower vine. The relationship between rain forest plants and animals is by no means one way, however. As we will see in the following chapters, most rain forest plants depend on animals to pollinate their flowers and disperse their seeds, and many have defensive or nutritional relationships with ants.

It would seem logical to suggest that if plant communities differ among rain forest regions, for historical or environmental reasons, we should expect to see corresponding differences in animal communities. Such differences between the plant communities do exist, both in the structure of the forest and in the plant species present. Among the most striking differences are the abundance of dipterocarp tree species in Asian forests and the proliferation of epiphytic bromeliad species in the Americas (Fig. 2.1; see Plate 2.1, opposite p. 150). Both dipterocarps and bromeliads have profound effects on the animal communities in their respective rain forests, some of which are discussed below and some in later chapters. We might also predict that more diverse plant communities would support more diverse animal communities. Here the correlation is less clear: the rain forests of central and western Amazonia are exceptionally rich in both plants and animals, but African rain forests combine a relatively low diversity of plant species with a high diversity in some groups of plant-dependent animals, including primates (see Chapter 3), squirrels, and terrestrial herbivores (see Chapter 4). Madagascar's rain forests, in contrast, are rich in plant species but relatively poor in animal species. In this chapter, we will describe the plant communities of each region, and in subsequent chapters, the role of these plant communities in structuring animal communities will be explored.

(a)

(b)

**Fig. 2.1** Two families that characterize a particular rain forest region. (a) Bromeliads (on the right), growing here on a tree branch, are a prominent feature of Neotropical forests. (Courtesy of Dale Morris, Ecuador.) (b) Giant dipterocarp trees, such as this one in Borneo, dominate the forests of Southeast Asia. (Courtesy of Richard Primack.)

## Plant distributions

All rain forests share the same basic growth forms of trees, shrubs, herbs, climbers, and epiphytes. In addition, they share many of the same dominant plant families, such as the legumes (Fabaceae) and the spurge family (Euphorbiaceae). The presence of the same dominant families in tropical forests throughout the world, despite continental separation for tens of millions of years, suggests that there is a conservative quality to these basic building blocks of the forest community, but the reasons for this are not at all obvious.

In general, the most ancient land plants have the most cosmopolitan distributions. Ancient plant families have not only had a longer time in which to spread around the world, but as discussed in the previous chapter, there also were overland dispersal routes between rain forest regions available in the Mesozoic and early Tertiary that did not exist in later epochs (Morley 2003). Among the major flowering plant families found in all the major rain forest regions, some, such as the Annonaceae (the soursop family), Arecaceae (the palm family), Lauraceae (the laurel family), and Myristicaceae (the nutmeg family), may be old enough to have spread to all the southern land masses while they still formed part of Gondwana, and to have ridden north to Asia on India (Chanderbali et al. 2001; Sauquet et al. 2003). Even in these families, however, some shared or related genera probably result from much later dispersal events (Pennington & Dick 2004).

The presence of plants on oceanic islands that have never had overland connections shows that many species can also disperse long distances, even over water. This is true for species with varying types of seeds, including light, fluffy, wind-dispersed seeds (such as members of the Asteraceae, the sunflower family), dust-like seeds (such as orchids, Orchidaceae), small seeds in fleshy fruits that could be retained inside the guts of long-distance seafaring birds (such as members of the Aquifoliaceae, the holly family), or floating seeds (such as some Arecaceae and Fabaceae). Sometimes whole groups of plants and animals may be dispersed when a tangled mass of trees and vines falls into a rain-swollen river and is swept out to sea. The ability of plants to disperse long distances over water is best illustrated by the Hawaiian Islands. These islands are 3900 km (2500 miles) from the nearest continent (North America), yet at least 272 flowering plant species have reached these specks of land in the vast Pacific Ocean, establishing new populations, and in many cases undergoing extensive speciation (Wagner et al. 1990).

Large expanses of water—as well as deserts and mountain ranges—are, however, barriers to the dispersal of many rain forest plant species. The rain forests of the Hawaiian Islands have no representatives at all from many important rain forest plant families, including tree families such as the Annonaceae, Bignoniaceae (the catalpa family), Meliaceae (the mahogany family), and Myristicaceae, and herb families such as the Araceae (the aroid family) and Zingiberaceae (the ginger family). Other major rain forest families, such as the Lauraceae and Sapotaceae (the chicle family), are represented by only one or two species. There are also no figs (*Ficus*), despite their small-seeded fleshy fruits, only three species of orchids, despite their tiny wind-blown seeds, and only a single genus of palms. Hawaii is an extreme case, but the distributions of several major rain forest families, many genera, and almost all species show that the dispersal of plants between rain forest regions has been limited. Most such dispersal seems to have occurred when there were overland connections with an appropriate climate (Morley 2003), but the role of long-distance dispersal across oceans may have been underestimated (Givnish & Renner 2004; Pennington & Dick 2004). Some of the restricted plant families have major ecological importance and affect many other species in the community, such as the bromeliads in New World forests. In other cases, such as the dipterocarps (Dipterocarpaceae), a family is widespread but has become important in only one region.

From a floristic point of view, the rain forests of the Neotropics are the most distinctive: a pattern that we will see repeated in several animal groups in the following chapters. Not only do Neotropical rain forests have several important plant families (e.g. Bromeliaceae, Cactaceae (cactus family), Cyclanthaceae (Panama hat family), and Vochysiaceae) and many important genera that have few or no representatives in rain forests elsewhere, but they also lack some families (e.g. Pandanaceae (screw pine family)) and genera that are important in most other rain forests. In comparison with the distinctiveness of the New World, the floras of Old World rain forests have more in common with each other. African rain forests are distinguished more by the relative poverty of their flora than by unique floristic elements. Madagascar has a very rich rain forest flora that has both African and Asian affinities, as well as unique features resulting from its prolonged isolation. The diverse flora of New Guinea's rain forest, in contrast, is relatively young and surprisingly similar at the generic and family levels to that of Asia. This similarity has encouraged botanists to treat the whole region from Southeast Asia to New Guinea as a single floristic unit, often called Malesia, with the main floristic boundary between New Guinea and Australia. This is in striking contrast to the vertebrate faunas considered in later chapters, where there are major differences between Asia and New Guinea and major similarities between New Guinea and Australia.

## Rain forest structure

An alternative to comparing the families, genera, and species of plants present in different rain forests is to compare their structure, including such characteristics as canopy height and the number of trees per hectare. An additional commonly used characteristic is the basal area of trees, that is the total area of the tree stems in a hectare, measured at 1.3 m (4.3 feet) height on the trees. Such comparisons are not as simple as might be expected, since different researchers have used different methods and made different measurements in various parts of the tropics. The most easily comparable data come from studies that describe the characteristics of a forest by censusing every tree in research plots, which are typically 0.1 or 1 ha (0.24 or 2.4 acres) in area. Large numbers of such studies have been published, and 46 have recently been gathered together and synthesized (de Gouvenain & Silander 2003). The main conclusions of this survey are that the typical dipterocarp forests of Southeast Asia are taller than rain forests elsewhere, followed in decreasing height by typical rain forests in South America, Africa, and Madagascar. There were not enough data available to include New Guinea and Australia in this comparison. The differences are substantial. Lowland rain forests in Southeast Asia typically have a tree canopy at 30–50 m (100–165 feet) above the ground, with emergent trees reaching 70 m or more; while, at the other extreme, most lowland rain forests in Madagascar are less than 30 m tall. African rain forests contrast with rain forests elsewhere in tending to have a lower density of trees. Across the range of densities, which vary from 300 to 1000 trees per hectare, African rain forests tend to be on the lower end of the scale, in the range of 300–600 trees per hectare. Despite the low density of trees, African forests do not have an especially low basal area of trees, suggesting that these forests are lowest in the density of small trees, which have less of an effect on total basal area.

The differences in canopy height between rain forests are largely attributed by de Gouvenain and Silander (2003) to the occurrence of tropical cyclones (see Chapter 1), which reduce the advantages that might otherwise accrue to a tree that is taller than its neighbors. All their Madagascan sites were subject to cyclones, as were particularly low forests in the Caribbean, Australia, and the Philippines. In contrast, the exceptionally tall forests of Malaysia and Indonesia are all outside the cyclone belt. Additional environmental factors are also likely to be important in some areas. Exceptionally poor or shallow soils reduce canopy height all over the tropics. It is also striking that the tallest forests are almost all dominated by trees in the family Dipterocarpaceae (see below), and it is possible that dipterocarps simply have an exceptional capacity for height growth.

Structural differences between forests would be expected to have an impact on the communities of animals that climb or fly through the forest, although there has been no systematic study of this in the tropics. Tall forests may provide more opportunities for animals to specialize on a particular layer in the forest, with distinct animal communities in the understorey, subcanopy, canopy, and emergent layers. The density of trees and the degree to which adjacent trees are connected by overlapping crowns and woody climbers may also be important. High density and connectivity would be expected to make it easier for climbing animals to move through the forest, while providing more obstacles for flying animals. The effects of differences in structure will be most easily observed when adjacent forests in the same area are compared, while comparisons between rain forest regions are confounded by many other differences. One possible consequence of the distinctive structure of Southeast Asian dipterocarp forests for animal movement is considered in Chapter 6.

## How many plant species?

Tropical rain forests are by far the most species-rich of all terrestrial ecosystems. Despite occupying only about 6% of the Earth's land surface area, it is thought that they support more than half of the total species of land plants and animals (Whitmore 1998). Turner (2001a) estimates that tropical rain forests worldwide support 175,000 species of vascular plants (flowering plants, conifers, and ferns), which is around two-thirds of the estimated global total. These rain forest plant species are divided in an approximately 3 : 2 : 1 ratio between the Neotropics, the Asia–Pacific region (Asia, New Guinea, Australia, and the Pacific islands), and Africa plus Madagascar (Table 2.1). Within the Asia–Pacific region, the greatest numbers of species are found in the wettest and least seasonal areas, centered on the Sundaland region of Southeast Asia and the island of New Guinea. New Guinea may well have the richest flora in the region, but it is also the least well-collected and new species continue to be added by every expedition. Outside these everwet areas, in drier and more isolated parts of the region, the richness of species declines dramatically.

**Table 2.1** The estimated numbers of plant species in the tropical rain forests of the world.

| Area | Number of species | Area of rain forest (millions of ha) | Comments |
|---|---|---|---|
| *Neotropics* | 93,500 | 400 | Mainly Amazon basin; also includes some other forest types |
| *African tropics* | 20,000 | 180 | |
| African mainland | 16,000 | | Includes West Africa, Congo region, and montane areas |
| Madagascar | 4,000 | | A preliminary estimate; the flora is not well known |
| *Asia–Pacific region* | 61,700 | 250 | |
| Malesia | 45,000 | | Malaysia and Indonesia, including New Guinea |
| Indo-China and adjacent areas | 10,000 | | An estimate as no flora has been completed for the region |
| Southern India | 4,000 | | Mainly the Western Ghats |
| Sri Lanka | 1,000 | | Sinharaja and neighboring localities |
| Australia | 700 | | Queensland rain forest |
| Pacific islands | 1,000 | | Isolated rain forests with low species richness |
| Total | 175,200 | | |

Modified from Turner (2001a) and Whitmore (1998).

On a smaller scale, we do not have a complete species list of all the plants and animals for any area of rain forest, and complete lists of just the plants are rare. A survey of a 1 ha plot (100 m × 100 m, somewhat larger than a typical soccer pitch) in the exceptionally diverse rain forest at Cuyabeno, Ecuador, in western Amazonia, found a total of 942 vascular plant species, of which about half were trees and the remainder were divided among shrubs, climbers, ground herbs, and epiphytes (Balslev et al. 1998). Most other studies have only looked at the trees. The largest set of comparable data from around the world concerns the number of trees species more than 10 cm (about 4 inches) in diameter found in a 1 ha plot (ter Steege et al. 2000, 2003; Turner 2001a; Ashton et al. 2004). There are still huge areas of tropical rain forest for which there are no plot data, so although some general patterns are clear, others will undoubtedly change as more data become available.

The most diverse tropical rain forest plots so far, with more than 250 trees species in a single hectare, are in central and western Amazonia, in the Pacific coast rain forests of Chocó Province, Colombia, in the Atlantic Coastal Forest of Brazil, and in Sarawak, Malaysia. Plots with more than 200 different tree species have also been found on the islands of New Guinea and Madagascar, but not—so far—in Africa (Fig. 2.2). The most species-rich African sites, as well many sites in Amazonia, Sundaland, and New Guinea, have 100–200 tree species per hectare.

Outside the core areas of each rain forest region, tree diversity is generally lower. Hence, in the Neotropics, forests in eastern Amazonia, the Guiana Shield (between the Amazon River and the mouth of the Orinoco), most of the Brazilian

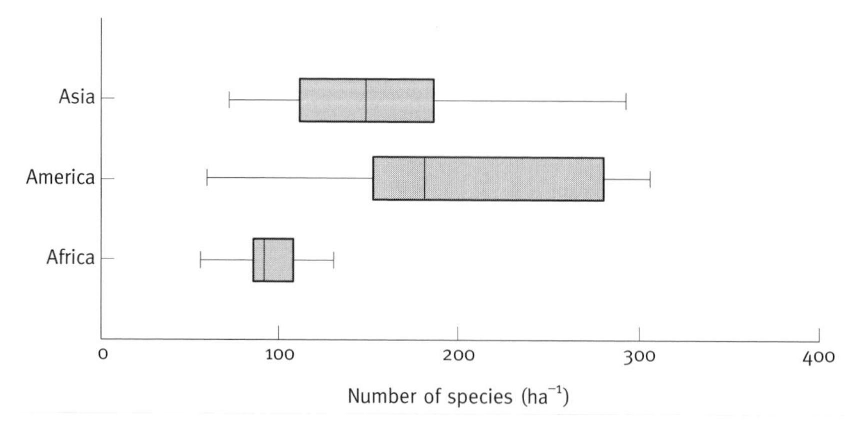


**Fig. 2.2** The number of tree species per hectare in rain forest sites in Asia, the American tropics, and Africa. The line in the box is the mean value, the horizontal line covers all values, and the box includes 50% of the values above and below the mean. (Modified from Turner 2001b, by extending the range from 270 to 290 species for Asian plots.)

Atlantic region, Bolivia, and Central America, usually have fewer than 100 species per hectare, although there are some exceptions. Similar declines in diversity with distance from the core area occur in other rain forest regions, but in tropical Asia even the northernmost rain forest sites studied—in the Himalayan foothills of northeast India (Proctor et al. 1998) and in southwest China (Zhu 1997)—have more than 100 species per hectare. The isolated Asian rain forest areas of the Western Ghats and Sri Lanka, in contrast, have fewer than 100 species in a hectare, while a plot in the northern Philippines has 99 (Ashton et al. 2004). An annual rainfall of less than 2000 mm (80 inches) or a severe dry season reduces tree diversity, as do periodic flooding and extreme soil types such as white sand, limestone, or peat. In the Amazon region, dry season length (i.e. the number of months with less than 100 mm of rainfall) is an excellent predictor of the maximum plot diversity within an area—with the highest diversity plots in areas with only 0–1 dry months—but a poor predictor of average plot diversity, since there are low diversity plots in all areas (ter Steege et al. 2003). The history of a rain forest area is probably as important as the current environment, particularly at the regional and global scales, with the highest diversities in areas where rain forest plant species have had a place to persist through the cool, dry glacial episodes.

A network of much larger, 16–52 ha, plots has now been established at 16 forest sites across the tropics by the Center for Tropical Forest Science (CTFS) of the Smithsonian Tropical Research Institute (Condit 1998; Losos & Leigh 2004). Within these plots, all plants more than 1 cm in diameter have been mapped, measured and identified: with more than 300,000 individual plants in the largest rain forest plots! Periodic re-censusing of the plots gives information on growth, mortality, and regeneration. Data from these plots are now being analyzed and will give greater insight into rain forest structure and dynamics. The results published so far show that species richness of large trees in single hectares plots is, in general, a good predictor of total tree species richness in these larger areas. African rain forests are an exception, with misleadingly low diversities at scales of 1 ha due to the tendency of one or several species to dominate a local area, but with more respectable diversities at scales of 10 ha and over. However, no African site approaches the very high diversities found in Amazonia and Southeast Asia, where the richest plots have more than 800 tree species greater than 10 cm diameter and more than 1000 species greater than 1 cm diameter in 25 ha.

To put the plant diversity of tropical rain forests in perspective, even the most species-rich temperate forests have fewer than 25 tree species in a 1 ha plot, and most have fewer than 10. In the richest tropical rain forests, on average every second tree is a new species, while many temperate forests are dominated over large areas by just one or two species. So even the "common" species in tropical rain forests may be "rare" by the standards of temperate forests, with important consequences for the animals that feed on them.

Low-diversity forests do occur in the wet tropics, but in most cases they reflect either extreme soil conditions, such as the "alan" peat swamp forests of Borneo dominated by *Shorea albida*, or recent disturbance. The young forests that develop after natural or human disturbances are sometimes composed almost entirely of a single successional species, often a species of *Macaranga* in Southeast Asia, *Cecropia* in tropical America, and *Musanga cecropiodes* or *Trema orientalis* in Africa. Large areas dominated by a single tree species do sometimes occur, however, on normal soils and with no evidence of recent disturbance. These seem to be particularly common in Africa (see below), particularly rare in Asia, and occasional in South America (Connell & Lowman 1989; Hart 1990). In South America, dominance by one or a few tree species is most common in the Guiana Shield region, where—as in Africa—tree species diversity is relatively low even in "normal" forests (Henkel 2003).

## Widespread plant families

### Pantropical families

Except on the most remote oceanic islands, the tree flora of any area of tropical rain forest is likely to include members of the following families: Annonaceae (the soursop family), Arecaceae (or Palmae, the palm family), Euphorbiaceae (the spurge family), Fabaceae (or Leguminosae, the legume family), Lauraceae (the laurel family), Meliaceae (the mahogany family), Moraceae (the fig family), Myristicaceae (the nutmeg family), Rubiaceae (the coffee family), and Sapotaceae (the chicle family) (Heywood 1993; Turner 2001a). There are also many genera that are shared by different continents, but very few species. One unusual example is the tree *Symphonia globulifera*, which is widely distributed in both African and American rain forests. The detailed fossil record of *Symphonia* has permitted a DNA-based reconstruction of its African origin, some 45 million years ago, and its transoceanic colonization of Central and South America 15–18 million years ago (Dick et al. 2003).

The continued success of the common rain forest families over tens of millions of years suggests that each has some special feature or features that gives it an advantage in competition with other plants in the forest. It is tempting to argue, for instance, that this advantage for the legumes is the nitrogen-fixing symbiosis with bacteria in the family Rhizobiaceae, but this symbiosis is neither universal in the family nor confined to it, and members of the most important rain forest subfamily, Caesalpinioideae, generally lack it (Perreijn 2002; Doyle & Luckow 2003). The features that give other successful families their ecological advantage are even less obvious, although most families have distinctive characteristics that are found in all or most species. The Lauraceae, for example, is an ancient, relatively unspecialized family, with small flowers and often small- to medium-sized berries or drupes. The family is notable for the presence of oil cavities throughout the leaves, bark, and other plant parts, which gives the plants a highly aromatic quality; well-known members of the family are laurel and cinnamon.

Other widely distributed tree families that are common in the rain forest include the Euphorbiaceae, Rubiaceae, Annonaceae, and Myristicaceae (Fig. 2.3). The Euphorbiaceae, in its traditional sense, has the greatest diversity of any tree family in Asian forest forests, and is also common in rain forests elsewhere. Recent classifications divide the family into several, separate, more or less closely related families, of which the largest are the Euphorbiaceae in a strict sense, with around 6000 species, and the Phyllanthaceae, with 1725 species. Trees and shrubs in the Euphorbiaceae can often be often recognized by their alternate, simple leaves with a long leaf stalk and often with a three-lobed fruit. Many members of the family have a milky sap, most notably the para rubber tree, *Hevea brasiliensis*. Species in the Phyllanthaceae, in contrast, lack latex.

The Rubiaceae is a large family with over 7000 species of mostly trees and shrubs. It is often the most diverse tree family in African forests, although it is also diverse and abundant elsewhere. The family is readily recognized by having opposite leaves with stipules (small triangular flaps) between the leaf stalks, and flowers often in dense inflorescences. Important members of the family include the cultivated coffee plant and many ornamental shrubs, such as *Gardenia*. The Annonaceae is a large family of around 2000 trees and shrubs, characterized by the metallic sheen of the leaves, the often sweet-smelling, stringy bark, and often aggregate fruits in which individual berries have fused together. The Myristicaceae is recognized in the field by its whorls of horizontal branches and the red sap that oozes from wounds in the bark. The fruit of the family is distinctive: when the fruit matures, the covering splits apart to reveal a single large seed covered by a brightly colored tissue, the aril, which is eaten by large birds.

### Legumes

Among the shared families, the legumes stand out because they typically dominate the rain forests of both Africa and the Neotropics, in terms of basal area and overall biomass. In Madagascar, legumes dominate the dry forest types but are less important in the rain forest. The legumes are generally less important in New Guinea and Australia, while in tropical Asia the dipterocarps dominate (see below), but there are many prominent rain forest legume species in all rain forest regions. As well as producing many of the biggest rain forest trees, the legumes are also an important family of climbers. In contrast to most of the widespread plant families, the legume family does not appear to be particularly old and did not become prominent in the fossil record until the Eocene (Doyle & Luckow 2003). This suggests that their current pantropical distribution may be a result of dispersal across northern land routes during the early Eocene thermal maximum or transoceanic dispersal (see Chapter 1) rather than an ancient Cretaceous origin on Gondwanaland.

(a)

**Fig. 2.3** Some tropical plant families. (a) Annonaceae fruits (*Polyalthia rumphii*) from Malaysia. This cluster of stalked fruits is the product of a single flower with many free pistils. (Courtesy of K.M. Wong.) (b) Nutmeg fruit (*Knema* sp., Myristicaceae), showing the arillate seed, from Gunung Palung National Park, Indonesian Borneo. (Courtesy of Tim Laman.)

(b)

Most legumes are immediately recognizable by the combination of pinnately compound leaves (though some have simple leaves) with stipules (small triangular green projections at the leaf base) and a one-chambered pod with one or

(a)

(b)

**Fig. 2.4** Members of the legume family are easily recognized by their pod-like fruits. (a) A legume tree covered with fruit. (Courtesy of Mark Brenner, Yucatan, Mexico.) (b) Pods of the leguminous vine (*Entada monostachya*), hanging over water, in Panama. (Courtesy of Chris Dick.)

more seeds. Legumes vary considerably in flower morphology: many species of tropical rain forest legume trees have radially symmetric flowers, while other subfamilies produce distinctive, bilaterally symmetric flowers (a "pea" flower), in which there are five petals: two pairs of two petals, and a single petal, "the standard" (see Plate 5.4, between pp. 150 and 151). In most cases, the fruits look like pea pods or bean pods, though they are often much larger. The pods of different species vary widely in shape, color, and size, with some opening at maturity and others remaining closed (Fig. 2.4). The pods may be brightly colored, some may be very fleshy and green, and some have sweet tissue, while others are dry, turning brown or black.

A special feature of the legume family is that symbiotic, nitrogen-fixing bacteria in the family Rhizobiaceae inhabit "nodules" on the root systems of many species. These legumes can absorb nitrogenous compounds from these bacteria, giving them a potential competitive advantage; the nitrogen absorbed from the bacteria allows legumes to build extra enzymes and proteins that can be used for rapid growth. These nitrogen supplies can also be used to produce poisonous chemicals, such as alkaloids, that deter herbivorous animals (particularly insects) from eating the leaves, bark, young fruits, and other plant parts. Despite the potential importance of symbiotic nitrogen fixation by rain forest legumes, few species have been studied and it is not yet clear how widespread this capability is among leguminous trees and climbers. A recent survey in the Neotropical rain forests of Guyana found that most species in the subfamily Caesalpinioideae—the most important subfamily of legume trees in rain forests—did not have nitrogen-fixing nodules, while most members of the other two subfamilies, Mimosoideae and Papilionoideae, did have them (Perreijn 2002). Among those species with nodules, the dependence on symbiotic fixation of atmospheric nitrogen varied from close to 0% to near 100%. Overall, symbiotic nitrogen fixation made a significant contribution to the nitrogen budget of the forest, but the great majority of this contribution came from a small number of common, nodulated legume species. The results of this study, and other similar ones, have potential implications for the management of logging operations, which could be designed to minimize damage to known nitrogen-fixers and to maximize their rate of recovery.

Legumes made up 50–75% of the biomass in the rain forest in Perreijn's study, which is high by pantropical standards. Given that nitrogen is the nutrient required in greatest quantity by plants and is likely to be limiting for growth in many situations, there is an urgent need for comparative studies in other rain forests. What, for instance, is the impact on the forest nitrogen budget of the apparent substitution of dipterocarps for legumes in Southeast Asian rain forests?

### Figs

The Moraceae (the fig family) (Fig. 2.5) is another important pantropical rain forest family, recognized by the latex that is found throughout the leaves, bark, and even immature fruit. In many tropical Moraceae genera, such as the breadfruit and its relatives (*Artocarpus*), individual fruits are embedded in a large, fleshy receptacle to form a complex "multiple fruit". By far the most important genus in the family is the fig genus, *Ficus*, in which both the flowers and fruits are completely enclosed inside a fleshy receptacle—the fig. Fig plants have a wider range of growth forms than any other genus of plants: trees, shrubs, climbers, and epiphytes, as well as hemiepiphytes, which start as epiphytes but then send roots down to the ground, and the bizarre "stranglers", in which the aerial roots eventually enclose and kill the host, leaving a free-standing fig tree.

Although the fig genus is pantropical—indeed, it is the only genus found in all the CTFS plots—the diversity of species and growth forms differs greatly between the major rain forest regions. The more than 750 fig species are currently classified into six subgenera and 20 sections (Berg 2003; Jousselin et al. 2003). Only two endemic fig sections occur in the Neotropics: around 120 species of epiphytes, hemiepiphytes, and stranglers in the section *Americana*, subgenus *Urostigma*, and 19 species of free-standing trees plus one strangler in the section *Pharmacosycea*, subgenus *Pharmacosycea*. In contrast, the islands of Borneo and New Guinea each support at least 140 species, representing all six subgenera and 10 or more sections, with the full range of growth forms found in the genus. Moreover, the local diversity of fig species in rain forests in Asia and New Guinea is extremely high, with 75 species (including 27 hemiepiphytes) recorded in the Lambir Hills National Park in Borneo (Harrison et al. 2003; R.D. Harrison, personal communication). As a continent, Africa is intermediate in fig diversity, with 81 species in seven sections, but only around 50 species occur in the rain forest, many in the near-endemic section *Galoglychia* in the subgenus *Urostigma* (Berg & Wiebes 1992). Madagascar has only 24 species, of which around two-thirds occur in the rain forest. Local fig diversities in Madagascan rain forest also

(a)
(b)
(c)

**Fig. 2.5** (a) Fig tree (*Fiscus stupenda*), with fruits produced among the leaves, in Gunung Palung National Park, Indonesian Borneo. (Courtesy of Tim Laman.) (b) Hemiepiphytic fig (*F. sumatrana*) that grew using a neighboring tree for support, from Gunung Palung National Park, Indonesian Borneo (note the roots of the fig growing around the trunk of the tree). This species starts as an epiphyte in the tree crown and then sends roots down to the ground. (Courtesy of Tim Laman.) (c) Figs produced on a tree trunk of *F. racemosa* in India. (Courtesy of David Lee, taken at Fairchild Tropical garden.)

appear to be low, with only 12 species known from the Ranomafana National Park (Goodman & Ganzhorn 1997).

These differences in fig diversity between rain forest regions do not necessarily translate into differences in the role that figs play in the ecology of the forest. In all the rain forest regions, the largest crops of fig fruits are produced by the giant hemiepiphytic and strangler species, and there is little evidence to suggest that the density of these plants—which is low everywhere—varies significantly between regions. Fig plants are very important in supporting the vertebrate communities of tropical rain forests because ripe fig fruits are available throughout the year (Shanahan et al. 2001). This contrasts with most other tropical forest plants, where all individuals of a species fruit together at one time. At some times of the year, figs may be the only fruit available and are eaten by many birds and mammals. A single hemiepiphytic tree of *Ficus pertusa* at Cocha Cashu in the Manu National Park had its figs eaten by 44 species of diurnal vertebrates, ranging in size from 10 g (0.5 oz) manakins to 10 kg (22 lb) spider monkeys, over a 21-day period (Tello 2003). For this reason, figs have been termed "keystone species"—species upon which many other species depend for sustenance or habitat. Some animals depend on figs for most of their diet year round. In areas of Sumatra and Sulawesi, the densities of hornbills and primates in the forest appear to be directly related to the abundance of fig fruits (M.F. Kinnaird, personal communication). On the rare occasions when no figs are available, hornbills will leave the area in search of other food sources.

Fig trees produce these successive crops of fruits year round as part of a coevolved relationship with tiny (*c.* 2 mm) fig wasps (Agaonidae). Fig flowers are pollinated by these highly specialized fig wasp pollinators, which must enter the figs and lay their eggs in fig flowers to complete their life cycle (Bronstein 1992). In general, each fig species is pollinated by a different wasp species, although an increasing number of exceptions to this strict one-to-one relationship have been found (Cook & Rasplus 2003). Each wasp species recognizes its own fig species by the volatile chemicals the figs release when ready for pollination. If a population of fig plants stopped producing crops of fig flowers, the short-lived pollinators would die out, and the fig plants would not be pollinated. This was dramatically illustrated with several fig species in northern Borneo after exceptionally severe droughts during the 1998 ENSO (El Niño–Southern Oscillation) event caused a break in fig production (Harrison 2001). It took up to 2 years for the fig wasps to recolonize from unaffected regions.

## Palms

The palms are one of the most characteristic families of tropical plants: indeed, in the popular imagination palms symbolize the tropics. There are more than 190 genera and 2000 species, most of which occur in rain forests. Palms are immediately recognizable by their unbranched woody trunks, either solitary or in clusters, a rosette of large, divided, feather- or fan-like leaves at the top of the trunk, and often complicated, hanging clusters of flowers followed by single-seeded fruits (Fig. 2.6; see Plate 2.2, between pp. 150 and 151). The most notable feature of the distribution of palms, like that of many other tropical families, is their relative impoverishment in Africa (Turner 2001a). Africa has only 16 genera and 116 palm species, many outside the rain forest, in contrast with 64 genera

**Fig. 2.6** A variety of palm species. (a) A cluster of fruits from the palm tree *Attalea tessmannii* from the Amazon forest of Peru. (Courtesy of Johanna Choo.) (b) Rattan stem, Gunung National Park, Indonesian Borneo. Note the spines along the stem that help the rattan attach to other plants for support. (Courtesy of Tim Laman.) (c) Edible palm fruit (*Salacca edulis*) for sale in a Malaysian market. (Courtesy of David Lee.)

(d)

**Fig. 2.6** *(cont'd)* (d) Fantail palm (*Licuala* sp.) in Lambir Hills National Park, Malaysia. (Courtesy of Tim Laman.)

and 857 species in the Neotropics. New Guinea, with more than 300 species, and Madagascar, with more than 170 species, both have more palm species than Africa, and even the tiny island of Singapore at the tip of the Malay Peninsula has more genera (18). Both fossil evidence and molecular studies suggest that the palm family is old enough for its pantropical distribution to reflect the break-up of Gondwanaland.

Palms are not only diverse in tropical rain forests, but also abundant, particularly in the understorey. Relatively few palm species reach the forest canopy, but where they do their elegant stems and feathery leaves give the rain forest a distinctive appearance. Shrub and tree palms are ubiquitous, but climbing palms have a more restricted distribution. By far the most diverse group of climbing palms is the spiny rattan palms (*Calamus* and related genera). Rattans are found in rain forests from Africa to New Guinea, Australia, and Fiji, but they attain their greatest abundance and diversity in Southeast Asia, where up to 30 species can coexist in the same area. Rattan palms climb into the forest canopy, aided by spines on the young stem and the underside of the leaf blade and leaf stalk. Climbing species also produce specialized climbing organs covered in hook-like spines, either as a whip-like extension of the leaf tip or, in some species of *Calamus*, in the form of a modified sterile inflorescence. The result of all of these thorns and spines is that rattans are highly effective at grappling onto surrounding vegetation and using these supports to grow towards canopy openings. Rattans wind upwards into the canopy, with the weight of the stem shifting from tree to tree, giving the forest an unusual appearance with climbing, feathery palms hanging onto the edge of the tree canopies. This ability to latch onto anything they come into contact with makes people avoid rattan clumps in the forest, and leads to their common name in Australia of "lawyer cane". Rattan species are particularly abundant in Borneo, although the great demands for rattan in furniture manufacture means that they are being ever more intensely harvested from the forest, with plantations being developed to grow them. In much of Southeast Asia, rattans are the most important forest product after timber.

Neotropical rain forests also have spiny climbing palms, in the unrelated genus *Desmoncus*, but these are smaller, far less diverse—only seven species are currently recognized—and usually less abundant than the Old World rattans. The Neotropical understorey palm genus *Chamaedorea* (*c.* 100 species) also has a single, nonspiny, climbing species, *C. elatior*, found in Central American rain forest. Madagascar was thought to have no climbing palms until the discovery of *Dypsis scandens*, which looks very like *C. elatior*, although it is in a different subfamily (Dransfield & Beentje 1995).

In addition to the economic importance of many rain forest palms, they play an important role in supporting communities of large vertebrates by producing large crops of medium to large, single-seeded, fruits, often with a fleshy layer inside the fruit covering. Many animals rely on eating the fleshy palm fruit or the seeds inside when there are no other fruits available, demonstrating their crucial role in maintaining populations of mammals and large birds throughout the year.

## Pandans and cyclanths

Both pandans (Pandanaceae) and cyclanths (Cyclanthaceae, the Panama hat family) are sometimes mistaken for palms, to which they are not closely related, because of their tufts of large leaves. Neither family is pantropical on its own, but they have complimentary distributions and both morphological and molecular evidence show that they are "sister groups", i.e. they are more closely related to each other than to any other family. This suggests that the common ancestor of the two families originated on Gondwana in the Cretaceous, with their descendents becoming separated as the supercontinent broke up, but the fossil record is too poor to confirm this. Today, the Cyclanthaceae are exclusively Neotropical, with at least 225 species of herbs, shrubs, climbers, and epiphytes. The leaves of cyclanths can look very like those of palms—some fossil "palm leaves" are probably cyclanths—but the plants are much less woody and the climbing species are attached by means of short roots, not grappling spines. The Pandanaceae occur in rain forests from West Africa and Madagascar throughout Asia to New Guinea, Australia, and the Pacific, i.e. in every rain forest outside the Neotropics. The family contains around 1000 species of small trees, shrubs, and climbers, with a few epiphytes. The 700 or so tree and shrub species in the genus *Pandanus* are often called "screw pines" because the long, spiny-toothed leaves are arranged in a spiral (see Fig. 2.13c). This genus is most diverse and abundant in the rain forests of Madagascar, Southeast Asia, and New Guinea. The other major genus, *Freycinetia*, with more than 200 species, ranges from Sri Lanka to the Pacific, and consists of woody root-climbers that are rather similar in appearance to some climbing cyclanths.

The fruits of pandans and cyclanths are common in rain forests, but little work has been done on what eats them and disperses the seeds. One of the few studies involved a hemiepiphytic cyclanth, *Asplundia peruviana*, in Peru (Knogge et al. 1998). For epiphytes, semiepiphytes, and branch parasites, such as mistletoes, seed dispersal is a particular problem because the seeds must be dispersed to a suitable branch or stem, or else be wasted. Hemiepiphytic figs seem to depend on massive seed production, so the wastage can be tolerated; mistletoes make use of specialist birds (see Chapter 5), and many epiphytes rely on ants (see Chapter 7). *Aspludia peruviana*, in contrast, exploits a small generalist primate, the saddle-back tamarin (*Saguinus fuscicollis*), which normally deposits most seeds on the ground. However, after consuming *Aspludia* fruits, the monkeys suffer severe diarrhea. The liquid feces run down the trunk, leaving *Aspludia* seeds stuck to the bark. No information is available on the chemistry of the fruits, but these observations strongly suggest that this cyclanth has evolved a strong laxative that increases the chance of its seeds being deposited on a suitable site for germination and growth.

## Climbers and epiphytes

Big woody climbers and epiphytic vascular plants are two of the most characteristic features of lowland tropical rain forests. Woody climbers—lianas—often form tangles in the tops of trees and are an important element of tropical rain forests (Putz & Mooney 1991; Schnitzer & Bongers 2002). Climbers are particularly abundant where forests are affected by natural disturbances, such as windstorms that knock down trees, or by selective logging. When the canopy is opened by these activities, woody vines often proliferate, sometimes forming dense green blankets that overtop and even kill smaller trees. Such dense tangles of vines seem to be more common in shorter African forests and in Neotropical forests affected by storms. Dense vine tangles also occur along rivers where flooding scours the banks. In all rain forest regions, many of the vines belong to such worldwide families as the Asclepiadaceae (the milkweed family), Convolvulaceae (the morning glory family), Fabaceae, Araceae (aroids), Apocynaceae (the dogbane family), Cucurbitaceae (the squash family), and Rubiaceae, but there are also families with climbers only or predominantly in one region, such as the Bignoniaceae and Sapindaceae in the Neotropics, the Dichapetalaceae in Africa, and the palms in Africa, Asia, and New Guinea (Turner 2001a). The rain forests of Southeast Asia appear to be relatively poor in climber species, apart from palms, though the species present are still sufficient to form vine tangles in disturbed areas.

Around 10% of all vascular plants are epiphytes and the great majority of these are restricted to tropical forests (Nieder et al. 2001). Although their diversity and abundance is greatest in montane forests, epiphytes are also an important component of all lowland rain forests. Frequent rain and mist supplies the water that they need, while their position in the tree canopy allows them to gain access to higher light levels than occur on the forest floor (Benzing 1990). Epiphyte communities play an important role in the food chains of the forest, providing nectar, fruit, leaves, and shelter for many of the insects and vertebrates that inhabit the canopy. Epiphyte communities are particularly well developed in Neotropical forests and partly account for the high overall number of plant species in these forests as compared to rain forests elsewhere (Gentry & Dodson 1987; Benzing 1990; Lowman & Nadkarni 1995). A 1 ha (2.5 acre) plot at Cuyabeno, Ecuador, in western Amazonia, contained 172 species of epiphytes (Balslev et al. 1998). Orchids (see below) and ferns are important as epiphytes throughout the tropics, but several large families with many epiphytes, including the Bromeliaceae (see below), Cactaceae, and Cyclanthaceae, are exclusively or almost exclusively Neotropical, while several others, such as epiphytic members of the Araceae, Gesneriaceae (the African violet family), and Piperaceae (the black pepper family), are concentrated in the Neotropics. Neotropical rain forests are also notable for the abundance of "ant gardens" (see Chapter 7), which have allowed a small number of specialist epiphyte species to become very common.

There are other epiphyte families that are primarily found in tropical Asia and New Guinea, but these families have fewer species. The most important of these families is the Asclepiadaceae (the milkweed family, now often included within the family Apocynaceae), with at least 150 epiphytic species, only two of which are found in the Neotropics. The two largest genera are *Dischidia* (at least 80 species) and *Hoya* (more than 100 species). Members of the Asclepiadaceae are recognized by their pairs of opposite leaves, abundant milky juice in all plant parts, waxy flowers, and pod-like fruits with fluff-covered seeds. Hoyas in particular are often cultivated for their glossy foliage and long-lasting flowers. Many epiphytic species of *Hoya* and *Dischidia* have associations with ants, including some that produce specialized leaves that house the ants and others that grow on arboreal ant nests (see Chapter 7).

Southeast Asian rain forests and, to some extent, other Old World rain forests are also notable for the abundance of litter-trapping "bird's nest" ferns, in the genera *Asplenium*, *Drynaria*, and *Platycerium*. One widespread species, *Asplenium nidus* (possibly a complex of many similar species), can attain both massive individual sizes (with a fresh weight of up to 200 kg) and a high density (30 large ferns per hectare) in lowland dipterocarp forest, providing an important habitat for canopy invertebrates (Ellwood et al. 2002; Ellwood & Foster 2004). The basket-shaped rosette of long, broad, leaves traps dead leaves and other organic matter falling from above, resulting in an estimated 3.5 tons per hectare of suspended soil and plant material at one site in Sabah. To some extent, these ferns can be seen as the equivalent of the New World bromeliads, but their diversity is much lower and they do not form water-filled tanks.

Rain forest epiphyte communities are impoverished in Africa, probably as a result of repeated past episodes of aridity, a relative lack of suitable forest refuges during glacial periods, and lower current rainfall in most forest areas. The relationship between rainfall and epiphytes is illustrated by a series of rain forest research plots; the diversity of epiphytes declines by half when annual rainfall drops below 2500 mm per year (Turner 2001a). Australian rain forests are similarly poor in epiphytes.

## Orchids

Orchids are the single largest angiosperm family, with approximately 20,000 species, and are also the largest family of rain forest epiphytes. Of the world's orchids, about 41% occur in tropical America and 34% are found in tropical Asia and New Guinea, but only 15% occur in Africa and 3% in Australia. Orchids

(a)

(b)

**Fig. 2.7** An epiphytic forest orchid (*Huntleya burtii*) from Braulio Carillo National Park, Costa Rica. (a) The whole plant. (b) Close up of a flower. (Courtesy of Catherine Cardelus.)

reach their greatest abundance at mid-elevations where fog and clouds provide ideal growing conditions for epiphytes. Orchids are well represented by species-rich genera in the New World and the Old World, but with different groups represented in each area (Fig. 2.7; see Plates 2.3 and 7.5, between pp. 150 and 151). In the Neotropics, the genus *Pleurothallus* and related genera have 4000–5000 species, mostly tiny epiphytic plants with whitish to yellowish to reddish flowers pollinated by small flies; the greatest diversity is found in mid-elevation cloud forests. In the Old World, the genus *Bulbophyllum* takes on a comparable role with around 4000 species. Species of *Bulbophyllum* occur throughout the world, but the genus reaches its peak of diversity in New Guinea and Southeast

Asia, with 200 species in Borneo alone. The flowers are usually tiny, and are often purple-colored and foul-smelling—the odor from flowers of *B. beccarii* has been likened to a herd of dead elephants (Pridgeon 1994)! Not surprisingly, these species are pollinated by flies. The flowers are often delicately shaped with elaborate fringes. *Bulbophyllum* species have thickened stems called pseudobulbs, which can aid in water and food storage.

In addition to these large genera of mostly tiny orchids, there are also examples of large but geographically restricted genera with larger flowers. In the Neotropics there are over 1000 species of *Epidendrum* orchids, species often with cane-like leaves, and the lip petals forming a nectar tube. Notable among the *Epidendrum* species are those, such as *E. ramosum*, with red and orange flowers that are hummingbird-pollinated. Overall, there is a bewildering array of flower colors, sizes, and shapes in the genus, with flowers ranging from white to yellow to pink to green to red, and with very delicate and elongate petals to other species with compact and thick flowers. The diversity of *Epidendrum* orchids in the Neotropics is paralleled by the genus *Dendrobium* in Asia and New Guinea, with over 1000 species with pseudobulb stems, with every possible color and shape of flower imaginable except for pure black.

Allied to these giant orchid genera are several other large genera with hundreds of species restricted to particular places, such as *Lepanthes*, *Maxillaria*, and *Oncidium* in the Americas and *Eria* in Asia. Many of these genera have such a diversity of flower colors, shapes, and sizes that it is difficult to identify them to genus except by technical characters. The only really distinctive genus is *Oncidium*, generally recognized by its lobed petals, often with an undulating margin and brown mottled coloration.

Africa is impoverished in epiphytic orchids, probably due to past episodes of drying and the relatively small area of upland forest. Madagascar is richer, presumably because it has been wetter for longer. One interesting—and historically important—example is the largely epiphytic genus *Angraecum*. Two-thirds of the 218 species in this genus occur in Madagascar, with the remainder in Africa or on islands in the Indian Ocean. Many species have night-scented white flowers with a very long spur, which secretes nectar at the base, and are apparently pollinated by nocturnal hawkmoths with very long tongues. In 1862, Charles Darwin predicted that the Madagascan star orchid, *A. sequipedale*, which has a floral spur an incredible 29 cm long, would be pollinated by a giant hawkmoth with a tongue long enough to reach the nectar (Darwin 1862). Such hawkmoths were subsequently found (Nilsson 1998). Flowers adapted for pollination by long-tongued moths are seen in other genera as well, such as *Aerangis*. The particular abundance of white, long-spurred orchids in Madagascar is striking and suggests that moth pollination may be more important on this island than elsewhere.

### Ground herbs

In comparison with the deciduous forests of the temperate zone, ground herbs are a relatively inconspicuous part of the rain forest flora. In the interior of closed-canopy rain forests, most of the small plants at ground level are seedlings and saplings of trees, shrubs, and climbers. Herbaceous plants have a patchy distribution, but they are most prominent in canopy gaps, on steep slopes, and in the montane forest. In contrast to the uniformity of most rain forest leaves, those of forest herbs display a wide range of distinctive textures and colors, including some species with variegated patterns and others that are iridescent (Lee 2001). All these peculiarities are probably adaptations to the continuous low light levels in the forest understorey, although the precise mechanisms by which they work are not always clear. The dominant herbaceous plants of all rain forests include ferns and the fern-like spike mosses (*Selaginella* spp.), plus several families of flowering plants: the Zingiberaceae (gingers), Cyperaceae (sedges), Araceae (aroids), Rubiaceae, Gesneriaceae (African violet family), and Orchidaceae.

## Neotropical rain forests

### Trees

In addition to the rain forest families found throughout the tropics, there are also certain families that are characteristic of one or two regions, giving them a distinctive quality. Four large, woody families form important components of the Neotropical rain forests and are much less common or absent elsewhere: the Vochysiaceae, with more than 200 species in the Neotropics and three in West Africa, and the Bignoniaceae, Lecythidaceae, and Chrysobalanaceae, which are pantropical but most diverse in Neotropical rain forests (Table 2.2).

The Vochysiaceae are mostly trees, plus a few shrubs and climbers, which produce abundant sprays of distinctive and delicate flowers often attracting large numbers of pollinators. The Bignoniaceae (the catalpa family) includes trees, shrubs, and numerous woody climbers, of which trumpet creepers and the *Catalpa* tree will be familiar to temperate gardeners. The family is usually readily

**Table 2.2** Some important plant families in different rain forests. Families found only (or almost only) in one rain forest region are in bold, while families in regular print are especially abundant in that region. Underlined families are absent or relatively rare in a region, but present in others. The families consist mostly of trees and other woody plants unless otherwise indicated: families especially common as epiphytes are indicated with an "**E**"; vines and climbers with a "**V**"; and large herbs with an "**H**".

| Neotropics | Africa and Madagascar | Asia and New Guinea |
|---|---|---|
| **Vochysiaceae** (*Vochysia*) | Dichapetalaceae | **Dipterocarpaceae** (dipterocarp) |
| Bignoniaceae (*Catalpa*) | Olacaceae (African walnut) | Fagaceae (oak and chestnut) |
| Chrysobalanaceae | <u>Lauraceae</u> (laurel) | Myrtaceae (myrtle) |
| Lecythidaceae (Brazil nut) | <u>Moraceae</u> (fig) | Palmae (palm) |
| Palmae (palm) | <u>Palmae</u> (palm) | Asclepiadaceae (milkweed) **E** |
| <u>Pandanaceae</u> (screw pine) | <u>Orchidaceae</u> (orchid) **E** | |
| **Bromeliaceae** (pineapple) **E** | | |
| **Cactaceae** (cactus) **E** | | |
| **Cyclanthaceae** (Panama hat) **V** | | |
| Passifloraceae (passionflower) **V** | | |
| Heliconiaceae (*Heliconia*) **H** | | |

Modified from Turner (2001a).

(a)

(b)

(c)

**Fig. 2.8** Some New World plant families. (a) Cannonball tree flower (*Couroupita guianensis*), with numerous stamens, from the family Lecythidaceae. (Courtesy of David Lee, Guyana.) (b) *Heliconia stricta* from Central America. Note the small flowers emerging from the upper lip of each of the cup-like bracts. (Courtesy of David Lee, taken at Fairchild Tropical Garden.) (c) Hanging *Heliconia* inflorescences, from Dominica. (Courtesy of Tim Laman.)

(d)

**Fig. 2.8** *(cont'd)* (d) Passionflower (*Passiflora ligularis*) grown for edible fruit, in cultivation in Costa Rica. The complex floral structure is described in the text. (Courtesy of Barry Hammel.)

recognized by its opposite, compound leaves and large tubular flowers. The tendency of many family members to lose leaves during the dry season and then to produce an entire crown of large brilliantly colored flowers, often yellow, pink, or orange, is one of the most spectacular sights of Neotropical forests. When these bignoniaceous trees flower, they attract large numbers of pollinators, including hummingbirds, butterflies, and bees. The Lecythidaceae (the Brazil nut family) is a family of small to large trees recognized by their large, showy flowers, with 100 or more stamens (Fig. 2.8a). The sweet-smelling flowers of Neotropical species are typically pollinated by bees. The fruits are distinctive; they are often woody pots ("monkey pots") with lids that must be gnawed off by rodents to get to the large, flesh-covered, angular seeds within. Brazil nuts are an example of this type of seed (Peres et al. 1997).

Neotropical rain forests are also notable for the abundance of small trees and shrubs that flower and fruit in the understorey (Gentry 1982), providing a food source for flower visitors and fruit-eaters. This is in particular contrast with the dipterocarp forests of Southeast Asia, in which the forest understorey is dominated by the nonflowering saplings of canopy trees.

## Hummingbird flowers

In the New World tropics, a number of large genera and families of plants are pollinated predominantly or exclusively by hummingbirds (Stiles 1981; Bawa et al. 1985). These genera and families present a diversity of typically red-flowered species and generate hummingbird activity throughout the rain forest. Hummingbird-pollinated species are usually red because this is a color that stands out against a green background, making it easier for birds to locate the flowers. In contrast, most bees and other insects see light in a spectrum from orange to ultraviolet, and cannot distinguish red from green. One genus of note is *Heliconia* (Heliconiaceae), which is centered in the Neotropics, with outlying species on Pacific islands as far west as New Guinea and the Moluccas (Fig. 2.8b,c; see Plate 5.3, between pp. 150 and 151). The 200 *Heliconia* hummingbird-pollinated species are giant herbs with banana-like leaves, found in tree fall gaps and other forest openings. Birds are attracted to the pendulous or erect red inflorescences and visit the succession of small flowers produced inside the red bracts. In contrast, the Old World outliers of the genus are apparently pollinated by bats (Kress 1985).

Another noteworthy family pollinated predominantly by hummingbirds is the Marcgraviaceae, an exclusively Neotropical family of 125 species of thick-leafed woody climbers and shrubs, often epiphytic. A unique feature of the family is the production of pendulous whorls of flowers below which are large, cup-shaped nectaries. These nectaries, which are highly modified sterile flowers, fill with sweet nectar and attract hummingbirds as feeders, pollinating the fertile flowers in the process.

An important climbing family, often pollinated by hummingbirds, is the Passifloraceae (the passionflower family). While the family is widely distributed in the tropics, the most important genus, *Passiflora*, is centered in the Americas; the genus has 400–500 species in the Americas but only a few species in Asia, New Guinea, and Australia, and only one species in Madagascar. *Passiflora* species are generally recognized by their palmately lobed leaves and tendrils, making them similar in appearance to climbing squashes. Their most distinctive feature is their elaborate flowers, with a radiating star of five sepals and five petals, numerous filaments forming a colorful necklace on the surface of the petals, and then a fused structure of five large stamens and three large, knobbed stigma lobes (Fig. 2.8d). This complex flower structure was useful to missionaries in the New World tropics to illustrate the story of Jesus Christ and the Crucifixion, a story often called the Passion, hence the name *Passiflora*.

## Bromeliads

Just as Asian rain forests are commonly called "dipterocarp forests" because of the prominence of this tree family, American rain forests could reasonably be called "bromeliad forests". Species of the family Bromeliaceae are the preeminent group of Neotropical epiphytes, with one outlying species, *Pitcairnia feliciana*, which grows on rocky outcrops in West Africa where it presumably arrived by long-distance wind dispersal (Givnish et al. 2004). The bromeliads give a special "look" to the forest that is unmistakable. This is an easily recognized family, known for the cultivated pineapple, which is native to arid northeastern Brazil, the Spanish moss that drapes trees in subtropical residential areas, and the numerous cultivated plants in the family. Bromeliads have adaptations for living in dry conditions, which have made them ideally suited for living perched on tree branches as epiphytes. Many bromeliad plants produce a distinctive rosette of elongated stiff leaves, often with sharp spines along the edges, sometimes with reddish, brown, or white coloring on the leaves, and with overlapping leaf bases, creating a basket-like appearance to the plant (Fig. 2.9). The overlapping leaf bases can hold water for days or even weeks or months following a rainstorm. In some bromeliads, the leaf bases are enlarged to form tanks, and specialized leaf hairs, called trichomes, absorb water and minerals from the enclosed reservoir (see Plate 2.4, between pp. 150 and 151). The reservoir in some species may hold up to 5 litres (over 1 gallon) of water.

Mature bromeliads produce an elongated inflorescence. The flowers and inflorescence structures are often brightly colored, in many cases bright red, and in many species are pollinated by hummingbirds. The flowers are typically tubular, with three petals and three sepals. The fruit is either a small berry eaten by birds, or a capsule, which opens to release the wind-dispersed seeds. Pineapples are the best-known members of the family, but bromeliads are also well known as ornamentals, because of their attractive variegated foliage, and often brilliantly colored inflorescences. Notable genera include *Aechmea*, *Vriesea*, *Bilbergia*, and *Guzmania*. Bromeliads are easy to grow as houseplants as they can grow with minimal soil, they can store water in their tanks so they can survive without being watered often, and they can grow in relatively dry conditions.

Bromeliads provide a wide range of benefits to canopy animals in the Neotropics (Benzing 2000). They provide pollen, nectar, and fruits to many birds and mammals, while other animals eat the leaves and young inflorescences. Many species of birds forage among bromeliads for the numerous arthropods and small vertebrates that live there. Mammals, such as the coati (Beisiegel 2001), also take advantage of this foraging opportunity. The water tanks of bromeliads provide sources of drinking water for monkeys and other canopy animals during periods of drought (Bennett 2000). Using these water tanks, animals do not have to descend to the ground to obtain drinking water. Many species of invertebrates, especially aquatic insects, as well as frogs and salamanders, have taken this a step further by completing the aquatic phase of their life cycle in these canopy water tanks (Fish 1983). These bromeliad tanks form self-contained aquarium-like communities, with herbivores, detritivores, and predators feeding, breeding, and hiding in the tank and water-filled spaces between the leaf bases. The significance of these bromeliad water tanks can be illustrated by one single, highly important example. Mosquito larvae need to breed in standing water, preferably away from insect-eating fish. Of 1500 species described throughout the world as of 40 years ago, 500 species are known from the Neotropics; and of 88 subgeneric groupings of mosquitoes, 23 are found only in the Neotropics, whereas 11 are unique to the Australia–New Guinea region, eight to the Southeast Asian region, and only four to Africa (Bates 1964). The great abundance of species in the Neotropics is very probably related in part to the presence of bromeliad tanks as breeding grounds. This was confirmed by a survey done 20 years ago; of the 962 mosquito species known at that time from south of the United States, 22% of the species had larva that used bromeliad tanks as breeding sites (Frank & Curtis 1981; Frank 1983). Many of these tropical mosquitoes are also restricted to the mid- or upper-tree canopy, where they also presumably deposit their eggs in the water trapped by epiphytic plants.

(a)

Fig. 2.9 Members of the family Bromeliaceae, an important group of New World epiphytes. (a) Epiphytic bromeliad in flower (*Aechmea* sp.) from French Guyana. (Courtesy of David Lee.) (b) Epiphytic bromeliad perched on a tree trunk, Costa Rica. (Courtesy of Tim Laman.)

(b)

### Other Neotropical epiphytes

A second noticeable family of Neotropical epiphytes is the Cactaceae (the cactus family). While cactus species are most common in arid desert environments, around 150 species grow as epiphytes, many of them in drier Amazonian rain forests. In many ways, growing in the canopy of trees has certain similarities to deserts; the plants there are exposed to full sun and often high temperatures and plants are cut off from the soil water supply and nutrient pool. Eighteen genera of cacti have epiphytic members, with 58 species in the genus *Rhipsalis*, 18 in the genus *Hylocereus*, and 13 in the genus *Selenicereus*. With their thick waxy stems and absence of leaves, cacti are eminently suited to conserving water and surviving long periods without rain. A single species, the mistletoe cactus (*Rhipsalis baccifera*), is also found in Africa and on several islands in the Indian Ocean, but this is an exception to the general pattern, perhaps resulting from long-distance dispersal by birds. The third important Neotropical epiphyte family is the Piperaceae (the black pepper family), a large group of 1000 succulent herbs or small shrubs, often with thick, rounded, beautiful leaves, and tiny flowers in long, greenish or white spikes. The largest genus is *Peperomia*, found throughout the world, but with species concentrated and often extremely abundant in the Neotropics. All of the epiphytic members are apparently found in the Neotropics, where they are readily recognized by their distinctive flower spikes.

## Asian rain forests

### Dipterocarps

The tree family Dipterocarpaceae (literally "two-winged fruits") plays a dominant role in the ecology and economics of Asian forests in a way that no comparable family plays in the other rain forest regions (see Table 2.2) (Ashton 1982; Blundell 1999; Turner 2001a). Dipterocarps dominate forests in Borneo, Sumatra, Java, and the Malay Peninsula, as well as the wetter parts of the Philippines, with the majority of the large trees being members of this one family and accounting for the majority of the biomass. First-time visitors to these forests will be amazed to slowly turn around and realize that virtually every giant tree is a dipterocarp, and yet that they belong to several separate genera and dozens of distinct species. Outside this core everwet area, dipterocarps gradually decline in diversity and abundance (Fig. 2.10). A secondary center of dipterocarp diversity exists in Sri Lanka. A few species of dipterocarps are found in the African tropics and Madagascar, though not in the rain forests, and in the highlands of South America, providing a testament to the family's southern Gondwana origin. The dipterocarps appear to have reached Southeast Asia from Africa via the Indian plate, and did not arrive until the middle Eocene, when a moist corridor between India and Southeast Asia resulted in a major influx of plants with Gondwanan affinities (Dayanandan et al. 1999; Morley 2003). Despite this relatively late arrival, the dipterocarps underwent a massive evolutionary radiation in Southeast Asia. Enumerations of research plots in Southeast Asian forests show a conspicuous proliferation of tree species within the dipterocarp genera *Shorea, Hopea, Dipterocarpus,* and *Vatica* (Manokaran et al. 1992; Turner 1997b). In any one forest in the Malay Peninsula, Sumatra, or Borneo it would be

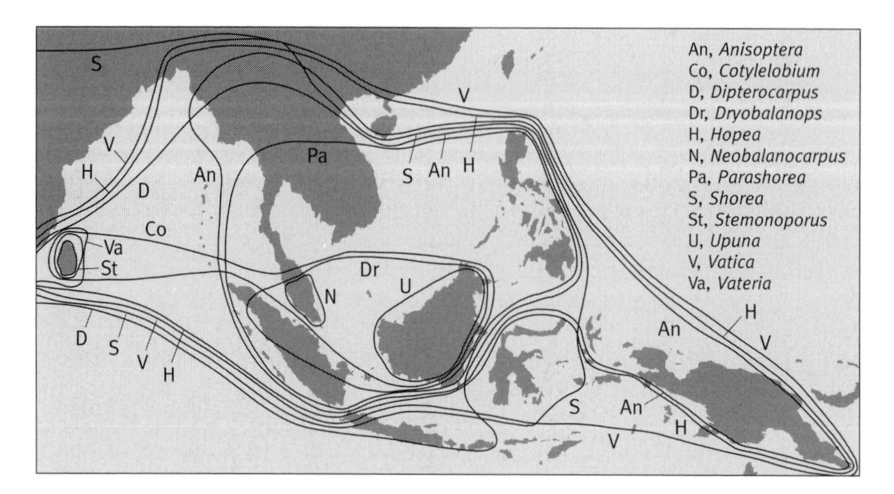


**Fig. 2.10** The ranges of dipterocarp genera, illustrating the center of distribution of this important tree family in the Sundaland region, with Sri Lanka being a secondary center. Each line encloses the distribution of all species in one genus. (From Whitmore 1998.)

common to find 25 species or more of *Shorea*, and six or more species of the other three genera. More recently, in geological terms, a few species have managed to disperse from island to island across the narrowing water gap to New Guinea, where they dominate in scattered patches.

Why should the dipterocarps be so dominant in Asian rain forests? There is no obvious single answer, but certain common features hint at the reasons behind their success. Dipterocarps tend to have smooth, straight trunks rising to great heights, without any side branches or forks until the canopy is reached (Fig. 2.11a). Dipterocarp forests often have canopy heights at and above 50 m (150 feet), which is higher than rain forests elsewhere (de Gouvenain & Silander 2003). The base of the tree is often buttressed. These growth characteristics emphasize the strength and stability of individual dipterocarp trees; trees do not typically fall over or get blown over as is often seen in Neotropical trees. Rather, dipterocarps often die standing, gradually losing their branches until only the trunk remains. As a result, the dipterocarp forest tends to be darker and more stable than forests in Africa (where the trees are shorter) and the Amazon (where trees may have a greater tendency to fall over and create large canopy gaps soon occupied by sun-loving trees and vines). Once dipterocarp trees reach the canopy and emerge from it, they produce a characteristic crown that is shaped like a cauliflower, with clusters of leafy branches evenly spaced around a dome. A tendency toward lower wind speeds in Southeast Asian rain forests than in the other regions (Thomas 2004) may favor this growth habit.

Another possible key to the dipterocarps' success and the long lives of individual trees is the presence in all plant parts of an oily, aromatic resin that presumably aids the plant in defense against attack by bacteria, fungi, and animals. This resin often accumulates where the bark is bruised and is encountered as hard, crusty, glass-like pieces on the trunk or on the ground. This resin, called dammar, is collected by the local people and used in varnishes or as boat caulking. The value of this resin is illustrated by the kapur tree, also known as the

Fig. 2.11 The dipterocarp family is extremely important in Asian forests. (a) Assortment of dipterocarp fruits, Borneo. Species vary widely in the size and shape of the wings. (Courtesy of Tim Laman.) (b) Looking up at dipterocarp trees, Borneo. Note their cauliflower-like branching pattern. (Courtesy of Tim Laman.) (c) Flowers of a dipterocarp species, *Hopea ponga* from India. (Courtesy of N.A. Aravind.)

Bornean camphor tree (*Dryobalanops aromatica*). Historically, this species was one of the main commercial sources of camphor, an essential oil of importance for its use in medicine and as a preservative. The crushed leaves have a distinctive camphor or kerosene-like smell. Dipterocarps also contain bitter-tasting tannins as a further deterrent to attack. Although non-dipterocarp trees also have chemical defenses in their foliage, dipterocarp leaves do seem to be peculiarly inedible, at least to vertebrates. This is illustrated by the fact that the colugo, a leaf-eating gliding mammal, lives in dipterocarp forests, and forages widely in the tree canopy for new leaves, but does not eat dipterocarp leaves (see Chapter 6). The orangutan and proboscis monkey, which also eat young leaves, again do not eat dipterocarp leaves. Dipterocarp leaves are simple in shape, with no teeth along the margins though sometimes with shallow scalloping, and are produced alternately along the twigs. The most distinctive feature of the leaves in many species is the strong pinnate pattern of raised small veins running between the secondary veins on the undersurface of the leaf.

The flowers of dipterocarps vary in size, some being small and others being relatively large and showy with five white, yellow, or pink petals and often with numerous stamens (Ashton 1982). The flowers are often scented and are adapted for pollination by a variety of insects—thrips, beetles, bees, or moths—depending on the species (Momose et al. 1998; Corlett 2004). Following flowering, a fruit is produced consisting of a single-seeded nut with a membranous wing-like calyx, looking like a badminton shuttlecock (Fig. 2.11b). The ratio of the fruit weight to the total wing area—known as the wing loading—is much higher in dipterocarps than in most other winged fruits, so they spin to the ground within a few meters of the parent tree (Osada et al. 2001). At least this is what usually happens. The fact that certain dipterocarp species have crossed major water barriers to reach New Guinea and the Philippines suggests that they can sometimes travel long distances. The key to their success must lie in occasional windstorms plucking the winged fruits off the tall trees and transporting them across rivers and seas.

A further reason for the success of the dipterocarps in the everwet areas of Southeast Asia may be the way most of the dipterocarp species over a wide region flower and fruit together only once every 2–7 years (van Schaik et al. 1993; Sakai et al. 1999; Sakai 2002). In an entire forest, only a few dipterocarp trees will flower in an ordinary year, but during a so-called "mast" year, almost every large tree reproduces (Fig. 2.12). Individual plant species have mast years in all rain forests, but only in Southeast Asia do the mast years of so many species coincide over such a large area. The trigger for the initiation of flower development may be a brief episode of low nighttime temperature caused by strong radiative cooling under cloudless conditions during a drought 2 months before flowering (Yasuda et al. 1999; Numata et al. 2003). However, these meteorological conditions do not always trigger mass flowering, showing that other factors are also important, presumably including the amount of resources the trees have accumulated since the last event. There are two main suggested advantages to this type of masting behavior. First, in the Asian everwet climate, with no distinct wet and dry seasons, plants need some cue to trigger the onset of reproduction. In this way, all the individuals of a species can flower at the same time and cross-pollination can occur. The multiyear seasonality of the El Niño cycle (see Chapter 1) provides the distinct set of conditions needed to coordinate reproduction. Second, and perhaps more important, mass fruiting at long intervals may prevent the build-up of populations of insects, birds, and mammals that would destroy the large and highly nutritious, oil-rich fruits (Janzen 1974; Kelly 1994). Synchronization of fruiting by many dipterocarp species across large areas is necessary for this to work, otherwise nomadic seed eaters, such as wild pigs, could simply move to wherever the trees were fruiting and destroy the whole crop (Curran & Leighton 2000). Thus, it is only in the mast years that any seeds survive long enough to germinate and grow into seedlings. Dipterocarps invest so much energy in reproduction during flowering years that they stop growing; in practice, they often have 5 or so years of growing without reproducing, followed by a heavy flowering year with no growth (Primack et al. 1989). This dipterocarp pattern is similar to the growth patterns seen in apple trees in temperate fruit orchards, with a year of heavy fruiting and little woody growth followed by a year of less fruit production and strong woody growth.

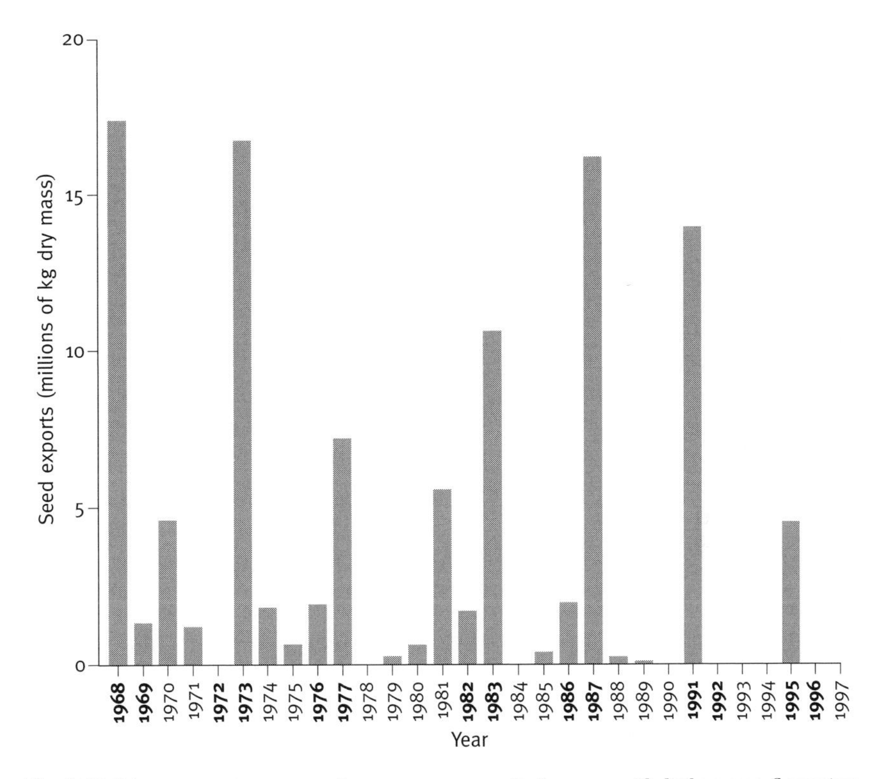


**Fig. 2.12** Dipterocarp trees mass flower once every 2–7 years, with little or no flowering in intervening years. This is illustrated by the export figures of illipe nuts (*Shorea* spp., section Pachycarpae), common dipterocarp species, from West Kalimantan, Indonesian Borneo, from 1968 to 1997. Years with El Niño–Southern Oscillation (ENSO) events are shown in bold; strong flowering years are associated with ENSO events. (From Curran & Leighton 2000.)

Producing successful crops of seedlings only every 2–7 years could be a major disadvantage in responding to the short-term recruitment opportunities that occur following the death of adult trees. Another important element in the success of dipterocarps is therefore the ability of dipterocarp seedlings of some species to survive for many years under the dense, shady canopy of established trees (Whitmore 1984). The resources provided by a large seed are an obvious advantage here. This effectively creates a "seedling bank" that can respond to opportunities created by an opening in the canopy overhead. In forests in Borneo, the seedlings of some dipterocarps species last for more than 15 years on the forest floor after a single fruiting event (Delissio et al. 2002). The variation among dipterocarp species in how well the seedlings can survive in deep shade and in how rapidly they can increase their growth rate in response to an increase in light levels allows the family as a whole to take advantage of a wide range of conditions (Brown et al. 1999).

Dipterocarp seedlings may also have an increased chance of survival as a result of a special form of the mutualistic relationship between roots and fungi called "mycorrhiza" (literally, "fungus root"), in which the plant receives mineral nutrients and water from the fungus in exchange for carbohydrates. Almost all plants form mycorrhizae, but unlike most other rain forest trees, dipterocarps are ectomycorrhizal—that is, the fungus forms a sheath over the outside of the roots. Ectomycorrhizal trees and their seedlings are linked by a network of fungal hyphae, which transfer nutrients from decaying organic matter to the plants. As soon as it germinates, a dipterocarp seedling can plug into the existing network and may obtain resources from its nearby parent, although this has not yet been convincingly demonstrated (Newbery et al. 2000). Whether this suggested ectomycorrhizal advantage exists or not, it is very striking that the same fungal association occurs in the oak family, which often dominates in Southeast Asian montane forests, and in legume trees in the subfamily Caesalpinioideae, which form extensive stands dominated by single species in parts of Africa (see below) and South America (Henkel 2003). Ectomycorrhizal associations are also the norm in low-diversity temperate zone forests. It is striking too that many of these ectomycorrhizal tree species have a pattern of heavy fruiting at multiyear intervals—mast fruiting—which is similar to that shown by the Southeast Asian dipterocarps (Green & Newbery 2002; Henkel 2003).

Many of these elements of the dipterocarp strategy for rain forest success seem to fit together. Wind-dispersed fruits are only practical in the rain forest for very tall trees that emerge from the forest canopy: there is too little air movement inside the forest. Large seeds produce seedlings that can establish and survive in deep shade. What is food for a seedling is also food for a beetle, rat, pheasant, or pig, but mast fruiting at long time intervals can satiate these seed predators so that many seeds escape consumption to grow into seedlings. Even in the rain forest, however, there are exceptions to some of these generalizations, including dipterocarp trees too small to emerge from the canopy and species that flower every year. Their strategies for survival must be different, and require further research. Moreover, in areas outside the everwet zone, such as in Thailand and the Western Ghats of India, dipterocarps flower and fruit on annual cycles in response to seasonal weather changes. In the rain forest at Sinharaja, in southwest Sri Lanka, which has a brief annual dry period, some dipterocarps have annual cycles while others show synchronized masting at multiyear intervals.

## Oaks and chestnuts

Walking in Southeast Asian rain forests, one may be surprised to find acorns from the genera *Lithocarpus* and *Quercus* (Fig. 2.13a) and chestnuts from the genus *Castanopsis* on the forest floor. The acorns come in a wide variety of nut shapes and cup shapes, but they are all immediately recognizable as acorns. Acorns are associated in most people's minds with temperate forests, but the Fagaceae (the oak family) is surprisingly well represented in lowland, tropical forests in Southeast Asia, and is one of the most dominant families at mid-elevations where the dipterocarps and other lowland rain forest families do not grow as well. Although rarely as conspicuous as in the northern temperate forests, the family attains its greatest diversity in tropical and subtropical Asia. The small equatorial island of Singapore, whose highest point is only 164 m above sea level, has 21 species (8 *Castanopsis*, 12 *Lithocarpus*, and 1 *Quercus*). A few species of *Lithocarpus* and one of *Castanopsis* extend to New Guinea, where they are prominent in montane forests, along with the Gondwanic, and largely southern temperate genus, *Nothofagus*, the southern beeches (Paijmans 1976).

**Fig. 2.13** A variety of Asian plant families. (a) Rain forest acorns (*Lithocarpus* sp.) in the family Fagaceae, from Malaysian Borneo. (Courtesy of K.M. Wong.) (b) Fruits of a Malaysian jambu tree (*Syzygium* sp.) from the family Myrtaceae, growing in cultivation in Brazil. (Courtesy of Chris Dick.)

Oaks (*Quercus* spp.) are also prominent in the montane forests of Central America, but only a single species, *Q. humboldtii*, has reached the northwestern margins of South America, apparently within the last 2 million years (Graham 1999).

(c)

(d)

**Fig. 2.13** (*cont'd*) (c) Screw pine stem with fruit (*Pandanus tectorius*) from the family Pandanaceae, growing in cultivation in Hong Kong. This widespread coastal species is used to illustrate the characteristics of the family. (Courtesy of David Lee.) (d) Wild durian (*Durio* sp.) from Borneo. (Courtesy of Tim Laman.) (See Plate 2.5, between pp. 150 and 151, for a further description of durians.)

Lowland oaks and chestnuts are almost entirely a Southeast Asian phenomenon, and there are no members of the oak family at all in southern India and Sri Lanka, in the whole of sub-Saharan Africa, or in most of tropical South America.

In the lowland forests of Southeast Asia, the acorns and nuts of the oak family represent food for squirrels, monkeys, pigs, and other vertebrates. The strong nutshell may help to protect the large seeds inside against the powerful bites of these animals. However, unlike the similar large nuts of the dipterocarps, acorns and chestnuts have no wings and depend for their dispersal on animals—probably rats and ground squirrels—that cache them in the ground for later consumption. Thus they must be edible, but not too edible!

### Speciose genera

The term "speciose" does not appear in most dictionaries, but is used by biologists to mean "with many species". It is widely believed that Southeast Asian rain forests are distinguished by the presence of plant genera with numerous coexisting species. There has been no systematic pantropical study of this question, but data from the CTFS rain forest plots shows that Lambir, Sarawak, has more species per genus than any other site (Lee et al. 2002; Ashton et al. 2004). The greater number of species per genus in Southeast Asia, if confirmed, could reflect relatively recent diversification of the rain forest flora there, but the pattern is also consistent with the alternative hypothesis that extinction has reduced diversity elsewhere.

The coexistence of multiple similar species of *Shorea* and other dipterocarps has already been mentioned. Another spectacular example is the genus *Syzygium*, in the Myrtaceae (the myrtle family). This huge genus of small to medium-sized trees occurs throughout the Old World tropics, from Africa to Australia, but is particularly diverse in Southeast Asia. In the 52 ha (130 acre) forest plot at Lambir, Sarawak, 49 species occur together and there are 45 species in a 50 ha plot at Pasoh, Malaysia. *Syzygium* species are readily recognized by their opposite, leathery leaves, fluffy flowers with numerous stamens, and round crunchy fruit with a characteristic cup of sepals at the top. These fruits are a major food source for fruit bats, other mammals, and birds. The Myrtaceae family is notable for its many cultivated fruit trees, among them the Neotropical guava tree, and a number of Asian *Syzygium* species have been brought into cultivation for their fruit known as jambus (Fig. 2.13b). Other genera with many coexisting species in Southeast Asian rain forests include *Aglaia* (Meliaceae, the mahogany family), *Diospyros* (Ebenaceae, the ebony family), *Ficus* (figs), and *Garcinia* (Clusiaceae, the mangosteen family).

## Rain forests in New Guinea and Australia

The lowland rain forests of New Guinea present an apparent paradox. The vertebrate fauna is very different from that of Southeast Asian rain forests, yet the composition of the flora is remarkably similar at the family and genus levels. The major floristic difference—one of considerable ecological importance—is that the dipterocarps are much less important, with the large Asian genera *Dipterocarpus* and *Shorea* entirely absent. Dipterocarps (in the genera *Anisoptera*, *Hopea*, and *Vatica*) are widespread in New Guinea and dominate over quite extensive areas (Paijmans 1976), but they are far less diverse than in Southeast Asia and are entirely absent from many areas. The dominant families in the few rain forest plots that have been enumerated in Papua New Guinea included the Lauraceae, Meliaceae, Moraceae, and Myristicaceae (Wright et al. 1997).

Many of the floristic similarities between New Guinea and Southeast Asia appear to reflect an influx of Asian rain forest plants into lowland New Guinea after the mid-Miocene collision between the Australian and Asian plates. The absence of a dry land connection between the two regions limited the exchange of vertebrates, but water gaps of a few tens of kilometers are much less of a barrier to many plants. Morley (2003) also suggests that elements of the Asian rain forest flora dispersed to the east on the land mass that is now the southwestern arm of the island of Sulawesi, but in the middle Eocene was attached to Borneo and supported an Asian flora. This flora therefore did not need to disperse across the Makassar Straits between Borneo and Sulawesi, which today forms the eastern boundary of the Asian faunal region. The predominance of Asian plants in New Guinea, rather than the other way round, seems to be a result of the relatively recent arrival of the Australian tectonic plate at tropical latitudes, which has not given time for a diverse indigenous tropical lowland flora to evolve. New Guinea's montane forests, in contrast, have a major component of southern origin, including the southern beech, *Nothofagus*.

In contrast to New Guinea, Gondwanic and Australian plants are more prominent in the smaller area of tropical rain forest in Australia, which has many endemic genera. Here, a much better fossil record shows little evidence for a post-Miocene influx of Asian plants, suggesting that the shared elements of the flora may reflect older connections and dispersal routes. The Australian rain forest is also noticeable for its concentration of primitive flowering plant families, including two woody climbers in the endemic family Austrobaileyaceae. These differences from New Guinea presumably reflect the very different history of the Australian rain forest, which, as discussed in Chapter 1, covered much of northern Australia during the early to middle Miocene, before becoming restricted to the northeast because of the drying of the continent. The lower representation of Asian lowland rain forest genera in Australia may have enabled the surviving southern genera to hold their own, while in New Guinea they were excluded from the lowlands by the Asian influx (Ashton 2003).

## African rain forests

One of the most notable of rain forest patterns is that African rain forests are relatively poor in plant species (Richards 1973; Turner 2001a), particularly in palms, orchids, and Lauraceae (the laurel family), as well as in epiphytes and woody vines in general. In many prominent tropical plant families, Africa only has 10–20% of the numbers of species in Asian and American forests. An extreme case is the Piperaceae (the black pepper family), where Africa only has three species in contrast to 500–1000 species in the Americas and Asia. In contrast with these absences, two less well-known families of rain forest trees without common names in English, the Dichapetalaceae and the Olacaceae, are more abundant in African forests than in rain forests elsewhere (see Table 2.2).

This relative poverty of species in African forests is probably due to multiple factors, but fossil evidence suggests that it is largely a result of extinctions in the 30 million years since a species-rich rain forest spanned the entire continent (see Chapter 1). Today, most African rain forests are drier and more seasonal than rain forests elsewhere, leading to the absence of climate-sensitive species, particularly among the epiphytes, most of which need abundant moisture to survive because their roots are not in the ground. Africa was also more strongly affected by drying during past glacial periods than other rain forest areas, resulting in forest loss and fragmentation (Maley 2001). In South America and Asia, species survived in the remaining wet areas. African forests have also been affected by human activity over a longer period. This contrasts with the relatively recent arrival of people in South America and the very low human population density in much of Southeast Asia's everwet forests until recently.

## Monodominance

African rain forests do not have a major distinctive element in their plant community, but are notable for the large expanses of forest dominated by a single tree species (Connell & Lowman 1989; Hart 1990; Newbery et al. 2000; Weber et al. 2001). Single-species dominance occurs in rain forests elsewhere, but not over such large areas. The most striking examples involve leguminous trees—all apparently lacking the nitrogen-fixing symbiosis. The mbau, *Gilbertiodendron dewevrei*, dominates the canopy of large patches from southeastern Nigeria and Cameroon across the Congo River basin (Torti et al. 2001). In the northeast Congo basin, some of these patches extend over hundreds of square kilometers, with mbau contributing 80–90% or more of the large trees. Mixed-species rain forests, with a diversity of trees more typical of rain forests elsewhere, grow adjacent to mbau forests. Monodominance does not mean that other tree species are absent, however, just that they are rare; the total number of tree species in 20 ha plots of mixed and monodominant forest in the Congo was very similar (Ashton et al. 2004).

Several factors seem to contribute to the persistent dominance of *Gilbertiodendron* (Hart et al. 1989; Hart 1995). The dense, continuous tree canopy casts a deep and uniform shade, in which only the most shade-tolerant plants can survive. Moreover, the litter layer is much deeper than in adjacent mixed forests, making it very difficult for the roots of germinating seeds to penetrate through to the soil below. These conditions favor *Gilbertiodendron*, which has large seeds that supply food reserves for the young seedling, as well as shade-tolerant saplings that can survive for years under an intact canopy.

The dominance of the forest by a single tree species probably reduces the diversity of insects, as there are fewer types of plants on which to feed. Mbau forests also seem to support a lower density and diversity of primates and other vertebrates than nearby mixed-species forests, although they may be an important habitat for some individual species. Colobus monkeys eat the new leaves and elephants uproot saplings to eat the roots, but the real bonanza occurs when the mbau trees fruit. As in Asian dipterocarp forests, fruiting is not an annual event, but occurs at intervals of several years. In mast years, these forests produce as much as 5 tons per hectare of large, edible, nutritious seeds (Blake & Fay 1997). Beetles destroy much of the crop, but enough is left for almost the entire forest mammal fauna to switch from their normal diets. Rodents, duikers, pigs, buffalo, elephants, gorillas, chimpanzees, and humans all find them irresistible. In non-fruiting years, the animals in the monodominant forests must either survive by eating figs, leaves, and fungi, or else migrate out of the area. In these periods, animals lose weight and have a greater chance of dying.

In the Korup National park in Cameroon, three species of giant legume trees—*Microberlinia bisulcata, Tetraberlinia bifoliolata,* and *T. moreliana*—codominate groves within an otherwise species-rich lowland rain forest (Green & Newbery 2002). These three species produce large fruit crops at 2–3-year intervals, although they flower heavily in some nonfruiting years. It would be very interesting to compare these and other low-diversity African forests, as well as similar forests in South America (Henkel 2003), with the much more diverse dipterocarp forests of Southeast Asia, with which they share single-family dominance of the canopy, mast fruiting at multiyear intervals, large, poorly dispersed seeds, and ectomycorrhizal associations with the roots.

Even the "mixed forests" in Africa are more heavily dominated by one or a few species—often of legumes—than rain forest elsewhere and, although most attention has been paid to monodominance in the canopy, the understories of African rain forests are also often dominated by a single species (Ashton et al. 2004). This further limits local plant species diversity and thus the variety of resources available for animals.

## Elephants in African rain forests

Forest elephants appear to play a uniquely important role in African rain forests, as a result of their size, movements, and feeding habits (see Chapter 4 and Plate 4.3, between pp. 150 and 151). No equivalent large mammals currently exist in the Neotropics and there is no direct evidence that any of the elephant-sized species that inhabited the Neotropics until the arrival of people lived in the rain forests. Neither Madagascar nor New Guinea ever had any animals this big. Forest elephants and rhinoceroses occur in Southeast Asian forests, but they appear to have much less influence on the vegetation. However, this impression may partly reflect the greatly reduced densities of elephants throughout most of Asia in historical times and it is possible that the importance of forest elephants in Asian rain forests has been underestimated.

African elephants damage or kill plants both accidentally, by trampling them in the course of other activities, and deliberately when feeding. During their feeding, they uproot or knock over small trees and debark or tear branches off larger ones. This damage can create gaps in the forest, as well as help maintain existing openings. The movements of elephants while foraging and over longer distances between favored sites create more or less permanent paths, with no vegetation at ground level. The importance of these activities to other forest herbivores is discussed in Chapter 4.

Fruit is a major item in the diet of African forest elephants and they are known to eat a huge range of fruit species. Many of these fruits are eaten by other forest mammals, such as gorillas, chimpanzees, or duikers, but African rain forests are also notable for the presence of very large, hard, fruits that fall to the ground when ripe. One of the most striking is the large round yellow fruit of *Strychnos aculeata,* which looks like a large billiard ball and is almost as hard. These fruits are formed by a vine in the tree canopy and come crashing down to the ground when ripe. The fruits are so smooth and tough that it is hard to imagine any animal being able to open them. However, elephants are capable of breaking through the 7 mm (0.25 inch) thick shell to get at the soapy slime inside, which surrounds the dozen or more round seeds. This slime is poisonous to many animals and is used both as a medicine and as a fish poison by local people. It has been speculated that elephants eat these fruits for the intoxicating effects of the slime rather than for nutrition (Martin 1991).

Elephants seem to be the primary dispersers of many large-seeded fruit trees in the African rain forest, such as the Guinea plum *Parinari excelsa*, with pits similar to a giant olive pit, the mango-like *Irvingia gabonensis*, and others such as *Panda oleosa*, *Balanites wilsoniana*, and the maskore tree *Tieghemella heckelii* (Fig. 2.14) (White et al. 1993; Yumoto et al. 1995). Elephant paths lead to fruiting individuals of favored species, such as the greenish-yellow maskore fruit. Fruits favored by elephants are typically at least 4–5 cm in diameter, dull-colored, and hard, with a shell protecting the strong-smelling flesh. Many of the

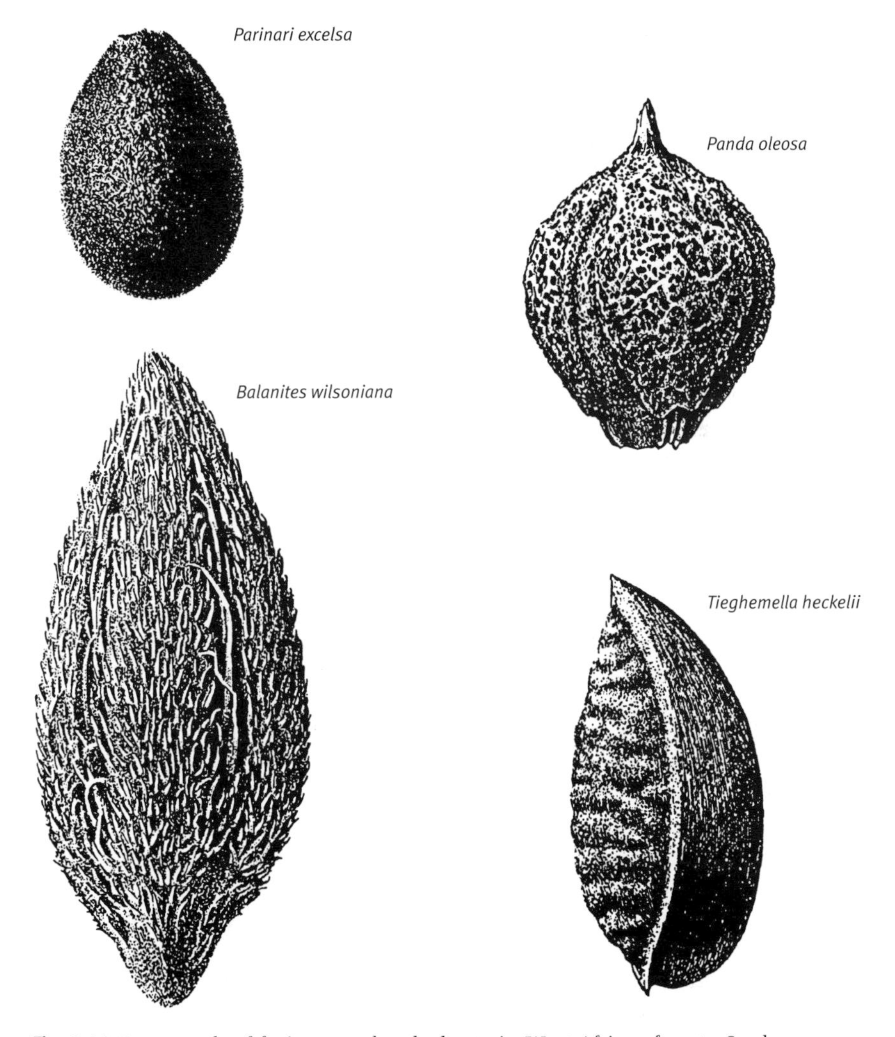


**Fig. 2.14** Large seeds of fruits eaten by elephants in West African forests. Seeds are shown at 80% of full size. (From Martin 1991.)

fruits contain large, nut-like pits, which protect the seeds as they pass through the intestines of the elephant. Fruit consumption by an elephant has been shown to provide multiple benefits in various plant species: reliable consumption of the entire fruit crop, long-distance seed dispersal over tens of kilometers, enhanced seed germination as a result of passage through the gut, and more vigorous seedling growth as a result of the fertilizing effect of the elephant dung (Cochrane 2003a; Nchanji & Plumptre 2003). Numerous forest trees and climbers benefit from seed dispersal by elephants and several, including most of those named above, seem to be entirely dependent on elephants. The elimination of elephant populations will eliminate these dispersal services and, in the long term, lead to the decline and eventual extinction of the plants on which they depend. Asian elephants also disperse seeds, but they seem to be less important in this role and there is no evidence that any plant species is exclusively dependent on them (Corlett 1998).

The hard shells of these very large, elephant-dispersed seeds not only shield them on their slow passage through the elephant, but also protect the nutritious interior from rodents and other seed-eaters. This protection is important because elephant dung attracts seed-eaters. The kernels of *Panda oleosa* are the toughest in Africa. One animal, however, has learned to overcome this barrier. Rain forest chimpanzees in several parts of West Africa use stone tools to crack open the nuts of *Panda* and other elephant-dispersed species (Mercader et al. 2002). The nuts are placed on stone or rock anvils and hit repeatedly with stone hammers. A skilled chimpanzee can crack up to 100 nuts in a day.

## Madagascan rain forests

Despite their small total area, the rain forests of Madagascar are surprisingly species-rich, probably because they were not subject to the severe drying during glacial episodes that occurred in Africa. As a result of the island's long history of isolation from other regions, they also have a very high proportion of endemic plant species—that is, species found nowhere else. An estimated 96% of the 4220 tree and large shrub species found on the island are endemic (Schatz 2001). In contrast to the vertebrate fauna considered in later chapters, this floristic distinctiveness does not run very deep, and although there are seven endemic families and many endemic genera, most species are in genera that are shared with Africa. There are some conspicuous exceptions, however. Notably, Madagascar has a diverse palm flora that includes genera with Asian affinities not represented in Africa (Dransfield & Beentje 1995). The largest endemic family is the Sarcolaenaceae, with around 60 species of trees and shrubs. The Sarcolaenaceae is the sister group of the dipterocarps and shares the ectomycorrhizal symbiosis of this family (Ducousso et al. 2004).

Madagascan rain forests are distinct in their generally low canopy heights, high tree densities, and the rarity of large-diameter trees (de Gouvenain & Silander 2003; Grubb 2003). Exposure to frequent cyclones is the simplest explanation for this structure, but low soil fertility may also contribute. Cyclone damage creates gaps in the canopy that benefit the famous "traveler's palm", *Ravenala madagascariensis* (Strelitziaceae), a giant relative of the bananas. Other distinctive features of Madagascan lowland rain forests include an abundance of palms, pandans, and bamboos (Grubb 2003).

One feature of the rain forest fauna also requires a botanical explanation. This is the striking poverty of the frugivore (fruit-eater) community in Madagascan rain forests (Goodman & Ganzhorn 1997; Rakotomanana et al. 2003). There are very few frugivorous birds compared with other rain forest regions, only three species of fruit bat, and, although many lemurs eat fruits when they are available, none depend on them year round. This seems to reflect a generally low density of plants that produce edible, fleshy fruits, relatively small fruit crops on these plants, and a strong seasonality in fruit production, with near zero fruit availability at certain times of the year. In many other rain forests, gaps in fruit production are filled by figs, which, as discussed earlier, produce ripe fruits throughout the year. Madagascan rain forests have relatively few rain forest fig species (fewer than 20) in comparison with other regions, but there is no published information on their abundance.

## Conclusions and future research directions

In contrast to the rain forest vertebrates, where very few families are pantropical, all the major rain forest regions share many plant families and a number of genera, but each region also has some distinctive botanical characteristics. Asian rain forests are notable for the dominance of dipterocarp trees and the presence of oaks and chestnuts in the lowlands. The tendency of dipterocarps and numerous other unrelated plant species to reproduce over a wide area at multiyear intervals results in long periods of flower and fruit poverty interspersed with brief and unpredictable periods of superabundance, with major consequences for the animal community. American rain forests have the greatest diversity and abundance of epiphytic species, most notably the water-filled bromeliads, providing canopy resources to many other species, as well as a number of distinctive tree, vine, and herb families. African rain forests are noted for their relative impoverishment of plant species due to past episodes of drought and forest contraction, the large areas of forest dominated by a single legume species, and the presence of plants with elephant-dispersed fruits. Madagascan rain forests are richer than might be expected from their small area and isolation. The lowland rain forests of New Guinea have a predominantly Asian flora—minus the dominance of dipterocarps—which, as we will see in subsequent chapters, interacts with a decidedly non-Asian fauna. Finally, the tiny area of tropical rain forest in Australia cannot be considered as simply an outlier of the more extensive rain forests of New Guinea, but shows evidence of its very different and much longer history. In each of these regions, a combination of biogeographical differences, past climates, and modern climate determine the types of plant communities that we see today. In later chapters, the impact of these plant communities on animal community structure and abundance will be developed.

The great diversity of rain forest floras makes rigorous comparisons between continents very difficult. In addition, the huge differences between years in plant reproductive activity—most strikingly in Southeast Asia but also, to some extent, in all rain forests—coupled with the very long life span of individual canopy trees, means that comparative studies must extend over many years or even decades to be meaningful. Understanding the causes and consequences of the observed differences will undoubtedly need an experimental approach, but experiments on the spatial and time scales required are beyond the capabilities of all but the largest and richest research organizations. Even impractical ideas for experiments can have great value, however, in stimulating thought and the design of more realistic experiments.

Although the rain forests of the Neotropics are the most distinctive in terms of the families and genera of plants that occur there, the lowland dipterocarp forests of Southeast Asia are undoubtedly the most distinctive in terms of their ecology. Both the family-level dominance by numerous species of dipterocarps and the supra-annual pattern of mass flowering and mast fruiting are unique to the region and have major consequences for the animals that live there. In comparisons between regions, dipterocarp dominance and supra-annual reproduction tend to be lumped together in the "food desert" view of dipterocarp forests. However, they are very different phenomena with different—and separate—consequences for other organisms in the forest. Dipterocarp dominance, for instance, is largely irrelevant to an understorey frugivore, but supra-annual fruiting cycles are highly relevant. Comparative studies are needed that separate the two phenomena.

There are no masting non-dipterocarp rain forests in Southeast Asia, but there are extensive areas of nonmasting dipterocarp forests. Dipterocarp trees mass flower and fruit every 2–7 years in the everwet zones of Southeast Asia, but change over to annual reproduction in the more seasonal areas to the north. It would be extremely valuable to investigate a series of forest sites along a transect from the southern tip of the Malay Peninsula up into northern Thailand to determine how the changeover from supra-annual to annual flowering affects the entire animal community. One hypothesis would be that the forest stands with an annual cycle of dipterocarp reproduction would support greater densities of flower-, fruit-, and seed-dependent insects and insect-eating frogs, reptiles, and birds. These seasonal forests in places such as Thailand would be predicted to have annual cycles of abundance in animal density and reproduction, reaching peaks of abundance when dipterocarp trees are reproducing and declining when trees are not reproducing. In contrast, everwet forests might have animal populations that are both lower in density and less variable over the course of a nonflowering normal year. Unfortunately, this comparison is unavoidably compounded by the direct effects of climatic differences on animal abundance. Studies on the few dipterocarp species that reproduce annually even in everwet forests would avoid this problem, but none of these species is abundant enough to have an impact on the entire animal community.

In order to investigate the effects of masting independently from dipterocarp dominance, comparisons need to be made with masting non-dipterocarp forests. Although mast fruiting at multiyear intervals involves many more species and much larger areas in Southeast Asian dipterocarp forests than elsewhere, masting also occurs on a smaller scale in some legume-dominated forests in Africa (Torti et al. 2001; Green & Newbery 2002) and the Neotropics (Henkel 2003). These mast-fruiting legumes show a number of other parallels with the dipterocarps, including large, poorly dispersed seeds and ectomycorrhizal associations with the roots. Studies of these various mast-fruiting tropical forests have so far used different techniques to test different hypotheses in each area. It would be instructive to do a pantropical comparison of mast-fruiting trees and forests, using standard methods, to look for common mechanisms behind their similarities. Parallel studies could contrast communities of vertebrates and invertebrates to determine how they respond to the masting cycle of the dominant plants.

An alternative, experimental, approach to understanding the role of dipterocarps in Asian rain forests would be to remove all the dipterocarps from a particular locality and observe the effects. This might seem hopelessly impractical, but selective logging already removes all the large dipterocarp trees from large areas of forest. This process could be completed by removing the remaining dipterocarp trees and letting the residual non-dipterocarp trees dominate the forest. Ideally, the resulting non-dipterocarp forest would be compared with a nearby control forest that had already been logged and from which a comparable number of non-dipterocarp trees had been removed. The hypothesis would be that the non-dipterocarp forest, because of its greater abundance of more edible plant families, would support higher densities of insects, leaf-eating vertebrates, and fruit-eating vertebrates. Such a project would need to be monitored for a period of decades as the trees grew to full size and the numerous pioneer trees gradually died off. Also, to be effective, the dipterocarps would have to be removed over a large enough area to encompass the dispersal distance of relevant animal species. The monodominant forests of Africa or South America could perhaps be investigated in a comparable manner, with all of the dominant species removed by repeated cutting over several years. However, the observation that adjacent monodominant and mixed forests have very similar floras—it is just the proportions of each species that are different—suggests that comparisons between the existing forest types in the same area would serve the same purpose. Southeast Asia is exceptional in that dipterocarp dominance is universal, so that non-dipterocarp forest can only be created artificially.

Bromeliads are the key distinguishing family of Neotropical forests. Experimental approaches might assist in understanding their role in forest ecology. An obvious experiment, that we think still has not been done, would be to remove all the bromeliads from an area of forest and compare this experimental forest with a nearby control forest with the bromeliads untouched. The hypothesis would be that the forest where the bromeliads had been removed would suffer a substantial loss of animal life, as insects, birds, lizards, and other animals of the canopy would no longer have a source of food, water, and breeding sites in the canopy. This decline in animal life would presumably be most severe during the dry season. It would also be valuable to check on microclimate conditions to determine if the removal of bromeliads had an impact on the humidity and temperature of the tree canopy.

Another method of investigating the importance of tank-forming bromeliads would be to plant large numbers of "artificial" bromeliads in the canopy of African or Asian rain forests. "Artificial bromeliads" could be made by folding together strips of stiff, green plastic to create a water tank similar in structure and water-holding capacity to a tank bromeliad. When these artificial bromeliads are placed in the canopy of trees would they be colonized by animals in the same way as living bromeliads? Is it simply the physical structure of the bromeliads, holding water in their overlapping leaf bases, that makes bromeliads ecologically important? Or do the bromeliad leaves have some active role in the process, such as the leaves releasing oxygen or chemicals into the water? Would the overall density of insects and other canopy animals increase as a result of the artificial bromeliads?

Elephants are keystone species in African forests. Where elephants have been removed from the forest by hunting, forest structure is expected to change and the fruits of many plant species to remain undispersed. A great deal of largely anecdotal evidence suggests that this is the case, but there is a need for more systematic studies. Africa now offers a range of rain forest sites from which elephants have been removed at different times in the past, providing an opportunity to investigate the long-term impact of elephant loss. Hopefully, the conclusions of these studies can be tested in future by the reintroduction of forest elephants to areas from which they had previously been removed. Comparative studies of other rain forests are needed to determine how the loss of large herbivores has affected the forest ecology in terms of tree density, the level of disturbance on the forest floor, and the types of plants growing on the ground. Reintroduction may be the only possible approach in Asian rain forests, where both rhinoceroses and elephants have been eliminated or greatly reduced in density in most areas. Indeed, large herbivores have been eliminated from most forests throughout the world (Janzen & Martin 1982); are there groups of large-seeded species that are now no longer being dispersed because their fruits are too big to be eaten by the remaining animals? The introduction of elephants to the Neotropics as a substitute for the extinct gomphotheres is almost certainly a bad idea, but exposing a domesticated elephant to "the fruits the gomphotheres ate" (Janzen & Martin 1982) in the forests of Costa Rica would be an alternative. Would the seeds survive passage through an elephant gut? Would germination be enhanced by gut passage and seedling growth by elephant dung?

This chapter has reviewed the topic of plant diversity and community structure. Certain differences among rain forests have been described, and in this conclusion some ideas for comparative and experimental investigation have been developed. In the next chapter, we will consider patterns in the primates, probably the most completely investigated group of animals.

## Further reading

Benzing D. (1990) *Vascular Epiphytes.* Cambridge University Press, Cambridge, UK.

Benzing D. (2000) *Bromeliaceae: Profile of an Adaptive Radiation.* Cambridge University Press, Cambridge, UK.

Gentry A.H. (1990) *Four Neotropical Rainforests.* Yale University Press, New Haven, CT.

Heywood V.H. (ed.) (1993) *Flowering Plants of the World.* Mayflower Books, New York.

Losos E.C. & Leigh E.G. Jr. (eds) (2004) *Tropical Forest Diversity and Dynamism: Findings from a Large-scale Plot Network.* University of Chicago Press, Chicago, IL.

Turner I.M. (2001a) Rain forest ecosystems, plant diversity. In *Encyclopedia of Biodiversity,* Vol. 5 (ed, Levin S.A.). Academic Press, San Diego, CA, pp. 13–23.

Turner I.M. (2001b) *The Ecology of Trees in the Tropical Rain Forest.* Cambridge University Press, Cambridge, UK.

Weber W., White L.J.T., Vedder A. & Naughton-Treves L. (eds) (2001) *African Rain Forest Ecology and Conservation: an Interdisciplinary Perspective.* Yale University Press, New Haven, CT.

Whitmore T.C. (1984) *Tropical Rain Forests of the Far East,* 2nd edn. Clarendon Press, Oxford.

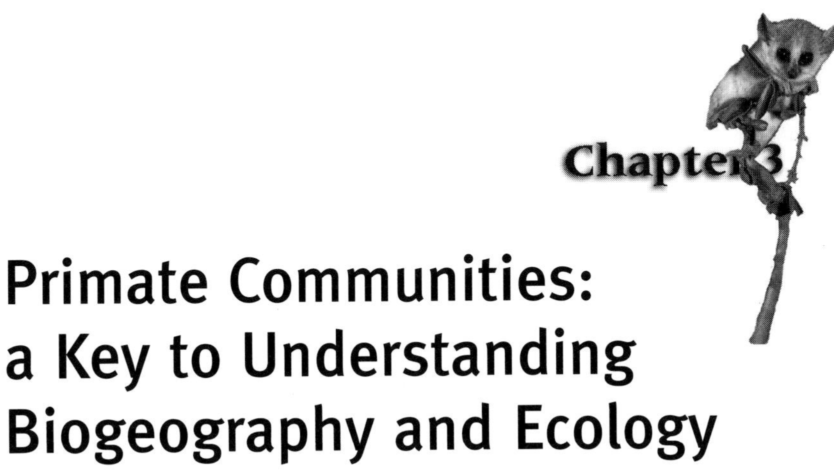


# Primate Communities: a Key to Understanding Biogeography and Ecology

No animal group illustrates the differences among the major rain forest regions better than the primates do. The unique biogeographical relationships that primates have with each of the five major rain forest areas sheds light on the forests' distinctive histories and features (Fleagle 1999; Fleagle et al. 1999). These relationships depend on the evolutionary history of primates in each area of the world, past and current environmental conditions, and human impacts both past and present. Each continent represents a separate set of evolutionary radiations and ecological adaptations to the rain forest environment. Thus, New World primates are evolutionarily distinct from the Old World primates of Africa and Asia, and although African and Asian primate communities have some similarities at the family level, the species in each region belong to separate radiations within each family and represent adaptations to the distinctive environments of their region. In particular, many African species represent adaptations to a drier, more open forest, while most Asian species are adapted to the tall, everwet dipterocarp forests. Neotropical species are adapted to an almost purely arboreal existence in the dense forests of South and Central America. The primates of Madagascar represent an entirely novel radiation of an early primate group, the lemurs, that expanded and persisted in the absence of later primates and without the constraints of competition with many of the more recent mammals of the large continents. Finally, the New Guinea–Australian region lacks native primates altogether, so it is possible to see how other rain forest animal groups have responded evolutionarily to this lack of a key mammalian group.

## What are primates?

Our closest living relative in the animal kingdom is a rain forest primate, the chimpanzee, so it is not surprising that these animals hold a great fascination for both biologists and the general public (Eimerl & DeVore 1965). But rain forest primates are far more than just a source of insight into human behavior and evolution. In most rain forests, primates represent a large proportion of the vertebrate biomass, are among the most abundant of large mammals, and play a key role in the ecology of the forest, particularly as important dispersers of seeds and as major consumers of leaves, fruits, and insects (Chapman &

Peres 2001). Moreover, 90% of all primates are associated with tropical forests, and the majority live in rain forests. Unfortunately, primates are also the most endangered group of rain forest animals, with many species threatened by hunting and habitat destruction. Primates as a group are valuable as indicator species for conservation purposes because they are easy to census, sensitive to hunting intensities, and typically are strongly affected by forest fragmentation.

Modern primates show an astonishing diversity of shapes, sizes, and behaviors, but at the same time, they are one of the most easily identified mammalian orders (Fig. 3.1). Most primates can be identified by various anatomical features, including a large head associated with an enlarged brain, a flatter face with shorter nose, and grasping hands and feet, often with opposable thumbs. Associated with a well-developed brain is a complex set of behaviors, often involving elaborate social interactions.

The principal divisions of the primates at the family level and below are associated with geographical areas, as primates are not able to easily cross oceanic barriers. Further, most species are primarily arboreal; even ground-dwelling species climb trees to obtain food or to take shelter when threatened, so that primates are only able to disperse across landscapes with at least some trees. The distribution at higher taxonomic levels, in contrast, is less easily explained by present-day geography. In modern systems of classification (Fleagle 1999; Groves 2001), the order Primates is divided into two suborders: the Strepsirhini (literally "inward turned nose", the wet-nosed primates) and the Haplorhini ("simple nose", the dry-nosed primates) (Fig. 3.2). The Strepsirhini includes the lemurs, which are found exclusively in Madagascar; and the lorises, pottos, and galagoes (also known as bush babies), which are found in Africa, India, Sri Lanka, and Southeast Asia. The Haplorhini includes the tarsiers, which are confined to Southeast Asia; the New World monkeys (Platyrrhini, literally, "flat-nosed"), found in South and Central America; and the Old World apes and monkeys (Catarrhini, "down-nosed"), widely distributed in Africa and Asia. Humans fall in the latter category, as *Homo sapiens* is an ape species of African origin. Note that the primate communities of African and Asian rain forests include species from several separate radiations, while those of the Neotropics and Madagascar are each composed of a single radiation.

The monkeys, apes, and humans—known collectively as anthropoids—differ from other primates in several respects: they are generally larger in body size but have a proportionally shorter torso, they have a larger brain in proportion to their body size, and they have nails rather than grooming claws. There are also many differences in their associated activity patterns. Anthropoids tend to be active during the day, while tarsiers and many strepsirhines are active at night. Anthropoids also have short snouts with less developed nasal regions, relying more on vision than smell.

## Old World versus New World primates

The main division of the anthropoids is between the Old World Catarrhini and the New World Platyrrhini. The anthropoids originated perhaps 50 million years ago in Africa, diversifying and spreading in the Old World over a period of more than 10 million years. During this time, South America was an island continent, far enough separated from Africa that primates could not easily cross the water

(a)

(b)

**Fig. 3.1** Some of the diversity of primates. (a) Mueller's gibbon (*Hylobates muelleri*), a lesser ape, from Borneo. (Courtesy of Tim Laman, taken in captivity.) (b) Potto (*Perodicticus potto*), a nocturnal strepsirhine from Kenya. (Courtesy of Harald Schuetz.)

(c)

(d)

**Fig. 3.1** (*cont'd*) (c) Emperor tamarin (*Saguinus imperator*), a small New World monkey. (Courtesy of Tim Laman, taken in captivity.) (d) Proboscis monkey (*Nasalis larvatus*) eating leaves, in Sabah, Malaysian Borneo. (Courtesy of Tim Laman.)

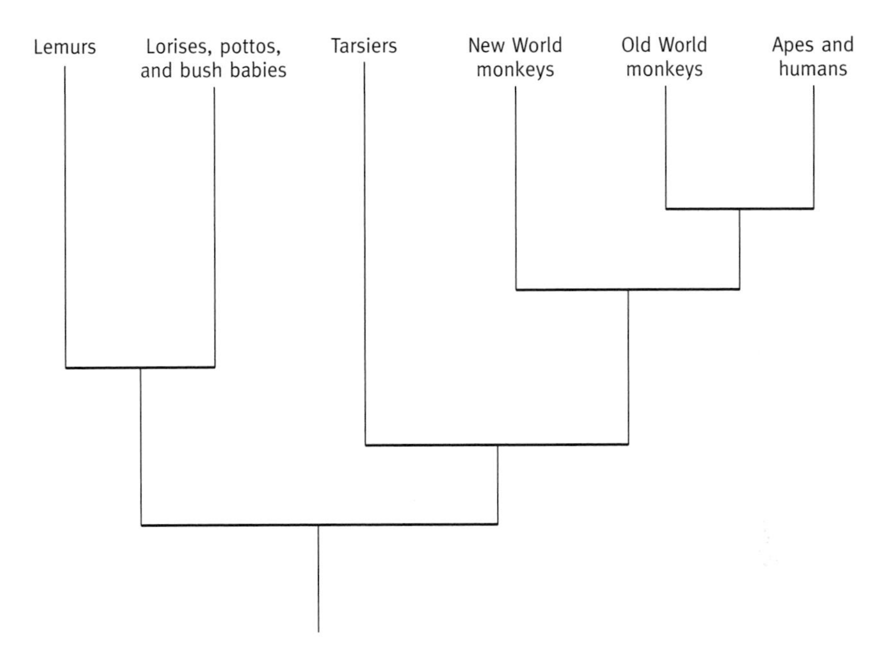


**Fig. 3.2** Our current understanding of the relationships between the major groups of primates mentioned in the text. (From multiple sources.)

barrier of the Atlantic Ocean. However, approximately 30–35 million years ago, primates did somehow manage to cross from Africa to South America, as evidenced by their appearance at that time in the fossil record (George & Lavocat 1993; Fleagle 1999).

The fact that all South American monkeys have similar characteristics to each other, which distinguish them from Old World monkeys, suggests that this dispersal across the water took place at a single time only in the entire history of primates (Fleagle & Kay 1997). The most likely, or rather least unlikely, scenario is that during a particular past period the continental shelves of Africa and South America were exposed due to lowered sea levels or elevated plate positions, and at the same time islands were exposed along the mid-Atlantic ridges between the continents. Oceanic currents sweeping northward along the western coast of Africa would be deflected from the bump of West Africa directly westward toward the South American coast. A small group of African monkeys, exploring a floating clump of trees in a river, or perhaps feeding on a tangle of trees and vines when the entire bank slumped into the river, could have been swept out into the Atlantic Ocean on a river current. The river current then joined the ocean current, which would have carried the tree raft and the monkeys perched on top directly to the South American coast. Alternatively, dispersal could have taken place over several generations as the oceanic islands of the Atlantic ridge were occupied by monkeys rafted from the African coast, and then served as dispersal points for further westward migrations. Modern Atlantic islands such as St Paul Rocks and Fernando de Noronha are today in this possible dispersal corridor. While this sequence may seem highly improbable, it only had to occur once over a period of millions of years to explain the arrival of monkeys in South America.

The primates that arrived in South America underwent an episode of adaptive radiation into a diversity of forms, which resemble various Old World monkeys, prosimians, apes, and even nonprimates, such as squirrels. Despite this radiation, they are still a relatively uniform group and share several unique characteristics that distinguish them as New World monkeys, separate from the Old World primates. One main difference is the shape of the nostrils. New World primates have broad nostrils, pointing to the sides, whereas Old World primates have narrow nostrils, directed forward. Also, the New World primates have a number of distinctive anatomical characteristics of the head, such as three premolars (the teeth between the molars and the canines), whereas the Old World primates have two premolars. (Check your own mouth to confirm that you are an Old World primate by ancestry, if not in modern residence.) In general, New World primates have relatively short forearms and lack an opposable thumb. Most obviously, all New World primates have a tail, whereas many Old World primates lack tails. In five of the New World genera, the muscular tail is even used like a prehensile fifth limb, which lets them hold on to branches while using their front limbs for picking up food. The prehensile tail allows for great acrobatics in species such as the spider monkey.

Another interesting contrast is in the area of color vision. Old World monkeys, apes, and humans are uniformly trichromatic—that is, all normal individuals have light receptors in the retina of the eye that are most sensitive to three different ranges of wavelengths (Jacobs 1995). In contrast, in New World monkeys, except for howler monkeys, most individuals are dichromatic—able to distinguish only two ranges of wavelength, like colorblind humans. The consequence is that Old World monkeys have a greater ability to distinguish the color red from the color green, allowing a more accurate determination of the first ripening of fruit and a greater ability to distinguish young leaves from older leaves on the basis of visual cues (Osorio & Vorobyev 1996; Dominy & Lucas 2001). It is noteworthy that the New World monkey that consumes most leaves, the howler monkey, is the only species currently known in which all individuals are trichromatic, allowing them to visually seek out young leaves to eat. In other New World monkeys in which color vision has been studied, some females are trichromats, while others, as well as all males, are dichromats. It has been suggested that the varied visual capabilities within a group of New World monkeys may allow different individuals to "specialize" on finding particular types of foods or detecting certain kinds of predators, with dichromats possibly having advantages in detecting some types of camouflage, as well as having better vision in low light intensities (Caine 2002). Another dramatic difference between Old and New World primates that may be related to the these differences in color vision is that several of the Old World primates, such as the drill and the mandrill, have naked red or pink rumps, presumably as a social cue, whereas this characteristic is absent in New World primates. Similarly, African guenons have dramatic variation in the color patterns of their faces, a characteristic less pronounced in New World primates.

The New World monkeys are relatively uniform in comparison with the older radiation of anthropoids in the Old World. The primary division among the Old World anthropoids is between the apes and the Old World monkeys, which are distinguished by many skeletal and dental characteristics. In addition, the apes have broad palates, broad nasal areas, and even larger brains. They are also extremely intelligent, and some species show complex tool use and the transmission of cultural behavior across generations. The 16–18 ape species that survive today are the remnants of a much more diverse radiation during the Miocene that not only dominated the rain forest primate fauna, but also occupied many other habitats. The Old World monkeys, by contrast, were less diverse during the Miocene and seem to have occupied mainly non-rain forest habitats, but have radiated dramatically since and now greatly outnumber the apes (Jablonski 2002). The living apes are divided into the Lesser Apes, composed of 11 species of gibbons from Asia, and the Great Apes, including orangutans from Asia, and gorillas, chimpanzees, and bonobos from Africa. The Great Apes have a longer time to first reproduction and a longer gestation period than Old World monkeys. Humans are classified with the apes and originated in Africa, but now are the most widely dispersed primate.

## Primate diversity

The total numbers of primate species found in the tropical rain forests of Asia, Africa, Madagascar, and tropical America are partly a function of the total area of rain forest available for them (Fig. 3.3). The Neotropics, with around 4 million km$^2$ (1.5 million sq. miles) of rain forest, has around 70 primate species, in contrast with Africa, which has around 50 primate species in 1.6 million km$^2$ of forest, and Southeast Asia with around 25 primate species in 1.1 million km$^2$ of forest (Fleagle 1999). However, Madagascar has at least 20 living species of forest primates in only about 50,000 km$^2$ of rain forest, and had several more as recently as a few thousand years ago. Africa and, in particular, Madagascar have somewhat more species of primates than one would expect given their current areas of rain forest, perhaps because the long period of occupation has allowed primates to undergo greater specialization than elsewhere. This is particularly

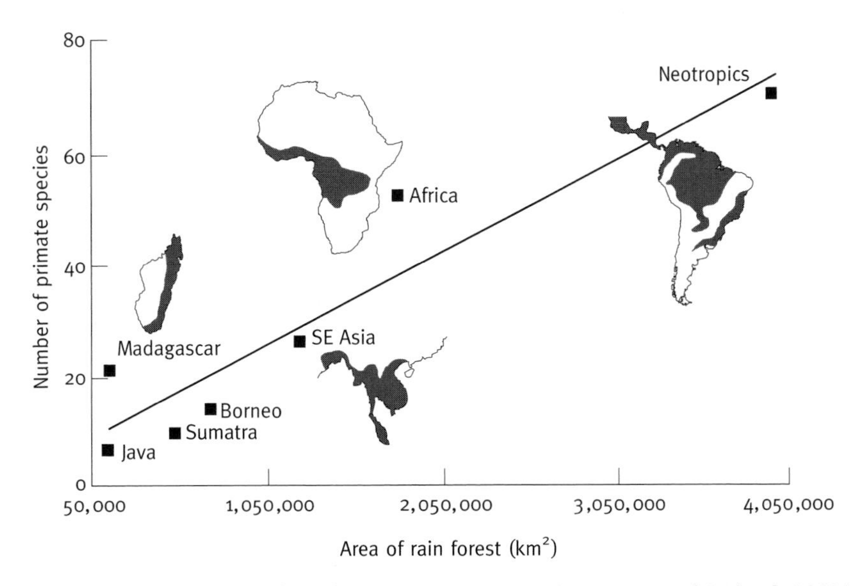


**Fig. 3.3** Larger areas of rain forest have more primate species. (From Reed & Fleagle 1995.)

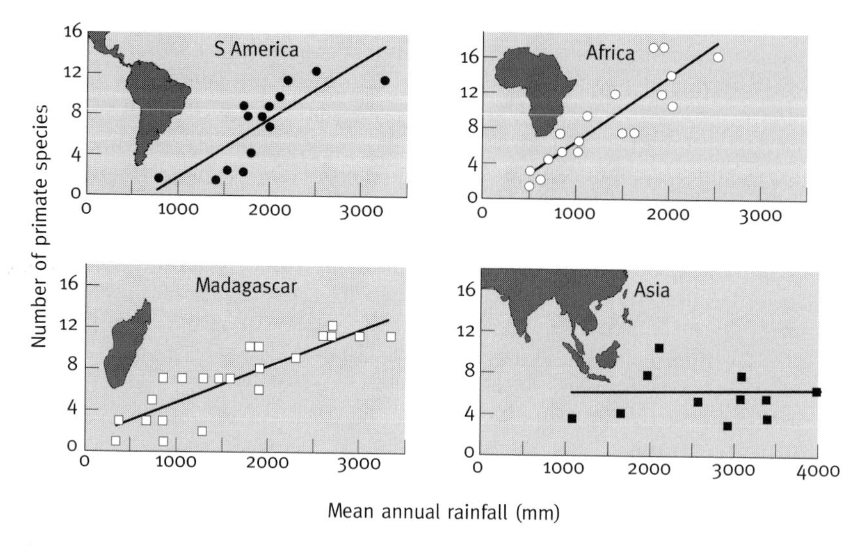


Fig. 3.4 The number of primate species at a site increases with rainfall in South America, Africa, and Madagascar. Such a relationship is not present for Asia. (From Fleagle 1999.)

likely to be true in Madagascar, where many other groups of birds and mammals that might compete with lemurs are absent (see Chapters 4 and 5).

The number of primate species found living together at the richest sites in each region is typically around a dozen, except in Asia, where seven or eight species is the usual maximum (Fleagle 1999), although some parts of Borneo have 10 or 11 species (Meijaard & Nijman 2003). Higher figures (up to 17) have been quoted for some African sites, but these all seem to include species using different parts of a habitat mosaic, rather than coexisting in the same forest. Primate diversity increases with the amount of rainfall at sites in the Neotropics, Africa, and Madagascar (Fig. 3.4). For example, in Africa, at sites with 1000 mm (40 inches) of annual rainfall, there are typically only five to eight primate species, while sites with 2000 mm of rainfall have more than 10. This pattern is not the same for each region, and there is also some indication that primate diversity may decrease at the highest rainfall levels (Kay et al. 1997). In general, African sites have more primate species than American sites with the same rainfall. In addition, Madagascar has more species at low rainfall sites than the Americas. There is no relationship between rainfall and number of species for the sites that have been studied in Southeast Asia, probably because most of these have abundant rain and many are on islands, where species numbers are usually lower than equivalent mainland sites.

## Primate diets

Primates are intelligent and adaptable animals, and many species are more or less omnivorous, taking whatever food is most available and nutritious at any one time. This tends to obscure neat categories of dietary specialization, although most species depend heavily on one or two types of food for a major part of each year. On each continent, there are species specializing on insects, fruits, leaves, and even plant saps, although the proportion of primate species with each diet differs greatly between continents. Many small primates feed on mixtures of fruit and insects, and many large primates feed on mixtures of leaves and fruits.

Dietary differences are sometimes reflected in the classification of primates, indicating the conservative nature of diet, linked in turn to a wide range of associated behavioral, morphological, and physiological adaptations. This is particularly obvious in the Old World monkey family, Cercopithecidae, which is divided into two very different subfamilies. The Cercopithecinae, including baboons, mandrills, mangabeys, guenons, and macaques, have cheek pouches and a simple stomach, and are omnivorous feeders, with a strong preference for ripe fruits when they are available. The Colobinae, including colobus monkeys, langurs, and the proboscis monkey, have no cheek pouches and a complex stomach, and are primarily leaf- and seed-eaters. Members of both subfamilies coexist in most African and Asian rain forests, but the Cercopithecinae are most diverse in Africa while the Colobinae are most diverse in Asia.

Intelligence and foraging adaptability are taken to the extreme in our primate cousins, the chimpanzees (*Pan troglodytes*). Although primarily fruit-eaters, there are big differences in diet between different groups, some of which seem to reflect differences in culture rather than simply resource availability (Whiten et al. 1999). In some areas, cooperative hunting of red colobus monkeys or other mammals by adult chimpanzee males has a significant impact on the populations of their prey. There are also striking differences between chimpanzee groups in the amount and type of tool use, including the use of sticks to "fish" for ants and termites, and of stones to break open hard nuts. Similar cultural differences have also been observed between groups of orangutans (*Pongo pygmaeus*), which are relatives of both humans and chimpanzees (van Schaik et al. 2003). As with the chimpanzees, some of these differences involve methods of foraging, such as the use of sticks to extract the highly nutritious seeds of *Neesia* trees from the irritant hairs that surround them, and the capture and eating of slow lorises (*Nycticebus coucang*) hiding in dense vegetation.

### Leaf-eaters

Some primates are predominantly leaf-eaters. The great advantage of a leaf diet is that leaves are readily available; the disadvantage is that leaves are of relatively low food value and often contain toxic chemicals. Thus, "good" leaves may not be as readily available as one might first think. Leaves take longer to digest than other foods, leading to a larger gut, slower processing time, a reduced metabolic rate, and lower activity levels. As a result, leaf-eaters are usually relatively big.

Each rain forest area that has primates has some species that eat leaves, but the extent to which these leaf-eating species rely exclusively on leaves differs from region to region, as do the physical adaptations that allow these species to survive on such a low-quality diet. The leaf specialists of African and Asian rain forests are all Old World monkeys in the subfamily Colobinae, which includes the colobus monkeys of Africa and the leaf monkeys (or langurs) of Asia (Davies & Oates 1994). These colobines vary in their dependence on leaves and are perhaps better categorized as feeders on "difficult" plant materials, since different species also take varying amounts of seeds and unripe fruits. The key colobine adaptation is their complex, ruminant-like stomach, which can hold a third of their body weight in food while bacteria detoxify plant defensive chemicals and digest the cellulose.

For sheer strangeness of appearance, it is hard to beat the Bornean proboscis monkey *Nasalis larvatus*, a specialized species of leaf monkey (see Fig. 3.1d). The most notable feature is the well-developed nose, which is large, drooping, and bulbous in adult males. Males weigh up to 22 kg (48 lb) and females up to 11 kg. Slender tails are about the same length as their body. Their fur is reddish brown with lighter buff areas on their underside. These animals have a pot-bellied appearance associated with the large stomachs and intestines needed to digest a diet consisting of leaves, and to a lesser extent flowers, fruits, and seeds. Proboscis monkeys are active during the day in their preferred habitat of mangroves and lowland rain forest near water. They are excellent swimmers, even leaping off trees into the water below.

The leaf-eating lemur species of Madagascar, such as the 600–900 g (20–30 oz) sportive lemurs (*Lepilemur*), are very different animals. These lemurs are exceptions to the rule that leaf-eaters are large and, unlike colobines, they have a relatively simple stomach. Instead, they are "hindgut fermenters", digesting leaves in the colon and an expanded cecum, excreting large particles rapidly while selectively retaining smaller, more digestible particles.

The New World lacks lemurs and colobine monkeys, and no species has developed such extreme adaptations for eating leaves as these Old World primates have. The howler monkeys (*Alouatta*) and woolly spider monkey (*Brachyteles*) are the closest to being leaf specialists in the Neotropics, but, unlike most colobines, they also eat substantial amounts of ripe fruit when it is available. Howlers and woolly spider monkeys are hindgut fermenters, with a long hindgut and expanded cecum. The nearest ecological equivalent to the Old World colobines are not primates but sloths (Megalonychidae and Bradypodidae), a group which is confined to the New World. Like colobines, sloths have a complex, ruminant-like forestomach for cellulose digestion and can reach high densities in Neotropical rain forests. The low metabolic rate, slow movements, long periods of inactivity, and general slothfulness of sloths are an extreme example of a trend seen in other arboreal leaf-eaters, none of which are particularly lively animals. The lack of high levels of activity is presumably an energy-saving mechanism, made possible by the abundance of their food supply and necessitated by the low nutritional quality of their diet.

### Insectivores

Other primates specialize in eating insects and other invertebrates. Insects are an important source of protein for many primates, including such large species as chimpanzees. However, the insect specialists are mostly small, active primates, such as the tarsiers (*Tarsius*) of Asia, the marmosets (*Callathrix*) of the Americas, the mouse lemurs (*Microcebus*) of Madagascar, and the dwarf galagos (*Galagoides*) of Africa. Small size is no disadvantage in overcoming insects, whereas a larger size would make it impossible to find enough insects to eat. The earliest ancestors of the modern primates may have been insectivores and such key primate adaptations as grasping hands and binocular vision may have evolved first as an aid to catching such small, active prey. The high quality of protein-rich insects as a diet is offset by the scarcity of insects at some times of the year in most rain forests, forcing insectivores to seek other food such as fruits, nectar, or plant sap.

### Frugivores

The third major diet type is frugivory—fruit eating. Fruit is conspicuous and often readily available, larger in size than insects, and more easily digestible than leaves, although it is typically low in protein. Frugivorous primates usually consume only the fruit pulp so the seeds, which are hard and often poisonous, are an unwanted waste product. Fruit-eating species form a major part of all rain forest primate communities, but there are striking differences in the ways that the different groups of primates deal with the seed problem. The fruit-eating Old World monkeys in the subfamily Cercopithecinae, which includes the African guenons and Asian macaques, all have well-developed cheek pouches. These are used to hold excess fruits, which are then returned one at a time to the mouth for processing, with the larger seeds being spat out (Corlett & Lucas 1990). This allows these monkeys to harvest many fruits quickly and process them more slowly as they move between fruiting trees. Fruit-eating apes, lemurs, and New World monkeys, in contrast, lack cheek pouches and often swallow fruits whole. This also permits rapid harvesting, but has the disadvantage that indigestible seeds can make up large proportion of the material in the gut at any one time. Both seed-processing strategies seem to work equally well for the primates, but may have very different consequences for the plants whose fruits they consume (see below).

## Primate communities

### African primates

One of the most noticeable differences among primate communities in different rain forest regions is in the range of body sizes (Fig. 3.5) (Kappeler & Heymann 1996; Fleagle 1999). Africa has the widest range of sizes among living rain forest primates, from the largely insectivorous dwarf galagos (also known as bush babies), weighing less than 100 g (3.5 oz), to the gorillas, which can attain 200 kg (440 lb) or more. The large size of gorillas confers a number of advantages, including safety from predators and dietary flexibility. Lowland gorillas (*Gorilla gorilla*) feed largely on fleshy fruits when they are available, but can also cope with fibrous alternative foods when fruit is scarce. Mountain gorillas (often now treated as members of a separate species, *Gorilla beringei*) have an almost entirely folivorous (leaf-eating) diet.

A typical African rain forest community will have several species of galagos, pottos, and their relatives—small (0.1–2 kg), nocturnal strepsirhines that feed on various combinations of insects, fruits, and gums—and one or two species of apes, of which the chimpanzee (up to 60 kg) is the most widespread. The other members of the community are all medium to large Old World monkeys. Typically, one to

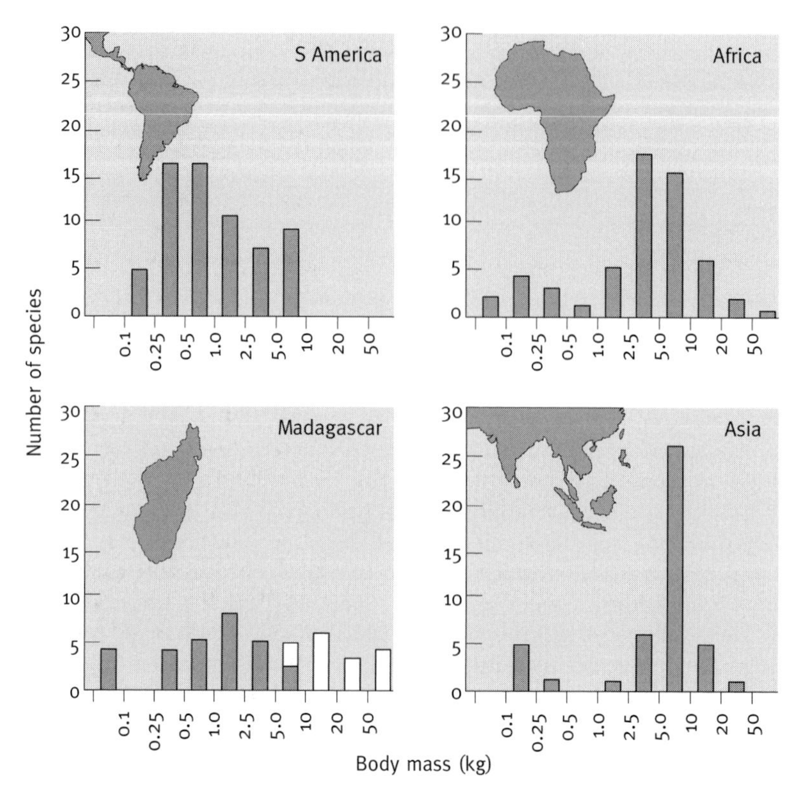


**Fig. 3.5** The range of primate body weights in four land areas. Recently extinct Madagascar species are indicated as white columns. Data include both rain forest and non-rain forest species. (From Fleagle 1999; courtesy of Peter Kappeler.)

three species of leaf- and seed-eating colobus monkeys (3–14 kg) are joined by up to eight species of frugivorous or omnivorous cercopithecines "cheek pouch" monkeys (1.5–50 kg), ranging from primarily terrestrial to entirely arboreal and active during the day. These include the approximately 20 species of guenons in the genus *Cercopithecus*, a genus that includes the Diana's, Mona, and de Brazza's monkeys (Glenn & Cords 2002). These are medium-sized monkeys, weighing from 4 to 12 kg, depending on the species, with a diverse diet. Most notable is their sleek, beautiful fur; often with contrasting white and darker patches on the face, throat, and body.

The mandrill (*Mandrillus sphinx*) is a large African cercopithecine monkey, found on the ground and at mid-level in the canopy in dense rain forests in Cameroon, Equatorial Guinea, Gabon, and the Congo. The mandrill is a substantial primate, weighing around 25 kg in males and 11 kg in females, with a diverse diet. Mandrills have dramatic colorings, a feature not found in New World primates. These include prominent nasal ridges, which are dramatically colored in streaks of blue and purple, and with scarlet also found in males. These animals also have exposed pink skin on their buttocks, which has some role in recognition in dense foliage and competitive signaling in their large foraging hordes of up to 800 members (Abernethy et al. 2002).

## American primates

The New World monkeys are the youngest and most uniform primate radiation, although local species diversities are similar to those in Africa (Fleagle 1999). American primate communities are unique in being dominated by arboreal small to medium-sized insect- and fruit-eating monkeys, with no living species weighing more than 12 kg (25 lb) (Fig. 3.5). The Americas lack the ecological equivalents to many of the primate species found elsewhere, with no terrestrial species, only one nocturnal genus, *Aotus*—with 10 species known as the night monkeys or douroucoulis—and few leaf-eating or suspensory species. There is, however, a distinct subfamily (Pitheciinae), containing the sakis (*Pithecia*), bearded sakis (*Chiropotes*), and uacaris (*Cacajao*), which are specialized feeders on fruits and seeds with hard coverings. The majority of New World primates weigh less than 2.5 kg, with the species-rich, squirrel-like marmosets and tamarins typically weighing less than 1 kg. A particularly widespread species is the squirrel monkey (*Saimiri sciureus*), a small monkey found over a large area of the Amazon, from northern Peru and Colombia to northeastern Brazil. These 1 kg monkeys are distinctive in appearance with soft, thick fur, large ears, nose, and lip region with exposed black skin, a contrasting reddish back and legs with white or buff underparts, and a long, black-tipped tail. Groups of 10–40 animals live in primary or secondary forest, often along streams, foraging mainly during the day in trees for a highly varied diet of fruits, berries, nuts, flowers, buds, seeds, leaves, gum, insects, spiders, and small vertebrates.

None of these small New World species readily descends to the ground, where their small size would make them highly vulnerable to predators. Perhaps because of their small size and arboreal habitats, marmosets are apparently unable to cross rivers, and many species have distributions restricted to one side of a river system. This is in marked contrast with African primates, such as gorillas, chimpanzees, bonobos, drills, mandrills, and mangabeys that are larger in size and are found more often on the ground. Such large animals are less vulnerable to predators.

Both the absence of large primates from tropical American forests and the low diversity of leaf-eaters may be explained by the prior presence and high density of sloths as arboreal leaf-eaters when the primates first arrived. There were, in fact, several, considerably larger, 20–25 kg species of primates in the Americas in the recent past, but they have since died out, perhaps because of hunting by early human inhabitants of the continent (MacPhee & Horovitz 2002). It has been suggested that these larger species may have been partly terrestrial in their habits (Heymann 1998), although there is little evidence to support this. The low diversity of nocturnal primates may be similarly explained by the prior presence and diversity of night-foraging marsupials. Thus, in striking contrast to the situation in Madagascar, considered below, the primate radiation in the New World was constrained by the diversity of preexisting mammal species that already occupied several potential primate niches.

## Asian primates

Although the New World rain forests have an abundance of small, active insect-eating primates (22 species), tropical Asia has only the big-eyed, acrobatic tarsiers (four or five *Tarsius* spp.) (see Plate 3.1, between pp. 150 and 151) and the slow-moving lorises (one species of *Loris*, two of *Nycticebus*). There is also a very large (40–100 kg) species of arboreal, frugivorous ape, the orangutan (*Pongo pygmaeus*), now restricted to Borneo and Sumatra, although it was found throughout Southeast Asia in prehistoric times (see Plate 3.2). The Bornean and Sumatran populations have been separated for a long time and are sometimes treated as two separate species, *P. pygmaeus* in Borneo and *P. abelii* in Sumatra. All the other primates in Asian rain forests, including the cercopithecine macaques, the colobine leaf monkeys (or langurs), and the gibbons, which are small apes, have adult weights in the 5–12 kg (10–25 lb) range (see Fig. 3.5). Gibbons are remarkable primates of Asian rain forests, with no equivalent in the other rain forest regions. Amazingly long forearms allow them to swing from branch to branch through the canopy (see Fig. 3.1a), with an agility matched only by the Neotropical spider monkeys.

One possible explanation for both the relatively low diversity of most Asian primate communities and the deficiency in small species is that Asian rain forests have a high diversity of potential competitors. Most striking is the diversity and abundance of squirrels (more than 30 species) that have been present since Miocene times, including ground squirrels, flying squirrels, and giant squirrels. As many as a dozen species can coexist in the same area. Asian rain forests also have the squirrel-like, but largely terrestrial, tree shrews (Tupaiidae, 19 species), which vary in weight from 20 to 400 g and feed largely on insects and small fruits (Table 3.1) (Emmons 2000). African forests are intermediate in both the numbers of small primates (14 species) and the numbers of squirrels (14 species). Africa also has the anomalurids, a group of seven species of flying rodents

**Table 3.1** Species richness of some small, canopy, vertebrate animals in rain forests.

| | Small primates* | Squirrels | Tree shrews | Parrots | Tree kangaroos | Other† |
|---|---|---|---|---|---|---|
| Australia | — | — | — | 13 | 2 | — |
| New Guinea | — | — | — | 43 | 8 | — |
| Java, Borneo, Sumatra | 2 | 37 | 10 | 7 | — | 1 |
| Southeast Asian mainland | 2 | 31 | 5 | 6 | — | 1 |
| Philippines | 1 | 2 | 1 | 12 | — | 1 |
| India | — | 7 | 1 | 6 | — | — |
| Sri Lanka | 1 | 6 | — | 5 | — | — |
| African rain forests | 14 | 14 | — | 8 | — | 7 |
| Madagascar | 35 | — | — | 3 | — | — |
| Amazon basin | 22 | 7 | — | 50 | — | 2 |
| Central America | 4 | 7 | — | 32 | — | 4 |

* Includes lorises, galagos, bushbabies, pottos, tarsiers, small lemurs, marmosets, and spider monkeys.

† Includes colugos (Dermoptera) in Asia, anomalurids in Africa, and kinkajous and olingos (Carnivora, Procyonidae) in the Americas.

Data from Nowak (1999) and Juniper & Parr (1998). Based on overlapping ranges between animal distribution and rain forest, supplemented by information on habitat. Due to an incomplete knowledge of ranges and habitat types, these values should be regarded as approximate.

ecologically similar to flying squirrels (Nowak 1999). The earlier diversification of squirrels and other squirrel-like mammals in Asian forests prior to the arrival of the primates probably put small primates at a competitive disadvantage in Asian forests, and to a lesser extent in African forests. The paucity of squirrels in tropical America (only seven species), and their recent arrival in the Pleistocene period, allowed American primates (as well as other groups such as parrots), to undergo a radiation of small-sized species that was closed to them elsewhere.

Another, perhaps complementary, explanation for the low primate diversity in Asian rain forests is the dominance of the forest canopy by the inedible leaves and branches of dipterocarps, coupled with the tendency of these and many other species to flower and fruit at intervals of 2–7 years. As described in Chapter 2, the everwet forests of tropical Asia are truly "fruit deserts" most of the time, and also appear to be impoverished in the insects that feed on flowers, fruits, and seeds, as well as the amphibians, reptiles, and birds that depend on those insects as basic food items (Duellman & Pianka 1990). For Asian primates, dietary specialization would be fatal, since fruits and insects are unreliable sources of food, even if the squirrels were not already there as competitors. Thus, the majority of Asian primates are either primarily leaf-eaters, such as some langurs and the proboscis monkey, or are at least capable of subsisting largely on leaves when fruit is rare. As discussed earlier, leaf-eating favors a large size. Presumably it is because good-quality food is so often scarce and in such small patches that orangutans are wide-ranging, largely solitary, foragers, contrasting with the social feeding of the African great apes. The orangutans are frugivores by preference, but will eat leaves, bark, and even insects when fruits cannot be found, and some populations regularly kill and eat slow lorises (van Schaik et al. 2003). Reproduction in orangutans and other Asian primates is enhanced during dipterocarp fruiting years, when a majority of rain forest trees follow the lead of the dipterocarp species and fruit at the same time, allowing the animals to feast on the rich abundance of fruit (Knott 1998).

## Madagascan primates

The island of Madagascar provides a useful natural experiment, in that there was an independent radiation of the lemurs, filling the ecological niches occupied by monkeys, apes, and other primates in the rain forest areas of other regions. The oldest lemur fossils in Madagascar are from the Pleistocene and there are no definite lemur fossils from anywhere else, so we have no direct evidence for their arrival and diversification. Molecular evidence suggests that all the Madagascan lemurs are descended from a single species that arrived as much as 60–65 million years ago (Yoder & Yang 2004). Madagascar has been separated from the African mainland for 100 million years, so they must have crossed a formidable water barrier to get there. No apes or monkeys have managed to cross the Mozambique Channel, allowing the lemurs to diversify in their own unique way. The channel has also kept out squirrels, terrestrial herbivores (see Chapter 4), and many important groups of rain forest birds (see Chapter 5). Indeed, when the lemurs first arrived on Madagascar 60 million years ago, mammalian carnivores would also have been absent (Yoder et al. 2003). In striking contrast to the Americas, therefore, the lemurs radiated into an environment with few competitors. The Madagascan primate radiation is also much

## Box 3.1 The lemur radiation

The lemurs have undergone a broad diversification, evolving at least 32 living species, in 14 genera, in five families, widely separated from each other and only distantly related to the other strepsirhines (Fleagle 1999). Three additional families have become extinct within the last thousand years (Godfrey & Jungers 2002). The eight families are:

1 The **Cheirogaleidae** is a family of nocturnal, nest-building, solitary animals, with a weight of less than 500 g (20 oz). Most notable are the seven species of mouse lemurs, which feed on a diverse diet of insects, small vertebrates, fruits, flowers, nectar, and gums. A recently discovered dry forest species, *Microcebus berthae*, has the distinction of being the world's smallest primate at a weight of around 30 g with a total length of around 20 cm, and the slightly larger brown mouse lemur (*M. rufus*) is the smallest rain forest primate (see Plate 3.3, between pp. 150 and 151). Mouse lemurs have soft fur, long limbs and tail, large eyes, and a relatively short snout. They are probably the most numerous of all lemur species.

2 The **Lemuridae** are the typical Madagascar lemurs of medium size (1–4 kg, 2–9 lb), organized in groups, active during the day or intermittently during the day and night, and feeding on leaves and fruit. A well-known species is the ring-tailed lemur. Also in this group are the bamboo lemurs that feed almost entirely on bamboo shoots (Fig. 3.6a).

3 The **Indriidae** are specialized leapers with an enlarged intestine and cecum, which aid in the digestion of leaves. A well-known group of species is the sifakas, which leap actively between trees, hang by their limbs when feeding, and hop on their back legs when moving on the ground. The indri (*Indri indri*) is the largest of all living lemurs (weighing up to 7.5 kg).

4 The **Lepilemuridae** consist of the sportive lemurs, small drab lemurs that are nocturnal and eat leaves.

5 The **Daubentoniidae** consist of a single living species, the aye-aye (*Daubentonia madagascariensis*), certainly the strangest looking primate (Fig. 3.6b). The aye-aye is a medium-sized black lemur with coarse fur, huge ears, a bushy tail, two large, rodent-like incisors, and long clawed digits, particularly the third digit on the hand. The aye-aye forages for insects at night, tapping branches with an elongated finger, and gnawing on bark with their teeth, sort of a primate equivalent of woodpeckers, which did not reach Madagascar. There was also a second, five times larger, species, the giant aye-aye (*D. robusta*), until its recent extinction (Godfrey & Jungers 2002).

6 The **Archaeolemuridae** is an extinct family of rather monkey-like lemurs with estimated body weights in the 14–27 kg range.

7 The **Megaladapidae** is an extinct family of large-bodied (38–75 kg) lemurs that were probably slow-moving leaf-eaters.

8 The **Palaeopropithecidae** is an extinct family of "sloth lemurs", which included the largest known lemur, the gorilla-sized *Archaeoindris*.

older than the American radiation. As a result of both these factors, they now dominate the rain forest fauna in a way not seen in any other rain forest region.

Using living species to explore the evolutionary radiation of the lemurs presents some problems. Many of the larger species and genera were driven to extinction when people first arrived on the island 2000 years ago (Godfrey et al. 1997; Goodman & Patterson 1997; Burney et al. 2003). Most of the at least 17 known extinct species were larger in size than any surviving species. Among the most notable species that have gone extinct is *Palaeopropithecus ingrens*, a 50 kg leaf-eating sloth lemur. Its forelimbs were much longer than the hindlimbs, suggesting it was able to hang in a manner comparable to a modern sloth or orangutan. Another large, extinct sloth lemur is *Archaeoindris*, a 200 kg leaf-eating ground-dweller, probably comparable to a modern gorilla. Two thousand years ago, therefore, the Madagascan lemurs spanned much the same size range as the apes, monkeys, and strepsirhines do together in Africa. However, there are no

**Fig. 3.6** Two species of lemurs from Madagascar. (a) Bamboo lemur (*Hapalemur* sp.). (Courtesy of R.A. Mittermeier, Conservation International.) (b) Aye-aye (*Daubentonia madagascariensis*). (Courtesy of Harald Schuetz.)

fossil records from eastern Madagascar (Godfrey & Jungers 2002), so we can only speculate on whether any of these extinct species lived in the rain forest.

There are several characteristics that distinguish the ecology of lemur communities from other primate communities (Ganzhorn & Sorg 1997; Wright 1997). Firstly, Madagascar is unique in the diversity and abundance of relatively small, leaf-eating species. Folivorous primates make up most of the biomass of primate communities elsewhere, but in Asia, Africa, and the Americas this biomass is made up of a few species of large (more than 5 kg) primates. In Madagascan rain forests, in contrast, most of the biomass comes from a very high density of 1–5 kg folivores (although the recently extinct, very large lemur species were probably also folivorous). Secondly, some species of lemur often descend to the ground to search for food, in the same way as many unrelated African species and a few species in Asia do; although none at all do in the Neotropics. This is probably partly due to the relatively open nature and short height of most Madagascan forests and the lack of competition for food on the ground. Also, being on the ground does not add substantially to the risk of attack from predators, since the greatest danger to lemurs is from hawks and the fossa, an endemic species of carnivore (see Chapter 4), both of which can attack lemurs in trees. Thirdly, lemurs use their sense of smell more than primates do elsewhere. They mark their bodies and their territories extensively with scents produced from scent glands on their face, wrists, and urogenital region, depending on the species. In the relatively dry forests where most lemurs live, these scents may be more lasting than in wetter rain forests elsewhere. Fourthly, many lemurs are active at night or during both day and night, in contrast to primates in other regions that are more specialized for being active at just one time period, often during the day. There are no other large Madagascar mammals active at night, so this niche is available to lemurs, such as the avahi (*Avahi laniger*), which eats leaves, bark, buds, and fruit, and the aye-aye, which eats seeds, fruits, and insect larvae.

Despite the apparently low fruit supply in Madagascan forests (see Chapter 2), a number of lemur species, ranging in size from the 40 g brown mouse lemur (*Microcebus rufus*) to the 6.5 kg diadem sifaka (*Propithecus diadema*) (see Plates 3.3 and 3.4, between pp. 150 and 151), are highly frugivorous. These frugivorous lemurs have various ways of surviving the long nonfruiting season. Sifakas rest more, have a low metabolic rate, and females stop lactating. Red-bellied lemurs (*Eulemur rubriventer*), in contrast, travel further in times of fruit shortage, eat a wide variety of alternative foods, and may continue feeding after dark. The most extreme adaptation to the strongly seasonal supply of high-quality food is found in the squirrel-sized greater dwarf lemur (*Cheirogaleus medius*), which accumulates large amounts of fat in its tail when sugar-rich fruits are available and then hibernates for several months during the nonfruiting season (Fietz & Ganzhorn 1999). Most female and some male brown mouse lemurs do the same. This ability to hibernate or go into torpor is just the sort of adaptation that may have enabled the ancestors of modern lemurs to survive the long water crossing to Madagascar 60 million years ago (Kappeler 2000).

In addition to the role of lemurs in seed dispersal, discussed later, there is also evidence that several plant species, including the endemic traveler's palm (*Ravenala madagascariensis*), may depend on lemurs for pollination (Kress et al. 1994). Thus the lemurs of Madagascar have not only occupied the niches filled by primates in other rain forest regions, but they have also taken over some of the roles fulfilled by fruit- and nectar-eating birds and bats elsewhere. The role of the aye-aye, as a sort of primate woodpecker (see Box 3.1), is another example of this phenomenon. The larger species of recently extinct lemurs, such as *Archaeoindris*, may have extended the dominance of the lemurs to ground level, making use of resources consumed by other groups of terrestrial mammals in other rain forests (see Chapter 4). It is no exaggeration to call the Madagascan rain forest the "lemur forest".

## Primate equivalents in Australia and New Guinea

The rain forests of Australia and New Guinea are places to examine how animal communities respond in the absence of primates. The example is not clear-cut because these areas also lack other major groups of mammals, such as carnivores, ungulates, and squirrels (see Table 3.1). In these rain forests, some of the activities of primates have been taken over by marsupials, including species of possum and tree kangaroo. Marsupials are also abundant in Neotropical rain forests and the fossil record shows that they were once much more diverse than they are today. However, the presence of other mammalian groups, such as sloths and, later, primates, has restricted their diversity in the Neotropics. The rain forest marsupials in Australia and New Guinea, in contrast, vary greatly in size, diet, and time of activity in a manner somewhat similar to the divisions found in primate communities elsewhere, particularly in Madagascar (Smith & Ganzhorn 1996).

The tree kangaroos (*Dendrolagus* spp.) are the largest arboreal mammals in the rain forests of New Guinea and Australia, with adult weights of 6–15 kg (13–33 lb). In contrast to ground-dwelling kangaroos, the front and hindlimbs of tree kangaroos have similar proportions. They are agile climbers, leaping from branch to branch or to the ground. Their diets are little known, but appear to consist largely of leaves, although fruits and flowers are also eaten. Tree kangaroos are foregut fermenters, like colobines and sloths, digesting cellulose in a modified forestomach (Hume 1999).

The possums are an extremely diverse group of marsupials, with at least 35 species in New Guinea, ranging in size from the 20 g long-tailed pygmy possum (*Cercartetus caudatus*) to the 6.5 kg black-spotted cuscus (*Spilocuscus rufoniger*) (Fig. 3.7) (Flannery 1995a). For reasons that are not fully understood, the possums are most diverse in the highlands and there are relatively few species in lowland rain forests. The smallest species seem to feed mostly on insects, while most of the larger species are apparently more or less omnivorous. The ringtail possums (Pseudocheiridae) are slow-moving arboreal leaf-eaters. They are much smaller animals (150–1500 g) than the tree kangaroos and are hindgut fermenters, digesting cellulose in a greatly expanded cecum like the sportive lemurs of Madagascar (Hume 1999). In the absence of squirrels, marsupials have also evolved squirrel-like forms, including the sugar glider (*Petaurus breviceps*), which has converged toward the form of a flying squirrel. There are even marsupial

**Fig. 3.7** A marsupial mammal from New Guinea that looks and acts somewhat like a primate: the spotted cuscus (*Spilocuscus* sp.). (Courtesy of R.A. Mittermeier, Conservation International.)

equivalents of the Madagascan aye-aye, the striped possums (*Dactylopsila*), with powerful jaws, specialized teeth, an elongated fourth finger, and a long tongue for extracting wood-boring insect larvae (Rawlins & Handasyde 2002). Another curious feature of these animals is their rather skunk-like smell, associated with their skunk-like, black-and-white striped, color patterns.

Old World primates and marsupials meet only on the island of Sulawesi, in the center of the Indonesian Archipelago. Here it is a marsupial, the bear cuscus (*Ailurops ursinus*), which fills the arboreal leaf-eater niche, while the only primates are omnivorous macaques (Dwiyahreni et al. 1999). The bear cuscus is unusual for a cuscus, both in its large size (*c.* 7 kg) and the fact that it is active during the day. There is also a nocturnal frugivorous species on Sulawesi, the dwarf cuscus (*Strigocuscus celebensis*).

## Primates as seed dispersal agents

Primates play multiple roles in the ecology of tropical rain forests, as predators of invertebrates and, in some cases, vertebrates, as leaf-eaters, as seed dispersal agents, and even occasionally as pollinators. Primates are also the most threatened of rain forest animals and there are few areas left with an intact primate community in which all species occur at their natural densities. It has been suggested that the decline or local extinction of primates will lead eventually to the loss of plant species that depend on them for seed dispersal (Chapman & Onderdonk 1998). Yet rain forests have survived without primates for millennia in New Guinea and Australia, as well as on many tropical islands. What is so special about seed dispersal by primates?

Primates make up a substantial proportion of the total frugivore biomass in all rain forests in which they occur and eat huge amounts of fruit, but frugivory does not necessarily result in seed dispersal. What matters is the fate of the seed. Most primates kill some seeds by breaking them in the mouth, drop or spit out others, and swallow and defecate the rest. Seeds are usually hard and often toxic, so only a minority of specialized primate species destroy large numbers of seeds. The colobine monkeys of Africa and Asia, the pitheciine monkeys of America, and the sifakas (*Propithecus*) of Madagascar destroy the seeds in most fruits that they consume. Some tiny seeds may be swallowed accidentally, but most of these "seed predators" eat only unripe fruits, so even these seeds may not be viable.

The proportion of seeds that are dropped or spat out also varies greatly from species to species, with larger seeds and those that are easily separated from the flesh most likely to suffer this fate. Most of these seeds end up directly under the fruiting tree, which is not a good place for a seed to be, since it must compete with its siblings and parent, as well as survive the pests and diseases that are concentrated there. The cercopithecine monkeys of Africa and Asia, however, are seed-spitters with a difference (Corlett & Lucas 1990; Lambert 1999). Many seeds are dropped while feeding, but these monkeys also have cheek pouches in which fruit is carried away from the fruiting tree. The seeds are then spat out one by one as the monkey moves through the forest.

Apes, most lemurs, and most New World monkeys swallow most of the seeds in the fruits they eat and defecate them later, unharmed. In most cases, these primates are probably high-quality dispersal agents, carrying the seeds far from the parent tree and scattering them through the forest. Larger primates, such as chimpanzees or howler monkeys, may provide a lower quality dispersal service, particularly if they defecate on the ground or from low branches, because large numbers of seeds are deposited in a clump. In contrast, some smaller primates, such as the Neotropical tamarins (Knogge & Heymann 2003), defecate large seeds singly and small seeds in small clumps, while the feces of canopy primates, such as gibbons, often shatter before they reach the ground. This is an oversimplification, however, and the postdispersal fate of the seeds depends on numerous factors, including the attractiveness of primate feces to seed-predating rodents and the abundance of feces-burying dung beetles (Andresen 2002; Feer & Forget 2002).

Are primates providing a unique seed dispersal service in rain forests, or could other animals compensate for their loss, as they presumably have compensated for their natural absence in New Guinea? Birds and fruit bats are the other major seed dispersal agents in tropical rain forests (see Chapters 5 and 6, respectively), followed by mammalian carnivores and various ground-dwelling large herbivores (see Chapter 4). Studies that have compared the fruit diets of primates with those of other dispersal agents have invariably shown some degree of overlap, but there are also always fruits that are consumed largely or only by primates. Most of these "primate fruits" are relatively large, with large seeds, and many have a thick, inedible husk, which is hard to remove without the coordinated use of hands and teeth. If primates were lost from a rain forest area, these plant species would lose their major seed dispersal agents.

Overall, it appears that rain forest primates provide an important service for many plants species and an essential service for some. Primates and rain forests are truly co-dependent (Chapman & Onderdonk 1998). The data currently available are insufficient for pantropical comparisons, but dependence on primates may be highest in Madagascar, where the primate radiation is particularly ancient and there are relatively few frugivorous birds, bats, and other mammals (Dew & Wright 1998).

## Conclusions and future research directions

Primates give valuable insight into differences among rain forest regions because of their diversity and the variety of their ecological roles. These differences are well known because primates have been intensely investigated for insights into human evolution. Humans are currently altering primate distribution and abundance in a way that might incidentally give additional insight into rain forest ecology.

Primates are being intensively hunted for meat in many parts of the world and there are now vast areas of rain forest without any primates or with a greatly reduced density (Chapman et al. 2000; Cowlishaw & Dunbar 2000; Chapman & Peres 2001). When primates are eliminated from the forest, how will that influence the ecology of the remaining species? As discussed above, the most important impact is likely to be on seed dispersal. Several studies have now looked at the consequences of primate removal for primate-dispersed plants (e.g. Chapman & Onderdonk 1998; Pacheco & Simonetti 2000), but there is an urgent need for broader, longer term studies and for pantropical comparisons. Although primates feed on fruits in all areas of the world, the importance of primates as seed dispersers may be greater in some areas than others. It may also be possible to look at the roles of particular primate species, where only one or a few have been eliminated. For instance, macaques in Asia and howler monkeys in the Neotropics often persist in rain forest fragments from which other primate species have been lost.

Unfortunately, comparisons between forests from which primates have been eliminated by human activities and forests with intact primate communities are not as straightforward as might at first appear. Because all large vertebrates are hunted in addition to primates, it may be difficult to separate the special ecological role of primates from the impact of other large animals if all the large vertebrates have been removed. A better comparison would be between a forest where primates have been hunted out and a forest in which a new population of primates has been established. Alternatively, a before/after comparison could be made at a single site where primates are being reintroduced. Primate reintroductions are rare at present—one exception being the successful reintroduction of the charismatic golden lion tamarin (*Leontopithecus rosalia*) to forest fragments in southeast Brazil—but their number is certain to increase, providing a range of opportunities for studying their impact.

There are also lessons to be learned from the introduction of primate species *outside* their natural ranges, although we emphasize that this is not a reason to encourage such introductions, which can potentially cause massive ecological damage. In this case, the main interest is not in the primates' role in seed dispersal, but in their impact as predators, particularly on vertebrates. A surprising number of primate species have been released, deliberately or accidentally, in other parts of the world, usually on islands, but most of the "successes" have been with Asian or African cercopithecines (Long 2003). Macaques, in particular, are now established on previously primate-free islands throughout the tropics. The most worrying of these "successes" has been the establishment of a wild population of the long-tailed macaque (*Macaca fascicularis*) on the island of New Guinea, where it is feared they may affect the endemic wildlife, such as the birds of paradise, by stealing eggs from their nests. This movement of macaques across one of the world's major biogeographical boundaries could be seen as analogous to the arrival of monkeys in South America 30 million years ago, or of lemurs in Madagascar even earlier.

The successful introductions have almost all been with widespread "weedy" species like the long-tailed macaques. Other possible movements of primates across natural biogeographical barriers should remain forever as thought experiments. What if the apes had reached Madagascar or South America? Is the Amazonian rain forest unsuitable for large primates or is their absence simply a biogeographical accident? Imagine releasing populations of chimpanzees, gorillas, or other large African primates into a suitable Amazonian rain forest. Would these primates survive or perhaps thrive? Would these primates alter the surrounding biological community? How would orangutans do in an African or Amazonian forest? Such experiments are certainly not appropriate to carry out because of the potential for introducing new diseases and other unforeseen changes. The best strategy for most species is to protect them where they are currently living. However, at some point in the future, such primate introductions might be a potentially valuable conservation option. If we are unable to protect gorillas in Africa and orangutans in Indonesia, could they be protected in a large national park in Brazil, or Costa Rica, or on a Caribbean island, such as Puerto Rico?

Primates have great interest for humans because of their close relationship to people. But they are also useful for understanding rain forest differences. Other mammalian groups provide further valuable insights, and these will be considered in the next chapter.

## Further reading


Chapman C.A. & Peres C.A. (2001) Primate conservation in the new millennium: the role of scientists. *Evolutionary Anthropology* **10**, 16–33.

Cowlishaw G. & Dunbar R.I.M. (2000) *Primate Conservation Biology*. University of Chicago Press, Chicago, IL.

Eimerl S. & DeVore I. (1965) *The Primates*. Time Incorporated, New York.

Fleagle J.G. (1999) *Primate Adaptation and Evolution*, 2nd edn. Academic Press, San Diego, CA.

Fleagle J.G., Janson C. & Reed K. (eds) (1999) *Primate Communities*. Cambridge University Press, Cambridge, UK.

Groves C.P. (2001) *Primate Taxonomy*. Smithsonian Series in Comparative Evolutionary Biology. Smithsonian Institution Press, Washington, DC.

Hartwig W.C. (ed.) (2002) *The Primate Fossil Record*. Cambridge University Press, Cambridge, UK.

Reed K.E. & Fleagle J.G. (1995) Geographical and climatic control of primate diversity. *Proceedings of the National Academy of Sciences of the United States of America* **92**, 7874–7876.

Terborgh J. (1983) *Five New World Primates: a Study in Comparative Ecology*. Princeton University Press, Princeton, NJ.

# Chapter 4

# Carnivores and Plant-eaters

Primates are not the only group of mammals that show major differences among rain forest regions. Few other mammals are as well studied as the primates, but many differences between rain forest regions are too obvious to overlook. In this chapter, we use two relatively well-known groups of mammals with very different diets and behaviors—the carnivores and the plant-eating mammals of the forest floor—to illustrate general differences between rain forests. These are both ecological, rather than taxonomic, groups, although a single taxonomic order, the Carnivora, accounts for most carnivorous mammals. Their inclusion in the same chapter is justified by their occupation of the same forest floor habitat —although a few carnivores are mainly arboreal—and the fact that the plant-eaters provide the major food for the carnivores. They have also had major reciprocal influences on each other over evolutionary time, so that many elements of the ecology and behavior of herbivores make sense only in relation to the carnivores that threaten them, while many features of carnivores are clearly adaptations to their preferred prey. In addition, most members of both groups are very poor at crossing even narrow ocean barriers, so their spread across the Earth's surface has been subject to similar constraints and opportunities. Finally, large carnivores and large terrestrial herbivores are the animals especially vulnerable to human impacts, so understanding their ecological roles and how these may differ between rain forest regions is particularly urgent.

## Carnivores

Carnivores are a key element in the rain forest animal community. They reduce the population numbers of their prey, including large terrestrial herbivores, primates, rodents, birds, reptiles, insects, and other, smaller, carnivores. Prey animals also avoid places and times of day associated with greater exposure to attack: the "ecology of fear" (Ripple & Beschta 2004). By influencing the population sizes and behaviors of herbivorous animals, carnivores have an indirect impact on the structure of plant communities. Equally important may be the influence of large carnivores on the abundance of the smaller carnivores, which are the main predators of birds and other small vertebrates. The removal of large carnivores can therefore have major consequences for the rest of the rain forest community as the effects propagate from level to level down the food web, from top carnivores to plants, in what is sometimes called a "trophic cascade" (Pace et al. 1999; Terborgh et al. 2001). At some isolated sites, the removal of carnivores has lead to substantial increases in rodent populations, and a consequent decline in seedling abundance. Unfortunately, we currently know far too little about the complexities of rain forest food webs to predict what the general impact of the loss of particular carnivore species will be in larger areas.

Paradoxically, many rain forest carnivores are also important seed dispersal agents. Despite their obvious adaptations for killing and eating other animals, most carnivores eat at least some fruit and some, including many civets, procyonids, and mustelids, are best described as opportunistic omnivores. The teeth and jaws of carnivores are not designed for chewing and the digestive system is unspecialized, so seeds pass through undamaged into the feces.

Although carnivores and Carnivora are practically synonyms today, the conquest of the rain forest by members of this mammalian order has been achieved in stages. By the beginning of the Miocene, modern families of Carnivora were already established in Asia, Africa, and North America (Nowak 1999). A single, mongoose-like species then crossed the substantial water gap from Africa to Madagascar around 20 million years ago, giving rise to a remarkable radiation of endemic carnivores (Yoder et al. 2003). This water-crossing feat was not repeated until around 9 million years ago, when a kind of raccoon (now extinct), from the North American family Procyonidae, traversed the narrowing water gap between Central and South America (Webb 1997). However, it was then not until the first definite land connection arose, around 3 million years ago, that more procyonids and four additional families of Carnivora—the cats (Felidae), dogs (Canidae), bears (Ursidae), and weasels (Mustelidae)—finally crossed from Central America into the southern continent. Among the major rain forest regions, only Australia and New Guinea remained free of Carnivora until recently, but this too is changing due to human influences. The Australian dingo is a descendent of domesticated dogs brought to Australia (and New Guinea) several thousand years ago, while the European fox and feral cats were brought by European settlers more recently. All three species have had a major impact on native wildlife but, so far, are not established within the rain forest.

The mammalian carnivore niche in the rain forests of both modern Australia–New Guinea and pre-Pliocene South America was occupied by marsupials. A variety of giant marsupial carnivores from several lineages appears in the Miocene fossil record in both regions. Descriptions of these animals as "marsupial lions", "marsupial wolves" and so on gives an idea of their size, but understates the bizarreness of their appearance. However, most of the larger species had disappeared before the end of the Tertiary. In South America, the large carnivore niche in open habitats was dominated by giant flightless "terror birds" immediately before the faunal interchange, although it is unlikely that these were present in the rain forest. In Australia and New Guinea, marsupials still dominate, but there are no longer any really large species.

It is also important to remember that mammals are not the only flesh-eating vertebrates in tropical forests. At La Selva, Costa Rica, more than 100 species of vertebrates eat other vertebrates, of which only 14 are mammalian carnivores (Greene 1988). In all the major rain forest regions, all but the largest prey species are shared with snakes and, in many cases, birds of prey (see Chapter 5). Both the Amazon and Congo rain forests also support small (< 2 m, 6.5 feet), stream-dwelling crocodiles: the dwarf crocodile (*Osteolaemus tetraspis*) in the Congo basin and the dwarf caiman (*Paleosuchus trigonatus*) in the Amazon basin. When not hunted, the dwarf caiman is reported to have by far the highest biomass of any vertebrate predator in the central Amazonian rain forest and must have an impact on the populations of the small mammals that it consumes (Magnusson & Lima 1991). Neotropical crocodile diversity was much higher in the Miocene and the forest fauna included large terrestrial species that may have survived until the arrival of placental carnivores in the Pliocene (Kay & Madden 1997). Old World rain forests, from Africa through Asia to Australia, also support monitor lizards (*Varanus* spp.) of various sizes, some of which prey on other vertebrates.

## Cats

Members of the cat family (Felidae) are the ultimate killing machines, with none of the compromise adaptations to an omnivorous diet shown in other rain forest carnivores (Sunquist & Sunquist 2002). Cats include the largest carnivores hunting in the forest, some of which prey upon even the largest forest herbivores (Fig. 4.1). Cats are present in the Americas, Asia, and Africa, but their diversity and size range in the rain forest differs markedly among these three regions. Many of these rain forest cats also occupy a range of other vegetation types and, in these species, the rain forest animals are usually considerably smaller than their cousins in other habitats, perhaps as a result of the generally smaller size of the available prey in rain forests. Some species are among the most widespread of all vertebrates, with the cougar (*Puma concolor*) distributed from the Canadian Yukon in the north to the Straits of Magellan in the south, while the leopard (*Panthera pardus*) used to occupy most available habitats in both Africa and Asia.

Asia is the richest in rain forest cats, with 10 species in Southeast Asia, ranging in size from the 200 kg (440 lb) tiger (*Panthera tigris*) and the 60 kg (130 lb)

(a)

**Fig. 4.1** Large cat species found in rain forests. (a) Sumatran tiger (*Panthera tigris* var. *sumatrae*). (Courtesy of Tim Laman, taken in captivity.)

**Fig. 4.1** *(cont'd)* (b) Leopard (*Panthera pardus*) in the wild, Kruger National Park, South Africa. (Courtesy of Tim Laman.) (c) Jaguar (*Panthera onca*). (Courtesy of Tim Laman, taken in captivity.)

leopard (*P. pardus*) (Fig. 4.1b), to the tiny flat-headed cat (*Felis planiceps*), at 2 kg (4.4 lb) weighing less than a housecat (Sunquist & Sunquist 2002). Most of these species can coexist in the same rain forest. An even smaller species, the 1–1.5 kg rusty-spotted cat (*Prionailurus rubiginosus*), is found mostly in dry forests in southern India, but also occurs in rain forest on the island of Sri Lanka, which supports only three other cat species. The tiger generally hunts on the ground, often at night, preying upon even very large animals such as the gaur, a relative of domestic cattle weighing up to 1000 kg, although pigs and deer are usually most important. Individual rain forest tigers can roam over tens of square kilometers. The leopard hunts mainly at night, taking a huge range of prey, including ungulates, monkeys, smaller vertebrates, and even insects. Where tigers and leopards coexist, the tiger takes bigger prey.

All of the other Asian rain forest cat species weigh less than 25 kg, and seven species weigh under 15 kg. The clouded leopard, *Neofelis nebulosa*, is intermediate in size, weighing up to 23 kg, but is notable as one of the most arboreal of the Asian cat species, able to hunt in trees and to leap onto prey from overhanging branches. Its relatively short legs, huge paws, and very long tail are all probably adaptations to an arboreal life. It also has the longest canine teeth in relation to head size of any living carnivore and the backs of the canines are very sharp, like the extinct saber-toothed cats. Many of the smaller Asian cat species typically hunt near water, catching prey such as birds, rodents, other small vertebrates, and insects, as they come to drink or feed near riverbanks or wetland margins. One species is known as the fishing cat, *Felis viverrinus*, for its readiness to swim and its ability to catch fish and crustaceans. The tiny flat-headed cat also lives along riverbanks where it catches fish and frogs. With its short legs and tail, long head, and tiny ears, this species looks, in some ways, more like a civet or mustelid than a cat.

Although Africa has many species of cats, including the familiar lion and cheetah, only two species live in the rain forest, the versatile leopard and the African golden cat (*Profelis aurata*), which is restricted to forest (Hart et al. 1996). Golden cats are large-pawed, lightly built, 8–16 kg predators that, like the leopard, mainly hunt on the ground, taking mostly small mammals and birds, but also duikers and monkeys. African rain forests have neither a giant tiger- or jaguar-sized cat nor any really small species. The ecological role of the small cats is perhaps taken over by the forest genets, the smallest members of the rather cat-like civet family (Viverridae), which is described in more detail later in this chapter. Overall, however, African rain forests have a considerably lower diversity of carnivorous mammals than rain forests in Asia and the Americas, despite the high diversity of potential prey species. Africa has no rain forest dogs, bears, or raccoons, and relatively few rain forest species of cats, civets, and mustelids (weasels, badgers, and their relatives).

Although cats did not enter South America until 3 million years ago, the American rain forest is intermediate in numbers of cat species, with six species that range in size from the 120 kg jaguar (*Panthera onca*) (Fig. 4.1c) and the 60 kg puma (*Puma concolor*, also known as the cougar or mountain lion), to four smaller species including the ocelot (*Leopardus pardalis*, 16 kg), the jaguarundi (*Herpailurus yaguarondi*, 7 kg), the margay (*L. weidii*, 4 kg), and the oncilla (*L. tigrinus*, 3 kg), also known as the tiger cat, tigrina, or little spotted cat (Sunquist & Sunquist 2002). All these species can coexist in the same area of rain forest, although the oncilla is less widespread than the other five species. Most, if not all, of these cat species appear to have originated in North or Central America before the interchange, and thus represent recent invaders, rather than a separate South American radiation of rain forest cats (Webb 1997).

The jaguar is more closely related to the Asian tiger than to other American cats, a fact that appears less surprising when it is remembered that the modern distribution of the tiger extends north to the Arctic circle, suggesting that its ancestors had no need for a tropical land route to the Americas. The jaguar hunts mainly on the ground, often near water, and primarily at dawn and dusk (Rabinowitz 2000). Like their Asian cousins, jaguars can easily kill the largest prey available in the rain forest and they are known to take peccaries, capybaras, agoutis, deer, tapirs, caimans, river turtles, and even fish. The leopard-sized pumas have a diet similar to the jaguar, but, like leopards, also include smaller vertebrate prey and even large insects. Ocelots appear to hunt opportunistically, taking small terrestrial prey (< 4 kg) more or less in proportion to their availability. The rather weasel-like jaguarundi is a largely terrestrial feeder on small vertebrates and arthropods, but it also climbs well. The margay, in contrast, seems to forage mostly in the trees, catching small mammals, birds, lizards, and tree frogs. Margays have the ability to rotate their hindfeet through 180°, allowing them to run straight down a tree trunk like a squirrel (de la Rosa & Nocke 2000). These are the only cats that include fruit as a regular part of their diet. Finally, the oncilla is a largely terrestrial consumer of small vertebrates and insects, which is reported to adapt well to disturbed and even suburban habitats (de la Rosa & Nocke 2000).

## Dogs

Unlike the cats, members of the dog family (Canidae) are relatively uncommon in rain forests. Dogs were confined to North America until the late Miocene, before spreading to Eurasia (*c.* 9 million years ago), Africa (4.5 million years ago), and finally South America (2 million years ago) (Hunt 1996). The Neotropics has two small rain forest canids, both of which are little known and apparently rare (Macdonald & Sillero-Zubiri 2004). The small-eared dog (*Atelocynus microtis*) is a medium-sized (8–10 kg, 18–22 lb), rather cat-like, predator, with a long slender muzzle, short ears, and bushy tail. It appears to be mostly solitary and at least partly aquatic, with most sightings in or near water. Its diet includes fish, frogs, small mammals, and fruits (Nowak 1999). Bush dogs (*Speothes venaticus*) are smaller (5–7 kg), with a very short muzzle, legs and tail, and are also often associated with water. In contrast to the small-eared dog, they live in social groups that hunt cooperatively. This allows them to prey on animals larger than themselves, up to and including the largest of all Amazonian land mammals, the lowland tapir (*Tapirus terrestris*) (Wallace et al. 2002). Most of their diet, however, seems to consist of large rodents. The crab-eating fox (*Cerdocyon thous*) also occurs widely in forests and savannas in South America often near riverbanks, where it subsists on small vertebrates, crabs, crayfish, insects, and fruit.

African rain forests lack dogs, but Southeast Asian forests have the widely distributed dhole (*Cuon alpinus*), a large-eared, 12–18 kg predator that hunts in packs of about 10 individuals (see Plate 4.1, between pp. 150 and 151). Cooperative hunting turns this medium-sized wild dog into a fearsome predator, able to overcome prey as large as deer, wild pigs, and wild cattle. There are even reports of dhole packs overcoming tigers. Another canid, the omnivorous raccoon dog (*Nyctereutes procyonoides*), is widespread in East Asia and enters the rain forest on the northern margins of the tropics.

## Weasels and their relatives

The weasel family (Mustelidae) has more species globally than any other group of carnivores, but relatively few occur in tropical rain forests (Nowak 1999). Most are long, slender animals such as weasels, but there are also stocky, badger-like forms, and intermediate species such as martens (Fig. 4.2a). Rain forest mustelids are most diverse in Southeast Asia, where they include the mainly terrestrial Malayan weasel (*Mustela nudipes*), the larger, tree-climbing yellow-throated marten (*Martes flavigula*), the large (up to 14 kg, 31 lb) terrestrial, badger-like hog-badger (*Arctonyx collaris*), ferret-badgers (*Melogale* spp.), and smaller (1–3 kg) stink badgers (*Mydaus* spp.), and the aquatic otters. African rain forests support only aquatic otters and the omnivorous honey badger or ratel (*Mellivora capensis*), which can kill animals up to the size of small forest duikers. Honey badgers get their common name from their habit of raiding the nests of wild bees. Mustelids are, again, relatively diverse in the Neotropics, where they include the large (5 kg), slender, arboreal tayra (*Eira barbata*), the predominantly terrestrial, short-legged and slender-bodied grison (*Galictis vittata*), the tropical weasel (*Mustela africana*), and several otters, including the giant river otter (*Pteronura brasiliensis*), which can exceed 2 m (6.5 feet) from head to tail and weigh 22–34 kg. The hog-nosed skunk (*Conepatus semistriatus*) occurs in rain forests in Central America and adjacent parts of South America. As with cats, this Neotropical diversity is surprising in view of the late arrival of mustelids, which took place only after the formation of the Panama land bridge 3 million years ago.

Mustelids in general are opportunistic carnivores, eating whatever invertebrates, small vertebrates, eggs, and fruit they can obtain. The slender, weasel-like species are fast, active hunters and can catch and kill vertebrates up to their own size or larger. For example, the Neotropical tayra can run, swim, and climb to catch prey, and although it eats mainly rodents, it also consumes fruit and honey and can kill animals as large as small deer (Nowak 1999). The stocky, badger-like species, in contrast, seem to concentrate on soil invertebrates, such as earthworms, although some species also take small vertebrates and plant material.

## Bears

One interesting contrast is found among the three species of tropical bears (Ward & Kynaston 1995). The spectacled bear (*Tremarctos ornatus*) of South American montane forests is the most herbivorous of the three and feeds extensively on fruit, as well as the leaves of bromeliads and cacti. Its choice of food is biogeographically revealing, as both bromeliads and cacti are strictly New World families. In contrast, the Malayan sun bear (*Ursus malayanus*), which is the smallest of all bears, feeds extensively on the nests of termites and wild bees, palm hearts, fruits, and small vertebrates, using its strong curved claws to rip apart trees in search of food. Its legs turn inward, giving it expert tree climbing

(a)

**Fig. 4.2** (a) Yellow-throated marten (*Martes flavigula*) in a tree in Gunung Palung National Park. (Courtesy of Tim Laman.) (b) Binturong (*Arctictis binturong*) in Gunung Palung National Park, Indonesia. (Courtesy of Tim Laman.)

(b)

ability but an ungainly gait on the ground. The sloth bear (*U. ursinus*) of the Indian subcontinent and Sri Lanka has a diet also composed of fruits and social insects. Protrusible lips and the absence of the inner pair of upper incisors allow it to suck up termites like a vacuum cleaner. A third Asian bear species, the omnivorous Asiatic black bear (*U. thibetanus*), inhabits mostly mountainous and temperate regions of Asia, but also occurred in the northern range of Asian tropical rain forests. Africa has no bears south of the Sahara.

## Civets and mongooses

A variety of other carnivores inhabits tropical forests in one or more regions. One of the most distinctive groups of small to medium-sized carnivores is the civets (Viverridae), including the genets, which are found from Africa through Asia, but are most diverse in the Sundaland region of Southeast Asia (Fig. 4.3) (Nowak 1999). Civet species vary in weight from 600 g to 15 kg (1.5–35 lb). Somewhat cat-like in appearance, civets typically have an elongate body, short legs, a long, bushy tail, and a pointed muzzle. Many species have striped or spotted fur, perhaps in some species as warning coloration to predators that they can produce foul-smelling secretions from their anal glands. These secretions also function in territorial marking and social interactions, and the secretions from several species are used commercially in the production of perfumes. In other species, the patterns may aid in camouflage while hunting. Their patterned fur, elongated body, and long tail give many of them a beautiful appearance. Most civets are agile and skillful climbers, and some are also skilled swimmers. Civets eat small vertebrates, invertebrates, and fruits, and some species are important seed dispersal agents in both Asian and African forests. Several

(a)

**Fig. 4.3** Members of the civet family. (a) Common palm civet (*Paradoxurus hermaphroditus*) from Borneo. (Courtesy of Hans Hazebroek.)

(b)

(c)

**Fig. 4.3** (*cont'd*) (b) African civet (*Civettictis civetta*) in Kenya. (Courtesy of Harald Schuetz.) (c) Common genet (*Genetta genetta*) in Kenya. This is not a rain forest species but it illustrates the appearance of the genus. (Courtesy of Harald Schuetz.)

civet species, such as the common palm civet (*Paradoxurus hermaphroditus*) in Asia, have been able to adapt to human-dominated landscapes, living in house rafters and drains. The commercial harvesting of the anal gland secretions for the perfume industry has resulted in the introduction of the Malay civet, *Viverra tangalunga* (see Plate 4.2, between pp. 150 and 151), to several islands between Southeast Asia and New Guinea that previously lacked mammalian carnivores (Nowak 1999).

Some civets, including the common palm civet, are largely arboreal, foraging in the canopy for small vertebrates, insects, and fruits. One of the most distinctive of the arboreal Asian civets is the binturong (*Arctictis binturong*), a very large species (up to 15 kg) with long coarse black hair and a prehensile tail that it uses during its nocturnal foraging (see Fig. 4.2b). The only other carnivore with a truly prehensile tail is the Neotropical kinkajou (*Potos flavus*, Procyonidae), which is unrelated but ecologically rather similar (see below). The African palm civet (*Nandinia binotata*) is an arboreal omnivore with similar habits to the Asian palm civets, although recent molecular evidence suggests that this animal may not be a civet at all, but rather an early branch from the group that gave rise to cats, civets, and mongooses (Yoder et al. 2003). Similar evidence has shown that the Asian linsangs (*Prionodon* spp.) are not cat-like civets, but civet-like relatives of the cats (Gaubert & Veron 2003). African rain forests support several species of partly arboreal genet: small, spotted or blotched, cat-like carnivores that are confined to Africa, except for one species that extends north into Europe. Other civet species in both regions, such as the Malay civet and the African civet (*Civettictis civetta*), live mostly on the ground. There are also civets in both regions that are semiaquatic, notably two Southeast Asian species in the genus *Cynogale* and the aquatic genet (*Osbornictis piscivora*) of the Democratic Republic of the Congo.

The mongooses (Herpestidae) are related to the civets and have a similar Old World distribution (Nowak 1999). The rain forest mongooses are small, long-bodied, ground-dwelling carnivores, and many species are found only near water. These animals are known for their agility and quick movements; so fast and agile are mongooses (*Herpestes* spp.) that they can attack and kill poisonous snakes without being bitten. Most mongooses have a diet consisting of insects, other invertebrates, and small vertebrates, although many also consume some fruits. Cusimanses (*Crossarchus* spp.) are dark, shaggy, social mongooses that inhabit the rain forests of West Africa, moving around in large, noisy groups (Kingdon 1997). They are reported to break the shells of snails and eggs by throwing them against a tree or other hard objects.

## Raccoons and their relatives

The Americas lack native civets and mongooses, though mongooses have been introduced into the region to control pest rats, most notably in islands of the West Indies, with disastrous consequences for the native wildlife. In the Neotropics, their ecological roles are, to some extent, played by a distinctive New World family of carnivores, the Procyonidae, which includes the familiar North American raccoon. The single Old World representative of this family is the red panda (*Ailurus fulgens*) of the Himalayan Mountains. The procyonids are another of the many animal groups that entered South America from the north only within the last few million years, after a land route became available (Brown & Lomolino 1998). In tropical American forests, the coatis (*Nasua* spp.) are long-tailed and longer snouted versions of the raccoon, weighing 3–6 kg (6.5–13 lb). Coatis are inquisitive diurnal omnivores, foraging on the ground but also going into trees. Males are solitary, except during the breeding season, but females and juveniles travel in stable groups. The long tail is often carried straight up while walking, giving a group of coatis an unmistakable appearance. The name "coatimundi"—often used in English for the species—is applied in South America only to the lone males. Other rain forest members of the family include the kinkajou (*Potos flavus*), the olingo (*Bassaricyon gabbii*), and, in Central America, the cacomistle (*Bassariscus sumichrasti*). The kinkajou is a nocturnal, arboreal, monkey-like forager, weighing 1.5–4.5 kg, with a prehensile tail, eating fruits, seeds, insects, and small vertebrates. The olingo also lives in trees, but without the prehensile tail and weighing less at 1–1.5 kg. The cacomistle is a small (< 1 kg) arboreal procyonid that looks remarkably similar to the forest genets of tropical Africa. It has been little studied, but is said to be an opportunistic omnivore (de la Rosa & Nocke 2000). The crab-eating raccoon (*Procyon cancrivorus*) thrives in all Neotropical lowland habitats, including rain forest, but apparently always near water. It is largely nocturnal and forages largely on the ground, eating an unusually wide range of foods.

### Madagascan carnivores

Until recently, the seven living members of Madagascar's carnivore fauna were assigned to two or three separate families, each believed to result from a separate colonization event. Molecular evidence has now shown that, despite their diversity of form, they are all descended from a single, mongoose-like ancestor that crossed the ocean from Africa, 18–24 million years ago (Yoder et al. 2003).

The largest Madagascan carnivore is the fossa (*Cryptoprocta ferox*) (Fig. 4.4a), which presents a cat-like appearance similar to a small puma, weighing 7–12 kg (15–26 lb) with a compact body, short smooth hairs, a very long and slender tail, and short retractile claws (Garbutt 1999). Other distinctive features are its reddish brown color, long facial whiskers, and somewhat elongate muzzle. The fossa is widely distributed across Madagascar, primarily in woodlands and forests. It eats a range of vertebrates and invertebrates, but in rain forests the diet is dominated by lemurs, including adults of the largest species, such as the sifakas, for which they are the major predator (Wright 1998). Fossas are superb climbers and apparently take lemurs at night while they are sleeping.

Two additional rain forest carnivores, the fanaloka (*Fossa fossana*) and falanouc (*Eupleres goudotii*), were previously considered to be civets (Garbutt 1999). The fanaloka (or "Malagasy striped civet") is a shy, nocturnal animal the size of a domestic cat (< 2 kg), but more fox-like in appearance (Fig. 4.4b). It apparently prefers streams and marshy areas, and its diet includes insects, small mammals, and a variety of aquatic animals (Goodman et al. 2003). The falanouc is a larger animal (2–4 kg), with an elongated snout and tiny conical teeth, like an insectivore. It feeds almost exclusively on earthworms and other invertebrates. There are also four mongoose-like species, three of which occur in the rain forest.

A true civet, the small Indian civet (*Viverricula indica*), has been introduced to Madagascar from tropical Asia some time in the last 2000 years, but is more common in degraded and agricultural habitats than in rain forest.

### Australia and New Guinea

The rich fossil record suggests that the vast Miocene rain forests of northern Australia supported a mammal fauna as diverse as in rain forests elsewhere,

(a)

Fig. 4.4 Carnivores in Madagascar. (a) Fossa (*Cryptoprocta ferox*). (Courtesy of Harald Schuetz.) (b) Fanaloka (*Fossa fossana*). (Courtesy of Harald Schuetz.)

(b)

including leopard-size marsupial lions and wolf-size carnivorous kangaroos (Long et al. 2002). The subsequent drying of the Australian continent eliminated most of this fauna and the extensive rain forests of New Guinea are apparently too young to have evolved substitutes. As a result, the absence of large, or even medium-sized, carnivores is one of the most distinctive features of the rain forests of New Guinea and Australia today (Flannery 1995a).

Australia and New Guinea lack native members of the order Carnivora and they have been only partly replaced by meat-eating marsupials in the order Dasyuromorphia. The largest species in New Guinea today is the New Guinea quoll (*Dasyurus albopunctatus*), which at around 700 g in weight (about the size of a squirrel) is no threat to anything larger than a rat. All other rain forest carnivores weigh less than 0.5 kg and feed mostly on insects. Australian rain forests have a larger (< 5 kg, 11 lb) species, the spotted-tailed or tiger quoll (*D. maculatus*), which is now the largest native mammalian carnivore on the Australian mainland, preying on animals as big as a small wallaby, as well as smaller mammals, birds, and insects.

Until recently, the largest mammalian predator in the region was the thylacine or Tasmanian wolf (*Thylacinus cynocephalus*), which weighed up to 35 kg. These marsupials bore a striking resemblance to wolves in their overall body shape, especially the shape of the head and the arrangement of teeth for stabbing and cutting meat. The marsupial ancestry, however, was shown by the presence in the female thylacine of a pouch for carrying her young. The thylacine occurred in New Guinea and throughout Australia, but underwent a dramatic decline following the introduction of the domestic dog by human settlers less than 5000 years ago. It was apparently eliminated from mainland Australia and New Guinea around 3000 years ago, but persisted in Tasmania until the 1930s. Records show that it could attack the largest native prey, including kangaroos and wallabies (Jones & Stoddart 1998). There is, however, no definite evidence that it occurred in tropical rain forests, and the fossil records from New Guinea are from alpine grassland.

The role of the thylacine as Australia's largest mammalian carnivore has now been assumed by the dingo, a large (up to 20 kg) wild dog descended from domesticated animals brought from Asia 3000–5000 years ago (Savolainen et al. 2004). Although generally not a forest animal, the dingo is an important predator in the margins of rain forest patches in the wet tropics.

## Herbivores of the forest floor

The tropical savannas and grasslands of Africa teem with highly visible herds of large mammals (Ripley 1964; Kingdon 1997), yet a naturalist is considered lucky to glimpse a small duiker tip-toeing alone through the African rain forest. Similarly, it is a rare event to catch the eye-shine of a solitary mouse deer by torchlight in Asia or to see plant-eating mammals in rain forests elsewhere. Rain forests contain far more plant biomass than savannas, so why are the animals that eat it so hard to see? Part of the answer is that the large mammals of many rain forests have been almost hunted out, with the few survivors so wary of humans as to be invisible (Wilkie & Carpenter 1999; Bennett & Robinson 2000). In the few well-protected reserves where there has been no hunting for several decades, the greater visibility of forest floor mammals is striking. Even in pristine rain forests, however, the density of forest floor herbivores is limited by the low availability of edible plant matter that can be reached from the ground. Much of the forest biomass is in tree trunks and branches, and flushes of new leaf material, along with flowers and fruits, are produced mainly in the canopy, far out of reach of the animals below. The dense canopy of the rain forest greatly reduces the amount of light energy reaching the understorey, so grasses are rare and all plants grow slowly. Slow-growing understorey plants cannot easily replace lost tissues, so they protect themselves from herbivores with high concentrations of tough fiber and toxic chemicals, such as bitter tannins.

The limited availability of nutrients from plants near ground level has had a major role in shaping the terrestrial plant-eating faunas of rain forests, but the very different taxonomic compositions of these faunas show that historical processes have also had an overwhelming influence. There are some striking examples of convergence on an "ungulate-like" body form in unrelated animals, but also many equally striking differences in the form, size, and diversity of species in the different faunas. Although comparative ecological studies are still largely lacking, these differences are likely to be very significant for ecological processes because of the major role that forest floor herbivores play in shaping vegetation, through their patterns of foraging, their dispersal of seeds, and the way in which the larger species trample plants as they move.

No single order of mammal has managed to dominate the terrestrial herbivore niche to the extent that the Carnivora dominate the carnivore niche, but the even-toed ungulates (Artiodactyla, or Cetartiodactyla if the whales are included) have been by far the most successful. Familiar members of this order include the pigs, deer, cattle, antelopes, giraffes, and hippopotamus. The most successful artiodactyls are the ruminants, represented in the rain forest by the mouse deer and chevrotains (Tragulidae), a giraffe relative, the okapi (Giraffidae), the deer (Cervidae), and the bovids (Bovidae: cattle, antelopes, and their relatives). These four families are not evenly distributed, with the okapi confined to Africa, deer to Asia and the Neotropics, and both bovids and tragulids to Asia and Africa. Ruminants are distinguished by their specialized adaptations to herbivory, including their dentition and a compartmentalized stomach where symbiotic microorganisms process plant material. The tragulids are small, forest-adapted animals while the other ruminants appear to be primarily adapted to open, grassy habitats, with some lineages having become secondarily adapted to forest.

The artiodactyls are related to a second, much smaller, group of hoofed mammals, the odd-toed ungulates, or Perissodactyla. The perissodactyls flourished in the early Tertiary, when they included giants such as the 20-ton *Indricotherium*, the largest known land mammal, but are they now represented by only the horses, rhinoceroses, and tapirs. Although traditionally combined as the "true ungulates", the Artiodactyla and Perissodactyla are now known to be part of a larger group (the Ferungulates) that includes not only the whales, but also the Carnivora (Lin et al. 2002).

Throughout the Tertiary, South America supported a diverse fauna of ungulates, assigned to several different orders (Cifelli 1985). The Notoungulata in particular radiated to fill many herbivore niches, from the rabbit-like typotheres to the rhinoceros-like *Toxodon*. Most of the fossil evidence for these animals comes from temperate and nonforest habitats, but a rich fossil fauna from La Venta, Colombia, dated at 12–14 million years ago (middle Miocene), appears to represent a lowland forest environment, along with more open riverine habitats (Kay & Madden 1997). The ungulates included species ranging in body size from 2 kg to more than 1000 kg (4.5–2200 lb), suggesting an analogy with the rich ungulate fauna of modern African rain forest, rather than modern South America. However, the relationship between these animals and the living ungulates is not clear. They declined in both diversity and abundance during the Pliocene and most were extinct before the faunal interchange brought new predators and competitors, although the last survivors were seen and hunted by the first humans to reach the continent.

Ungulate-like forms also arose independently in other mammalian lineages, including relatives of the hyraxes that dominated the small to medium-sized herbivore niche in Oligocene Africa (around 25–30 million years ago) before the Miocene influx of true ungulates (Kingdon 1997), and the giant caviomorph rodents that are still such a prominent part of the Neotropical rain forest community. The ancestral stock of caviomorphs arrived in South America at the end of the Eocene, around 31–37 million years ago, at about the same time as the New World monkeys (see Chapter 3). A variety of alternative source areas and routes have been proposed, but the molecular evidence argues strongly for a single colonization event (Huchon & Douzery 2001). The caviomorph invaders subsequently radiated into a wide range of ecological niches, apparently displacing marsupials and endemic ungulates from some of them.

There are also terrestrial plant-eaters with very different body forms, including the elephants in African and Asian rain forests, the now extinct ground sloths of the Neotropics, and the kangaroos and wallabies of Australia and New Guinea. Remember also that several of the Old World apes and monkeys, considered in the previous chapter, are either partly or largely terrestrial plant-eaters, as are some lemurs, whereas the New World monkeys are all arboreal.

## Feeding strategies

Food scarcity is a problem for an animal that is limited to foraging on the forest floor. Although arboreal and flying animals may exploit fruits, flowers, and leaves in the forest canopy (see Chapters 3, 5, and 6), terrestrial animals must wait for these foods to fall from above (or to be dropped by monkeys, birds, and fruit bats) or make the most of what is available growing close to the ground. There are, however, some plant species in all the rain forest regions that make use of terrestrial animals to disperse their seeds; these species provide a more regular supply of fruits, which are often green in color and fibrous in texture. In rain forests around the world, fallen fruits and seeds are important to forest herbivores. Many animals also seek out the newly sprouted leaves of forest herbs, tree seedlings, and shrubs in tree-fall gaps, at riverbanks, in manmade clearings, and at the forest edge—anywhere a break in the forest canopy lets in enough light to stimulate shoot growth at ground level.

Earning such a hard living has a number of consequences that unite the terrestrial herbivore faunas in different rain forest regions. Several feeding strategies are possible. Being small is one solution to the problem of resource scarcity. In general, smaller animals need less food, but it must be of higher quality, whereas larger animals can tolerate less nutritious foods but need larger quantities. This is because, while metabolic rate declines with increasing body weight, gut capacity remains a constant proportion of body weight. Larger animals can therefore afford to retain low-quality food in their guts for a longer time in order to digest it, while small animals must extract nutrients rapidly. Terrestrial rain forest mammals are often small, less than 25 kg (55 lb), especially if they include a substantial component of fruit in the diet. Fruits, though relatively nutritious, are scarce, so larger species must either spend the day foraging just to get enough to eat or include other plant material besides fruit in the diet. Leaves, shoots, and grasses are much lower in nutritive value than fruits, and large volumes need to be processed to provide sufficient energy and nutrients. Very large species like the Asian guar (*Bos gaurus*), which weighs up to 900 kg, can accommodate that extra bulk, and eat fruit only opportunistically. Size, however, is not a perfect predictor of diet, and Africa's smallest forest ungulates, the rabbit-sized dwarf and pygmy antelopes (*Neotragus* spp.), feed largely on leaves and shoots. Bates' pygmy antelope (*N. batesi*) of Africa, weighing only 2–5 kg, must eat a great diversity of plant species (> 200 species) to avoid being poisoned by any one species and to find enough food to eat. It must also forage nonstop throughout the night.

Browsing on the leaves and twigs of shrubs and small trees is possible in continuous forest, but species that graze on grasses and other herbaceous plants are usually associated with water, such as rivers, coastal areas, and marshes, or they depend on forest gaps and the forest edge. Gap and edge habitats have been extended greatly by recent human activities, such as farming and logging, but were created in the past by landslides, windstorms, floods, and natural fires. Both browsers and grazers frequently depend on natural salt licks, indicating just how tight their nutrient balance can be. In the Sumatran rhinoceros, *Dicerorhinus sumatrensis*, the availability of salt licks can influence local population density (Payne 1995). The bongo (*Tragelaphus eurycerus*), a large African forest antelope, is even known to swallow pieces of burned wood from lightening-killed forest trees, apparently to obtain salt.

The resource base in the rain forest is too limited and widely dispersed to support the extensive herds of grazers seen in tropical savannas, so rain forest herbivores are mostly solitary or occur in pairs, and they frequently defend a territory against others of their species. There are some exceptions to this, most notably the pigs, including the peccaries in the New World, true pigs in the Southeast Asia, and the red river hog and forest hog in Africa, where an omnivorous diet, including roots, shoots, fallen fruits, and small animals, permits larger group sizes. The very large group sizes (50–300 individuals) of the white-lipped peccary, *Tayassu pecari*, may enable the most efficient utilization of clumped hard nuts and seeds (their preferred food) as well as defense against predators (Kiltie & Terborgh 1983). The effect of resource limitations is still felt in these groups; the price of group living is that they must travel very large distances in search of food. The bearded pig of Malaysia, *Sus barbatus*, may move hundreds of kilometers in the course of a year to keep pace with fruiting patterns in the forest (Medway 1983). This usually solitary feeder is often found in montane areas, foraging for fallen acorns and chestnuts from trees in the Fagaceae family (Fig. 4.5). In dipterocarp masting years, pigs come together in large, loosely structured aggregations and descend into the lowlands to feast on the oil-rich dipterocarp nuts (Curran & Leighton 2000). In such years, pigs breed more actively and put on extra fat. In Africa, the red river hog (*Potamochoerus porcus*) also feeds in groups, though these groups tend to be somewhat smaller in size than those of Asian pigs, generally 4–12 individuals, with occasional temporary groups of up to 60 (Kingdon 1997). These pigs travel up to 4 km per day to reach feeding sites, where they eat fruits and use their snouts to plow up the ground in search of roots. They also eat small animals. The forest hog (*Hylochoerus meinertzhageni*) is a relatively sedentary browser and grazer.

**Fig. 4.5** Bearded pig (*Sus barbatus*) with an assertive attitude from Borneo. (Courtesy of Hans Hazebroek.)

Thus, it seems that to survive on the limited amount of plant material available at ground level, mammals foraging on the forest floor have had to adopt one or several of the following strategies: small body size; solitary or small social groups; continuous foraging; feeding on fallen fruit; extensive travel; or dependence on the lush and renewable plant material in swamps, gaps, or the forest edge. The same constraints apply in each of the five major tropical rain forest regions, and there has been some convergence in the foraging strategies adopted to deal with them. However, the actual animals involved, as well as the patterns of species richness on a local scale, vary enormously.

## Frugivorous ungulates

The most common foraging strategy, seen today in all the rain forests except Madagascar, involves a combination of small body size and feeding on fallen fruit. Other plant material, such as fresh fallen leaves or new leaves, shoots, and flowers are commonly eaten as well in most species, but fruit is a large component of the diet.

In African and Asian rain forests, the artiodactyls (even-toed ungulates) predominate as terrestrial frugivores (Happold 1996). African assemblages typically comprise species of duikers (*Cephalophus* spp., Bovidae, the cattle family) and the water chevrotain (*Hyemoschus aquaticus*, Tragulidae). Duikers often eat fruits that are hard and green, with a low sugar content (Emmons et al. 1983; Dubost 1984). They often spit up the seeds from their rumen or pass out the undamaged seeds in their droppings. The duikers are small, shy forest antelopes that look

(a)

(b)

**Fig. 4.6** Old World herbivores. (a) Bush duiker (*Sylvicapra grimmia*) in Kenya. While not normally a rain forest species, it illustrates the appearance of duikers. (Courtesy of Harald Schuetz.) (b) Greater mouse deer (*Tragulus napu*) from Malaysia. (Courtesy of Tim Laman, taken in captivity.)

rather like squat, short-legged miniature deer (Fig. 4.6a) (Kingdon 1989; Martin 1991). In horned species, both sexes have short and backward-facing horns to help them slip through the forest understorey. Most duikers have brown- or buff-colored upper parts with a pale stripe along the middle of the back as an aid to camouflage in the forest interior, but the striped-back duiker (*Cephalophus zebra*) lives up to its name with a bright orange coat marked with dark vertical stripes. Most duikers forage at night, although a few species feed during the day, and they forage alone or in pairs. Their name, "duiker", means "diving buck" and refers to their habit of quickly diving into dense vegetation when they feel threatened. In addition to plants, duikers will sometimes supplement their diet with animals such as frogs, birds, insects, and carrion. The diversity of duiker species is greatest in West Africa and the Congo basin, and is lower in the isolated forests east of the Congo basin. In the most westerly African rain forest region, centered around Liberia, there may be as many as seven species of duiker occurring in a single forest, and there is some suggestion that they avoid excessive dietary overlap by including different fruits in their diet (Happold 1996). Duikers occur at a lower density in forests dominated by one species of tree (monodominant forests, see Chapter 2) in comparison with mixed species forests, where their diet is more diverse and consistent.

Throughout most of the African rain forest, duikers are joined by the water chevrotain (*Hyemoschus aquaticus*, Tragulidae), a solidly built ungulate with a short, thick neck and a brown coat, with rows of white spots running along the body and forming lines along the flanks. Although, as its name suggests, it lives within a few hundred meters of water, it only enters water to avoid danger and its diet is largely fallen fruit. Two species of chevrotains, or mouse deer, also make up the small, fruit-eating guild of species in the forests of Asia, where there are no small species of bovid. They rival the dwarf antelope as the smallest artiodactyls; the lesser mouse deer *Tragulus javanicus* weighs only 2.0–2.5 kg (4.5–5.5 lb) and the greater mouse deer *T. napu* is not much larger at 3.5–4.5 kg (Fig. 4.6b). Like their African relative, the water chevrotain, mouse deer have no antlers or horns, and they are usually solitary. Although usually referred to as "deer", these tragulids represent an ancient, forest-adapted group of ungulates, which split from the other ruminants in the Eocene, before the origin and diversification of the deer and bovids (Hassanin & Douzery 2003).

### Frugivorous rodents

Although Neotropical rain forests appear to have supported small, fruit-eating ungulates in the past (Kay & Madden 1997), the terrestrial frugivore niche is today occupied by a very different group of mammals: large caviomorph rodents in the families Dasyproctidae and Agoutidae. Old World rain forests also have frugivorous rodents, but these are mostly small, rat-like animals. The larger caviomorphs, in contrast, in some ways resemble their Old World ungulate counterparts more than rats (Fig. 4.7) (Bates 1964; Goldblatt 1993). Adapted for a more mobile life style than most rodents, they have long legs and the hindfoot has a reduced number of weight-bearing toes. The high hindquarters give an arched appearance to the back that rounds into a much-reduced tail. Their heads are large and retain a rodent-like appearance, with strong jaw muscles and blunt noses. There are seven species of rain forest agoutis (*Dasyprocta*), and although they are very large for rodents (1.3–4 kg, 3–9 lb), they are much the same size as the small forest ungulates of Africa and Asia. All the Neotropical terrestrial rodent frugivore species tend to be rather similar in their habits, being fruit- and seed-eaters that take succulent plants when available. Unlike ungulates, but like

**Fig. 4.7** Certain pairs of rain forest mammals from tropical America (left) and Africa (right) are very similar in appearance and feeding habits. From top to bottom, left to right, the pairs are: capybara and pygmy hippopotamus; paca and water chevrotain; agouti and dwarf antelope; brocket deer and yellow-backed duiker; and giant armadillo and pangolin (which both feed on termites and ants). (From Bourliere 1973.)

many other rodents, they usually sit on their haunches to eat and manipulate the food with their forefeet (Fig. 4.8). This dexterity also enables them to bury seeds and nuts for use when fruit is not in season, and they are important seed dispersers as a result (Forget et al. 2002). The social unit is a territorial mated pair that remains together until death.

In contrast to the local species richness observed in African duikers, it is rare to find more than one species of agouti in any one locality in Neotropical forests

**Fig. 4.8** Many of the terrestrial herbivores in New World rain forests are rodents. Here a South American agouti (*Dasyprocta punctata*) eats a seed. (Courtesy of P.M. Forget.)

(Emmons 1997). Their ranges overlap very little, and the similarity in their habits suggest that the different species are ecological replacements for one another that may have arisen as populations became separated by geographical barriers such as major river systems. In most places, the agoutis are joined by the larger paca (*Agouti paca*), which are most common near water, and one of the two species of acouchi (*Myoprocta*). The acouchies are smaller than the agoutis (0.6–3.1 kg) but look very much like them and have similar diets. Like the agoutis, the two acouchi species overlap very little in their distribution. The less-well known pacarana (*Dinomys branickii*) is added to the group in the Amazon highlands. The pacarana looks like an overgrown guinea pig at 10–15 kg in weight, and is the last survivor of a once diverse caviomorph family, the Dynomyidae. The genus name, *Dinomys*, means "terrible mouse", and reflects its reputation as a fighter when cornered.

Although no other rain forest region has anything like an agouti or acouchi, frugivorous ground squirrels and murid rodents, such as the long-tailed giant rat (*Leopoldamys sabanus*) in Malaysia and the white-tailed rat (*Uromys caudimaculatus*) in Australia, also "scatter-hoard" seeds for future consumption and may make a similar contribution to seed dispersal if they fail to retrieve them all (Forget & Vander Wall 2001).

## African browsers

Most fruit-eating species are small. Larger species must either include more browse in their diet, or—as in the case of the large African duiker (*Cephalophus sylvicultor*) and the predominantly frugivorous red river hog (*Potomochoerus porcus*)—they must forage almost continuously. We can identify a loose foraging strategy of medium- to large-size browsers. Most of the medium-sized species will also tend to take seeds of woody legumes, fungi, or other accessible plant matter, and many of the larger species also graze, but the principal tendency, often associated with morphological adaptations, is toward browsing—that is, eating the shoots and leaves from woody plants. The largest browsers are the largest animals in the forest and can have a major impact on the structure and species composition of the forest, not only by what they eat but also by what they push over and tread on (Owen-Smith 1992).

The browsers are a variable group, both morphologically and taxonomically. In Africa, the bushbuck (*Tragelaphus scriptus*) and bongo (*T. eurycerus*) (family Bovidae) have varied diets but are predominantly forest-edge browsers of leaves, twigs, shoot tips, and vines (Kingdon 1997). These species look like crosses between an antelope and a small cow, with large eyes and ears, spiraled horns, and a red-brown coat marked with white stripes (Fig. 4.9a). The bushbuck is much the smaller of the two, just over 1 m (3 feet) at the shoulder and weighing 24–42 kg (53–93 lb), and it tends to be associated with water and the edges of swamps. The bongo is much bigger, standing 1.4 m at the shoulder and weighing up to 220 kg. A third species, the sitatunga (*T. spekei*), is intermediate in size and predominantly a grazer (see Plate 4.3, between pp. 150 and 151). These species avoid large areas of dense, closed-canopy forest, where little fresh greenery is available within reach, and instead make use of areas disturbed by landslides, floods, tree falls, and the activities of elephants (Kingdon 1997). Even then, they must range over huge areas to obtain enough nutritious browse.

One of the most unusual animals in the African rain forests is the okapi (*Okapi johnstoni*), confined to the equatorial forests of northern and east-central Democratic Republic of Congo (Kingdon 1997). The okapi is not a bovid, but a deep-forest relative of the familiar giraffe of African savannas. Like its larger relative, its body is short and compact and the neck and legs are long, though not as pronounced as the giraffe. It can reach up to 1.8 m at the shoulder and weighs 250 kg, and has a beautiful short, shiny coat that is the purplish-maroon color of an eggplant, with white horizontal stripes at the top of the legs. The long tongue is prehensile and is used to pluck leaves and buds and even small branches. Adequate food is hard to come by, and the okapi is usually solitary as a result.

The suite of African browsers is completed by the African elephant, found in the tropical rain forest zone of West and Central Africa (Happold 1996). Forest elephants are smaller, darker, and more solitary than their familiar savanna relatives, and have usually been considered a separate subspecies, *Loxondonta africana cyclotis* (Fig. 4.9b; see Plate 4.3). Recently, DNA sequence analysis has shown that forest elephants are, in fact, so different from savanna elephants that they deserve recognition as a separate species, *Loxodonta cyclotis* (Roca et al. 2001). The forest elephant feeds largely on shoots, leaves, and bark, but also takes fruit whenever available. Some trees with very large fruits depend entirely on elephants for seed dispersal, as described in Chapter 2. In the process of feeding, forest elephants knock down many small trees and bushes, opening up the forest and creating feeding opportunities for many other animals, both immediately on the fallen plants and later when plants begin to grow up in the gaps.

(a)

(b)

**Fig. 4.9** Some large African browsers. (a) Bongo (*Tragelaphus eurycerus*) from Kenya. (Courtesy of Harald Schuetz.) (b) African elephants (*Loxodonta cyclotis*) at a water hole in the Congo. (Courtesy of Roger Le Guen.)

(a)

**Fig. 4.10** Asian browsers. (a) Muntjac or barking deer (*Muntiacus muntjak*) from Bhutan. (Courtesy of Harald Schuetz.)

## Asian browsers

Although several species of bovids are found in Asian rain forests, the main browsers are deer (Cervidae) and two groups of odd-toed ungulates (Perissodactyla), the tapir and the rhinoceroses (Fig. 4.10). The smaller deer species (at 14–50 kg, 30–110 lb) are known as muntjacs (*Muntiacus* spp.) or barking deer. Their diets include a fair amount of fruit as well as browse. Barking deer are named for their habit of emitting deep, fox-like barks, particularly in dense vegetation and when predators are detected. The larger species of sambar or red deer (*Cervus* spp.) are larger browsers weighing up to 260 kg and as tall as 160 cm (5 feet) at the shoulder. They feed on leaves, shoots, and fallen fruits in tree-fall gaps, riverside forests, and other areas of lush growth.

A browsing habit is also typical of the Malay tapir *Tapirus indicus*, which has a prehensile proboscis comprising the snout and upper lips with which it reaches out and pulls leaves and shoots into the mouth. Tapirs also include fallen fruit in their diet, when available. In Asia, two species of rhinoceros browse in the forests, the Javan one-horned rhinoceros (*Rhinoceros sondaicus*) and the Sumatran rhinoceros (*Dicerorhinus sumatrensis*). These large Asian animals may have to travel long distances to fulfill their daily dietary requirements. In fact, the Javan rhino can travel 15–20 km (9–12.5 miles) in 24 h in search of adequate shoots, twigs, young foliage, and fallen fruit, and also needs a plentiful supply of water (Nowak 1999). The forest habitat of these Asian rhinos is a striking contrast to the savanna habitat of the African rhinos. The diverse diet that leads to this browsing style of foraging in the rhinoceroses is taken to extremes in the Asian elephant (*Elephus*

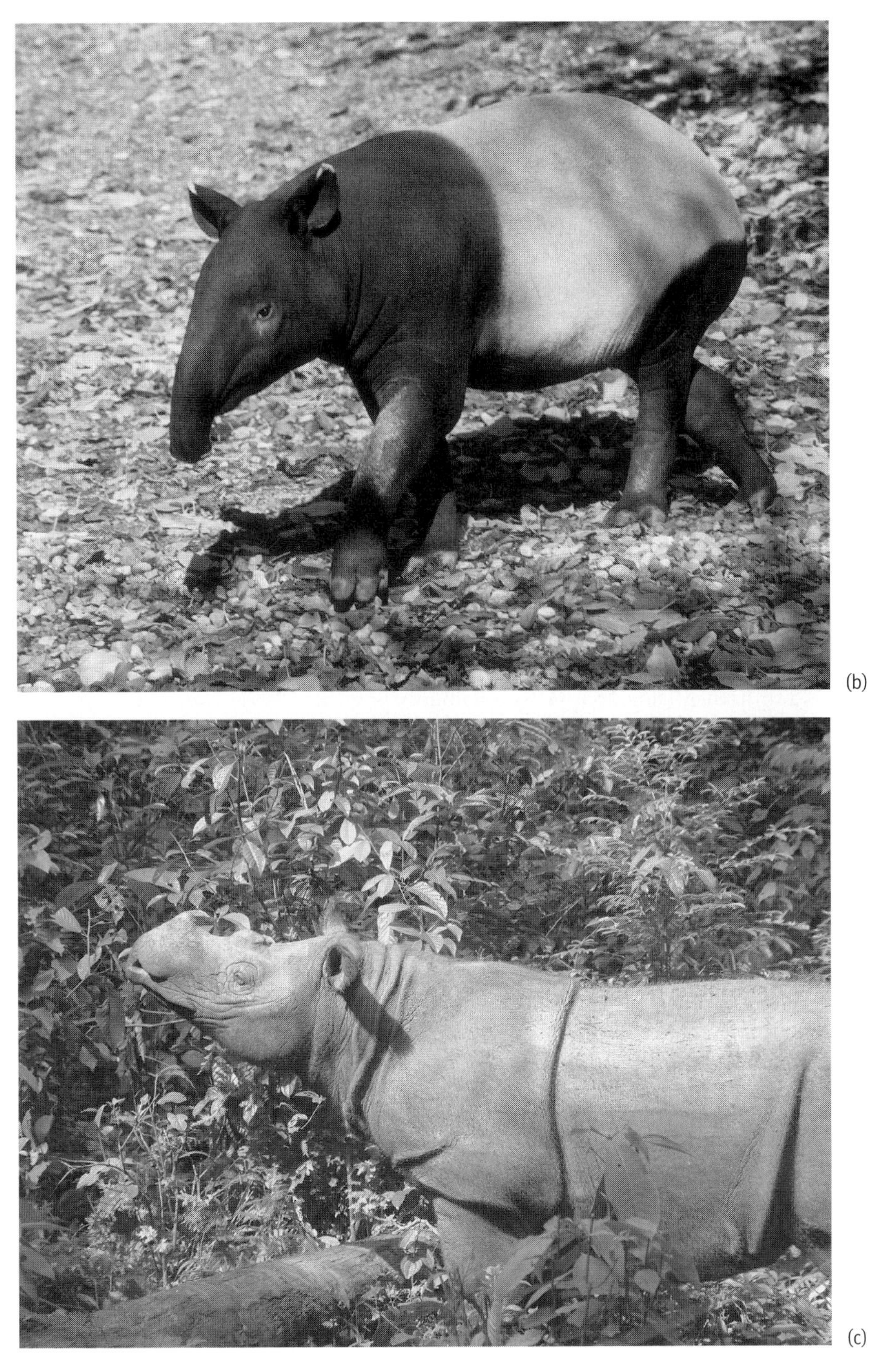

Fig. 4.10 *(cont'd)* (b) Malayan tapir (*Tapirus indicus*). (Courtesy of Jessie Cohen, taken at the US National Zoo.) (c) Sumatran rhinoceros (*Dicerorhinus sumatrensis*) browsing on leaves. (Courtesy of Nico van Strien, International Rhino Foundation.)

*maximus*), which may feed on the fruits, foliage, twigs, bark, and roots of up to 60 plant species, including trees, shrubs, climbers, herbs, and grasses (Medway 1983). Although Asian elephants are smaller than the African savanna elephants, the rain forest elephants of Asia and Africa have a similar size range.

When the okapi was first described from the forests of the Congo in 1901, there were suggestions that this would be the last new large mammal to be found (MacKinnon 2000). Of course this turned out not to be true, but the discovery of completely new mammal species, as opposed to the splitting of known forms (such as the African elephants, mentioned above), is a very rare event. The description of three new artiodactyls species within 4 years from the rain forests of the Annamite Mountains that border Laos and Vietnam is therefore particularly exciting. The strangest of these new species is the saola or Vu Quang ox (*Pseudoryx nghetinhensis*), found first in 1992 in Vietnam and then in 1993 in adjacent regions of Laos (Groves & Schaller 2000). Although clearly a bovid, the long, thin, backward-curving horns and conspicuous white markings on the face give the saola a strikingly distinct appearance and make its late scientific recognition even more surprising. Less than 1 m at the shoulder, and weighing about 100 kg, the incisors suggest a browsing habit, but this animal probably also eats grasses and herbs. Both DNA and morphological data suggest that it is most closely related to the cattle and buffaloes (Bovini). In the same region as the saola, at least two new muntjac species (*Muntiacus*) have been found, a very large species (*M. vuquangensis*) in 1994 and a small species (*M. truongsonensis*) in 1995 (Groves & Schaller 2000). Yet another new muntjac (*M. putaoensis*) was discovered in northern Myanmar in 1998 (Rabinowitz et al. 1999). The new discoveries emphasize how little we know about these shy and elusive rain forest browsers.

### South American browsers

The Neotropics lacks bovids, rhinos, and elephants. Deer and tapirs are the primary browsers in South America, although in contrast to Asia both groups are relatively recent (less than 3 million years ago) arrivals from the north over the Panama land bridge. There are two rain forest species of tapir, the South American tapir (*Tapirus terrestris*) and Baird's tapir (*T. bairdii*), found in Central America and west of the Andes in South America. Baird's tapir, at up to 300 kg (660 lb), is the largest land mammal in the rain forests of the Neotropics. Tapirs are found near water and often wear paths to the water's edge. They browse mainly on shoots of plants (especially aquatic species), as well as twigs, leaves, and fruit. Brocket deer (*Mazama* spp.) are shy, solitary animals of the forest weighing up to 25 kg. Their diet includes leaves, shoots, and grass, but they also feed heavily on fallen fruits when available (Gayot et al. 2004). They appear to lack the endurance of other deer and can be overtaken and killed by domestic dogs (Nowak 1999).

Although the Neotropics today has no equivalent of the elephants and rhinoceroses of Africa and Asia, this was not always true. The middle Miocene tropical forests of La Venta, Columbia, supported a variety of very large (> 500 kg) terrestrial herbivores, including ground sloths and a variety of endemic "ungulates" which appear to have been the ecological equivalent of the living Old World megaherbivores (Kay & Madden 1997). As recently as 10,000–12,000 years ago, at the end of the Pleistocene, Central and South America supported mastodons and gomphotheres (both relatives of elephants), as well as giant sloths

(Owen-Smith 1992; Martin & Steadman 1999). These animals may have played a similar role in maintaining open habitats within the forest and dispersing large seeds (Howe 1985). However, there is no direct evidence that these Pleistocene species occurred in rain forests.

## Grazers

Grazers cannot survive in closed-canopy rain forests because the lack of light restricts the availability of grasses, so grazing species tend to be associated with aquatic habitats or with large forest gaps and the forest edge. Some grazers, such as the African buffalo (*Syncerus caffer*, Bovidae), are so dependent on the existence of swamp grassland patches that they could reasonably be said to occupy the region despite, not because of, the rain forest. Forest buffaloes are very different from their larger, more robust savanna relatives and might be considered a separate species—like the forest elephants—were it not for the existence of intermediate forms (Kingdon 1997). Another African bovid, the sitatunga (*Tragelaphus spekei*), a spiral-horned relative of the bushbuck and bongo, is semiaquatic, grazing on reeds, sedges, and grasses in swamps (see Plate 4.3, between pp. 150 and 151). The pygmy hippopotamus (*Hexaprotodon liberiensis*) of West Africa, though less aquatic than the more familiar hippopotamus (*Hippopotamus amphibius*), relies heavily on water plants and grasses, as well as taking tender shoots, leaves, and fruit, and is found along streams and in wet forests and swamps (Fig. 4.11a) (Kingdon 1997). Hippos are artiodactyls, but not ruminants.

As in Africa, Asian grazers tend to use the forest for shelter and limited feeding, and move into more open areas to feed intensively. The tamaraw (*Bubulus mindorensis*) of the Philippines and the lowland anoa (*B. depressicornis*) of Sulawesi are mid-sized (up to 300 kg, 660 lb) buffalo-like bovids that eat grasses, ferns, saplings, and fallen fruits. The most impressive grazer is the gaur (*Bos gaurus*), a massive ox that can reach 2.2 m (*c.* 7 feet) at the shoulder and weighs 1000 kg (Fig. 4.11b). Herds of these huge beasts are led by a dominant bull through the forests of Asia, grazing on grasses and browsing shrubs and climbers in clearings, at the forest edge or open riverbanks. Only slightly smaller, the banteng (*B. javanicus*) prefers slightly drier habitats than the guar, but can also be found in the forest, occasionally in association with its larger relative (Medway 1983).

As a result of Pleistocene extinctions, such large grazers are now completely absent from South America. The largest surviving grazer is the semiaquatic capybara (*Hydrochaeris hydrochaeris*), the largest rodent in the world, weighing up to 65 kg (Fig. 4.11c). This stocky animal has a large head and rectangular muzzle and shows several adaptations to its semiaquatic life style: the ears and eyes are small and set high on the head, and the feet are partly webbed. It feeds on grass and browse, particularly aquatic plants, and is always found near water. Ecologically and morphologically it is most similar to the pygmy hippopotamus of Africa. Fossils of an even larger rodent, the 700 kg *Phoberomys pattersoni*, have been found in the upper Miocene of Venezuela (Sánchez-Villagra et al. 2003), showing that the absence of large grazing herbivores in modern Neotropical forests is not simply a reflection of size limits on rodent evolution. The largest herbivore still remaining in the Amazon is not even terrestrial; it is the Amazon manatee (*Trichechus innuguis*), a completely aquatic animal that feeds on water plants and weighs up to 500 kg.

(a)

(b)

**Fig. 4.11** Grazers from different continents. (a) Pygmy hippos (*Hexaprotodon liberiensis*) from Africa. (Courtesy of Tim Laman, taken in captivity.) (b) Gaur (*Bos gaurus*) from Malaysia. (Courtesy of James Elder.)

(c)

**Fig. 4.11** (*cont'd*) (c) Adult capybara (*Hydrochaeris hydrochaeris*) and juvenile from Brazil. (Courtesy of Harald Schuetz.)

## Australia and New Guinea

As we move to Australia and New Guinea, there is another dramatic shift in the dominant mammal group on the rain forest floor. Here, it is the members of the marsupial family Macropodidae, the kangaroos and wallabies, which slip through the understorey in search of edible plant materials (Fig. 4.12) (Flannery 1995a). These animals do not fit quite as readily into our classification of frugivores, browsers, and grazers, and indeed, for many species the diet is virtually unknown. The rain forests of New Guinea and Australia also support giant flightless birds, the cassowaries, which share the terrestrial frugivore niche filled by mammals elsewhere (see Chapter 5).

The best-studied rain forest macropodid is also the most peculiar: the musky rat-kangaroo (*Hypsiprymnodon moschatus*) is the smallest living macropodid, at around 0.5 kg (1.1 lb) (Nowak 1999), and it apparently diverged from the rest of the family before the evolution of such characteristic kangaroo features as bipedal hopping and the complex stomach needed to digest a fiber-rich diet. Musky rat-kangaroos are confined to the wet rain forests of northeast Queensland, where they occur at high densities and feed largely on easily digested fruit and seeds. This species has been shown to be a key dispersal agent (with cassowaries and white-tailed rats) for large seeds in these forests by burying them for future consumption, just like the agoutis of South American rain forests (Dennis 2003). A more recent radiation of rain forest wallabies in the montane forest of New Guinea has produced several species in the genera *Dorcopsis* and *Dorcupsulus*, ranging in size from around 2 kg—the smallest New Guinea macropodid—to around 5 kg (Flannery 1995a). The diets of these small forest wallabies are assumed to be largely vegetarian, but little else is known. Another group of

(a)

(b)

**Fig. 4.12** Herbivores from New Guinea and Australia. (a) The musky rat-kangaroo (*Hypsiprymnodon moschatus*) from northeast Queensland, Australia, eats mostly fruit but also eats invertebrates and fungi. (Courtesy of Belinda Wright.) (b) The forest wallaby or pademelon (*Thylogale stigmatica*) from northeast Australia and New Guinea is mainly a browser, but also grazes at the forest edge at night. (Courtesy of Andrew Dennis.)

small wallabies, the pademelons (*Thylogale* spp.), occur in both New Guinea and Australia. Most species seem to prefer more open habitats, where they can graze as well as browse.

As in South America, the fauna of browsing marsupials has been affected by recent megafaunal extinctions associated with the arrival of humans. New Guinea had at least one extinct 75–200 kg browsing diprotodon, *Hulitherium thomasetti*—an animal like a giant wombat or hairy hippopotamus—which overlapped with humans. It may have fed on bamboo: a sort of marsupial giant panda (Long et al. 2002). Related species occurred in the Miocene rain forest of Australia, although it is not clear if any of the more recent Australian diprotodons lived in the rain forest.

### Madagascar

Modern Madagascan rain forests seem to be an exception to the general rule that the mammals in each rain forest region have evolved species to utilize the sparse but diverse resources available near ground level. Part of this reflects Madagascar's long isolation, which means that many groups of mammals, including the artiodactyls, failed to reach it. Bush pigs now forage in Madagascan rain forests, but these were a recent introduction from Africa (Adrianjakarivelo 2003). Another factor is the mass extinctions of the last few thousand years, which have removed a large proportion of Madagascar's bigger native mammals as well as the giant, flightless elephant bird (Godfrey et al. 1997). Some of the extinct lemurs were probably at least partly terrestrial, and the largest species, the gorilla-sized *Archaeoindris*, must have been largely so, since only the biggest tree branches would support its weight. It is not unreasonable to suggest that some of these lemur species made use of the resources that support other terrestrial mammals in other rain forest regions. Three species of hippopotamus also survived until recently in Madagascar, including one that was very similar to the African pygmy hippopotamus and that may have had similar grazing habits.

## Conclusions and future research directions

In contrast to the primate communities discussed in Chapter 3, research and analysis of carnivore and plant-eater biology in tropical rain forests is still in an early stage of development. Certain patterns do emerge, but there is a compelling need for more research to verify and expand what has been learned to date. Our knowledge of primates has come largely from the habituation of unhunted groups to the presence of humans, so they can be observed continuously from close quarters. The nocturnal, secretive habitats and low population densities of most rain forest carnivores make this approach impossible and it is only slightly more practical with most forest floor herbivores. Much of what we know about these two groups of mammals, therefore, is based on fleeting glimpses of fleeing animals or on the interpretation of tracks, signs, and the contents of their droppings. New techniques are changing this, however. Automatic cameras triggered by animal movements can give more reliable estimates of densities and radiotracking gives information on how the available habitat is used. The application of molecular techniques to the analysis of feces can give far more information than the traditional "sieve and count" methods. The tools are now available to check what we think we know already and to answer questions that have previously not been possible to ask.

Carnivores may be the least understood of rain forest mammals, not only because of the difficulties of studying them, but also because their opportunistic and variable behavior makes it hard to interpret the limited data we have (Rabinowitz 1991, 2000). Comparative studies of ecologically similar groups across continents should be undertaken, making use of the few remaining sites where carnivores exist at near natural densities, as well as forests disturbed by logging and other human activities. There is no shortage of questions. For example, why should Asian and American rain forests have more carnivore species than African forests when the density and diversity of potential prey is at least as high in Africa? Why is there no jaguar- or tiger-sized carnivore in African rain forests? To what extent are the procyonids of Neotropical forests the ecological equivalents of the civets and mongooses of Africa and Southeast Asia?

Some questions about carnivores are best addressed in places where particular species have been removed, or exotic species introduced. It would be valuable to know how the removal of carnivore species affects the abundance of prey species and how this impact then propagates through the food web. When carnivores are eliminated by hunting, do their prey species increase, and if so, what are the consequences for the forest community? And when prey species are eliminated by human hunting, what impact does that have on carnivore behavior and numbers? Does that force the carnivores to leave the forest and prey on domestic animals?

Other questions cannot be addressed by considering mammalian carnivores alone. A monkey is just as dead if it is killed by a leopard, an eagle, or a python, and smaller vertebrates are prey to dozens of competing mammals, birds, and reptiles. How can so many species of carnivores coexist in one rain forest area? To what extent can an increased abundance of one species "compensate" for the absence of another as a result of natural factors (e.g. on islands) or human impacts?

The role of browsers and grazers also needs further investigation. In some cases, the exclusion of herbivores by fencing off an area of forest may be the most practical way to identify their impact; even elephants will avoid electric fences. Rodents are important understorey herbivores in New World forests, whereas artiodactyls (deer and/or bovids) are more important in Old World forests. These animals have very different teeth and digestive systems, so their supposed ecological equivalence needs more detailed study. It remains to be determined why Old World grazer and browser communities are richer in species than New World communities. Is this due to greater ecological specialization in Old World species? In many cases, African forest species appear to have evolved from savanna relatives. Why has this not happened to the same extent in New World forests, which also border extensive savanna and woodland zones?

There is increasing evidence that terrestrial herbivores play a key role in seed dispersal in all the major rain forest regions (except Madagascar), but also that this role may be very different in different forests. In the Neotropics, caviomorph rodents disperse seeds by caching them for future consumption. This puts the plant and rodent in direct competition for the seed reserves, since only unretrieved caches can give rise to seedlings. In Africa and Asia, in contrast, terrestrial herbivores ranging in size from mouse deer to elephants consume specialized fleshy or fibrous fruits that fall to the ground when ripe, dispersing the seeds when they are later defecated of regurgitated. In one case the seed itself is the reward, while in the other the seed is just ballast to be disposed of as quickly as possible. The ecological and evolutionary consequences of these two mechanisms of seed dispersal are likely to be very different. However, there is also some evidence of scatter-hoarding by Old World rodents, while both deer and tapirs disperse seeds in fallen fruits in the Neotropics. Is it possible that the contrast is not as great as the current literature implies? Pantropical studies with standard methods are needed to resolve this question.

Finally, we urgently need to understand the role of grazing and browsing herbivores in creating the very structure of the vegetation. While the role of African elephants in creating the openings in the forest required by many other species of animals and plants has received considerable attention, the role of the other grazers and browsers needs to be investigated. Asian elephants are a similar size to African forest elephants: do they have a similar impact? It would be valuable to learn how elephant activity affects the full range of species in the forest, such as birds, insects, and small rodents. We also need to know what happens to the vegetation when the browsers and grazers have been removed by hunting. Does the amount of ground-level vegetation increase, and how does the plant species composition change?

The extinction of all the megaherbivores in the Neotropics at the end of the Pleistocene must have had a major impact on both the structure and species composition of the rain forest, if—and this is still uncertain—they occurred in rain forests. Could we introduce forest elephants from Africa or Asia into Neotropical forests as a substitute? Should we do so? What would be the consequences? Would elephants eat the same fruits and browse the same plants as the extinct gomphotheres? The idea of bringing elephants to Neotropical rain forests may seem both impractical and potentially dangerous, since the ecological impact of elephant introduction is unpredictable, but just thinking about it highlights the fact that even the most "pristine" rain forests of the Neotropics have already been transformed by past human impacts. Unfortunately, this is also increasingly true of rain forests elsewhere, not just in the Neotropics. Forest floor vertebrates are particularly vulnerable to hunting and the densities of both large carnivores and large herbivores are almost everywhere lower than they would have been before the arrival of humans. In most rain forests, humans have now replaced large carnivores as the top predators and, in many areas, they have played this role for so long that we can only speculate on the nature of the "natural" predator–prey relationships. Understanding the role of past and present human hunters in rain forest ecology may be our biggest challenge.

## Further reading


Cranbrook, Earl of (1991) *Mammals of South-east Asia*, 2nd edn. Oxford University Press, Oxford.

Flannery T. (1995) *Mammals of New Guinea*. Cornell University Press, Ithaca, NY.

Garbutt N. (1999) *Mammals of Madagascar*. Pica Press, East Sussex, UK.

Kingdon J. (1989) *Island Africa: the Evolution of Africa's Rare Animals and Plants*. Princeton University Press, Princeton, NJ.

Kingdon J. (1997) *The Kingdon Field Guide to African Mammals*. Princeton University Press, Princeton, NJ.

Kricher J. (1997) *A Neotropical Companion*, 2nd edn. Princeton University Press, Princeton, NJ.

Nowak R.M. (1999) *Walker's Mammals of the World*, 6th edn. Johns Hopkins University Press, Baltimore, MD.

Rabinowitz A. (2000) *Jaguar: One Man's Struggle to Establish the World's First Jaguar Preserve.* Island Press, Washington, DC.

Chapter 5

# Birds: Linkages in the Rain Forest Community

Birds have important roles in the ecology of all tropical forest communities: they eat fruits and disperse seeds, pollinate the flowers they visit for nectar, and prey upon insects, spiders, and small vertebrates living on the leaves, twigs, and trunks of trees. Raptors are major predators of many larger forest animals, including frogs, reptiles, rats, squirrels, other birds, and even primates. Birds have received more attention from biologists than any other animal group, except perhaps the primates, so they provide a rich source of material for assessing the relative importance of past biogeographical histories and current ecological conditions in shaping rain forest communities (Karr 1989, 1990). Birds are particularly instructive in this regard because they can be readily divided into ecological guilds, such as frugivores, nectarivores, insectivores, and ground-feeders, making possible comparisons between bird communities that have no species in common.

The laws of physics place strict constraints on flying animals, so we might expect the birds (and bats) of the different rain forest regions to show more convergence in both form and function than the nonflying mammals considered in the previous two chapters. Rain forest bird communities in different regions do indeed show some striking examples of convergent evolution—one good example is the strong physical resemblance of New World toucans and Old World hornbills, which are completely unrelated families—but there are also many real differences in bird sizes, foraging behavior, and community organization. In some cases, a superficial resemblance between geographically separated families masks significant ecological differences. Seemingly minor differences in the foraging behaviors of flower-visiting birds, for example, have had significant impacts on the way that plants display their flowers and fruits. New World hummingbirds hover while they feed on flowers, whereas Old World sunbirds and Australian honeyeaters usually perch while feeding (Westerkamp 1990). As a result, the flowers—and subsequent fruits—of the many New World shrubs, epiphytes, herbs, vines, and small trees that are pollinated by hummingbirds are produced outside of the foliage, where the hummingbirds can get at them more easily. In contrast, Old World bird-pollinated plants typically produce flowers (and fruits) inside the foliage, where birds perched on twigs can reach the flowers readily.

## Biogeography

Most birds can fly, so we might expect that their distributions would be less affected by the geographical barriers between major areas of rain forest than those of other vertebrates. However, many rain forest birds—particularly those that live in the forest understorey—are either unable or unwilling to cross large open spaces, never mind the sea. Moreover, most modern groups of birds arose in the Cretaceous and radiated in the Tertiary, 35–65 million years ago, at a time when both South America and Australia were more isolated than they are today, and Madagascar was already separated from Africa by hundreds of kilometers of open sea (Cracraft 2001; Edwards & Boles 2002; Ericson et al. 2003). The tropics of Africa and Asia have been connected by dry land since at least the Miocene 20 million years ago, but the connecting habitat has usually been savanna or desert. Thus, forest-dependent species have rarely been able to disperse between Asia and Africa. A large proportion of rain forest bird diversity therefore is due to independent radiations of bird species in South America (Ricklefs 2002), New Guinea and Australia, Madagascar, and, to a lesser extent, Asia and Africa (Fig. 5.1). Although differences among these radiations were reduced after North and South America became connected through Central America, and Southeast Asia and the New Guinea–Australia region were linked by the stepping-stones of the Indonesian Archipelago, they are still very noticeable today. Only a few bird families—the swifts (Apodidae), swallows (Hirundinidae), pigeons (Columbidae), parrots (Psittacidae), cuckoos (Cuculidae),

(a)

**Fig. 5.1** Many bird groups are restricted to different areas of the world. (a) Hartlaub's turaco (*Tauraco hartlaubi*), a montane member of the African turaco family. (Courtesy of Tim Laman, taken in captivity.)

(b)

(c)

**Fig. 5.1** *(cont'd)* (b) Greater leafbird (*Chloropsis sonnerati*), a member of the Asian leafbird family, from Gunung Palung National Park, Indonesia. (Courtesy of Tim Laman.) (c) Sickle-billed vanga (*Falculea palliata*), a member of the vanga family, from Madagascar. (Courtesy of Harald Schuetz.)

hawks (Accipitridae), falcons (Falconidae), owls (Strigidae), and nightjars (Caprimulgidae)—are found in all major rain forest areas and, even in these families, the species in one region are almost always more closely related to each other than to species in other regions.

These separate radiations in different rain forest regions have sometimes produced birds that are so similar in appearance that only the DNA studies of the last 20 years have shown that these similarities result from parallel evolution, rather than from a shared ancestry (Sibley & Ahlquist 1990; Cibois et al. 2001). Past misconceptions are still reflected in the English common names, with terms like warbler, wren, treecreeper, and babbler applied to unrelated birds on different continents. Additional confusion arises from the tendency for the common names of tropical birds, particularly in South America and Australia, to reflect their supposed resemblance to more familiar species, so the antpittas, antshrikes, antthrushes, and antwrens are closely related to each other, but not to the birds after which they are named. The antthrushes of the Neotropics are thrush-like antbirds of the family Formicariidae, while the ant-thrushes of African rain forests are ant-following true thrushes in the family Turdidae. Ecological convergence between unrelated birds in different regions is most obvious for seed-eaters and for birds that catch insects in mid-air, known generally as flycatchers. Small, foliage-gleaning insectivores from different regions are also very similar. In contrast, fruit-eaters, flower visitors, scavengers, and other bird groups vary considerably among the rain forest regions and some have no obvious counterparts on other continents.

DNA studies are rapidly sorting out the evolutionary relationships between the major groups of birds, but the poor fossil record, due to their thin bones and lack of teeth, makes it difficult to relate this increased understanding of avian phylogeny to present-day biogeography (Ericson et al. 2003). The bird fauna of American rain forests is the most distinctive and also the richest in terms of species by a great degree (Stotz et al. 1996). Nearly 1000 bird species have been recorded in the enormous, near-pristine Manu Biosphere Reserve in Peru—roughly 11% of the 9500 bird species found on the entire planet (Wilson & Sandoval 1996). Despite the richness at the species level, fewer families of birds are represented in New World rain forests than in other rain forest areas, because each of a small number of large New World families occupies the ecological roles of several Old World families. Thus the extraordinarily diverse ovenbird family (Funariidae) includes species that look and behave somewhat like Old World larks, jays, tits, nuthatches, wrens, thrushes, dippers, treecreepers, and warblers. The ovenbirds, along with the antbirds (Formicariidae and Thamnophilidae), woodcreepers (Dendrocolaptidae), tyrant flycatchers (Tyrannidae), cotingas (Cotingidae), manakins (Pipridae), and several smaller groups, form part of an exclusively New World radiation of around 1100 suboscine passerines: a group of perching birds distinguished from the familiar oscine songbirds by their more simple vocal apparatus and relatively limited singing ability (Ridgely & Tudor 1994). Only around 50 species of suboscines occur in the Old World, representing a separate radiation that includes the pittas, broadbills, and asities. The suboscines may have entered South America from the south more than 40 million years ago, while the continent was connected to Africa through Antarctica (Ericson et al. 2003). Suboscines still dominate the rain forest understorey, where they have been joined more recently by a variety of oscine groups, the most important of which are the wrens (Troglodytidae) (Ricklefs 2002).

In the canopy of South American rain forests and in nonforest habitats, oscines now dominate, although the suboscine tyrant flycatchers are also prominent. Many of the Neotropical rain forest oscines belong to another major New World radiation, the "nine-primaried oscines" (Fringillidae, in the broad sense), which apparently entered from the north more recently. However, many representatives of this group are strong fliers, so it is likely that they spread to South America well before the formation of the Panama land bridge 2–3 million years ago (Ericson et al. 2003). These birds are distinguished by having only nine primary flight feathers (the feathers on the outer part of each wing), compared with 10 in most other passerines. Rain forest members of this radiation include the tanagers and their relatives (Thraupini) (see Plate 5.1, between pp. 150 and 151), the oropendolas, caciques, and orioles (Icterini), and also the woodwarblers (Parulini), many of which migrate in the spring to temperate North America to breed. Other major bird groups confined to the New World include the toucans (Ramphastidae) and hummingbirds (Trochilidae) (Table 5.1).

In contrast to the distinctiveness of Neotropical birds, the rain forest bird faunas of African and Asian rain forests are similar to each other at the family level because of dispersal via intermittent forest connections. Africa and Asia consequently share such important rain forest families as the hornbills (Bucerotidae), bulbuls (Pycnonotidae), and sunbirds (Nectariniidae). But both areas also have endemic families—the turacos (Musophagidae) in Africa and the leafbirds (Irenidae) in Asia—and the relative importance of the families they do share differs considerably. The pittas (Pittidae), for example, are a largely Asian group, with only two species in African rain forests, whereas the weavers (Ploceidae) are a largely African group, with many rain forest species in Africa but only a few species in Asia, none of which inhabits the rain forest.

Many major Old World bird families, including the woodpeckers, barbets, shrikes, tits, babblers, and hornbills, failed to colonize Madagascar, but two endemic groups of passerines have radiated to fill an amazing diversity of foraging niches (Morris & Hawkins 1998; Schulenberg 2003). The vangas (Vangidae) are a small family of around 17 species found only on the island of Madagascar. Although the relationship of the vangas to other bird families is unclear, they appear to represent an evolutionary radiation from a single ancestral type that colonized Madagascar in the past. Despite their common ancestry, the vangas have an exceptional diversity of morphologies, particularly in bill size and shape, but also in wing and tail size and shape, reflecting differences in foraging behavior and prey choice (Yamagishi & Eguchi 1996). The helmet vanga (*Euryceros prevostii*) has the appearance of a barbet or toucan, using its massive beak to grasp large insects, lizards, and frogs. The nuthatch vanga (*Hypositta corallirostris*) forages like a nuthatch or treecreeper, climbing up tree trunks and catching insects with its short, sharp beak. The sickle-billed vanga (*Falculea palliata*) has a long, downward curving bill that it uses for probing into tree holes and other cavities, like a woodpecker, in search of insects and other animals (see Fig. 5.1c). Recent DNA studies have resulted in the addition of four small, warbler-like species in the endemic genus *Newtonia* to the vanga family (Yamagishi et al. 2001), and have also identified a previously overlooked second radiation of endemic forest passerines, including at least 10 species so diverse in appearance that they were previously classified as bulbuls, warblers, or babblers (Cibois et al. 2001). Several smaller endemic radiations are also found among the nonpasserines.

In New Guinea, there are many species from separate New Guinea–Australian radiations of passerine birds, including the bowerbirds (Ptilonorhynchidae), honeyeaters (Meliphagidae), birds of paradise (Paradisaeidae) and several smaller families (Coates & Peckover 2001). The species diversity of pigeons and parrots, two widespread families, is also very high. There are also a few representatives

**Table 5.1** Endemic and dominant families of rain forest birds in each major tropical region.

| Region | Endemic families | Dominant families |
| --- | --- | --- |
| Tropical America | Tinamidae (tinamous)<br>Cracidae (curassows)<br>Psophiidae (trumpeters)<br>Momotidae (motmots)<br>Ramphastidae (toucans)<br>Galbulidae (jacamars)<br>Bucconidae (puffbirds)<br>Dendrocolaptidae (woodcreepers)<br>Formicariidae (ground antbirds)<br>Thamnophilidae (typical antbirds)<br>Furnariidae (ovenbirds)<br>Cotingidae (cotingas)<br>Pipridae (manakins) | Columbidae (pigeons)<br>Trochilidae (hummingbirds)<br>Ramphastidae (toucans)<br>Dendrocolaptidae (woodcreepers)<br>Formicariidae (ground antbirds)<br>Thamnophilidae (typical antbirds)<br>Cotingidae (cotingas)<br>Pipridae (manakins)<br>Tyrannidae (tyrant flycatchers)<br>Thraupini (tanagers)*<br>Parulini (woodwarblers)* |
| Africa | Musophagidae (turacos) | Cuculidae (cockoos)<br>Alcedinidae (kingfishers)<br>Bucerotidae (hornbills)<br>Pycnonotidae (bulbuls)<br>Laniidae (shrikes)<br>Sylviidae (Old World warblers)<br>Muscicapidae (Old World flycatchers)<br>Nectariniidae (sunbirds)<br>Ploceidae (weavers) |
| Southeast Asia | Irenidae (leafbirds) | Columbidae (pigeons)<br>Cuculidae (cockoos)<br>Alcedinidae (kingfishers)<br>Picidae (woodpeckers)<br>Pycnonotidae (bulbuls)<br>Timaliidae (babbers)<br>Sylviidae (Old World warblers)<br>Muscicapidae (Old World flycatchers)<br>Nectariniidae (sunbirds) |
| Australia/ New Guinea | Casuariidae (cassowaries)<br>Megapodidae (megapodes)<br>Menuridae (lyrebirds)<br>Ptilonorhynchidae (bowerbirds)<br>Maluridae (fairy-wrens)<br>Meliphagidae (honeyeaters)<br>Acanthizidae (Australian warblers)<br>Pomatostomidae (Australian babblers)<br>Orthonychidae (logrunners)<br>Paradisaeidae (birds of paradise) | Columbidae (pigeons)<br>Psittacidae (parrots)<br>Cuculidae (cockoos)<br>Muscicapidae (Old World flycatchers)<br>Ptilonorhynchidae (bower birds)<br>Paradisaeidae (birds of paradise)<br>Meliphagidae (honeyeaters)<br>Campephagidae (caterpillar birds)<br>Pachycephalidae (pitohuis) |
| Madagascar | Mesitornithidae (mesites)<br>Brachypteraciidae (ground-rollers)<br>Philepittidae (asities)<br>Vangidae (vangas) | |

* Treated as tribes of the family Fringillidae in this book.
Modified from Karr (1990).

from many Old World bird families, such as the hornbills, pittas, sunbirds, white-eyes, and starlings, which have all presumably entered New Guinea relatively recently from the west. However, many other major bird families of Old World rain forests, including the bulbuls, babblers, barbets, and woodpeckers, are entirely absent. Some Old World families, such as the hornbills and treeswifts, have reached New Guinea but not Australia.

## Little, brown, insect-eating birds

The best-known birds of tropical rain forests are large or brightly colored, or, more rarely, both. In reality, however, the majority of rain forest bird species are small, brown, and relatively inconspicuous. Small brown birds are particularly common in the understorey and lower canopy. These species actively search for insects and other small arthropods on living foliage, twigs, and tree trunks, in vine tangles, and in bunches of dead leaves. An indirect impact on plants has been demonstrated by experimentally excluding birds from tree branches in semideciduous rain forest in Panama (Van Bael et al. 2003). In the canopy, both arthropod densities and leaf damage greatly increased when birds were excluded, but there was no significant impact in the understorey. In Neotropical rain forests, the majority of these birds belong to the endemic radiation of suboscine passerines, while in the Old World these niches are occupied by many different families of oscine passerines.

### Mixed-species flocks

Insectivorous suboscines, including ovenbirds, antbirds, woodcreepers, and tyrant flycatchers, form a major part of the mixed-species flocks that are such a conspicuous feature of the New World understorey bird fauna (Munn 1985; Powell 1985). The suboscines are joined in these flocks by both nonpasserines, such as woodpeckers, and a variety of oscine passerines, including vireos (Vireonidae) and tanagers (Thraupini). Tanagers, tyrant flycatchers, trogons, nunbirds (Bucconidae), and cotingas dominate the more variable mixed-species flocks of the upper canopy. For the rain forest birdwatcher, long periods with no birds visible alternate with short bursts of frantic activity as a flock passes through. The understorey flocks consist of a dozen or more core species, each species represented by one breeding pair, which remains with the flock for several years. A variable number of more or less regular or occasional participants join the flock, resulting in up to 50 species foraging together as a group. The communal territory of the flock is defended by all the core species where it borders with the territories of other, similar flocks (Jullien & Thiollay 1998). At least in Amazonia, flock size, composition, and home range are highly stable between seasons and between years (Powell 1985).

Two basic hypotheses have been proposed to explain the advantages accruing to members of these mixed-species flocks (Thiollay 1999). One hypothesis is that the main advantage is increased safety from predators, such as hawks, cats, and snakes, because individual birds cannot keep an adequate lookout by themselves while they are actively searching for insects. This "many eyes" hypothesis has been supported by studies that compare the foraging behavior of flocking and solitary insectivores in the Neotropics (Thiollay 2003). Obligate flock members are more likely than solitary species to use active foraging techniques that make them conspicuous and are incompatible with sustained vigilance. An alternative, or additional, hypothesis is that flock members benefit from the prey found or flushed out by the flock as a whole.

Whatever the explanation, mixed flocks of understorey insectivores and omnivores are found in rain forests all over the world, although the families of birds involved show almost no overlap with those in the Neotropics. In Africa, mixed flocks include bulbuls, babblers (Timaliidae), Old World warblers (Sylviidae), Old World flycatchers (Muscicapidae), drongos (Corvidae), woodpeckers, white-eyes (Zosteropidae), sunbirds, and weavers. Woodpeckers are absent from Madagascar, where the endemic vangas join, and sometimes dominate, the mixed flock party. In Southeast Asia, mixed-species flocks are similar to those in Africa, but lack weavers and can contain up to 15 species of babblers—the most diverse group of birds in the region—as well as the endemic leafbirds. In Australia and New Guinea, these flocks attract many species from groups endemic to the region, including Australian babblers (Pomatostomidae), Australian warblers (Acanthizidae), pitohuis (Pachycephalidae), fairy-wrens (Maluridae), and birds of paradise. Although superficially very similar, it is not yet clear if any of these flocks are as highly organized and stable as those in the Neotropics.

## Birds and army ants

Mixed-species flocks also form around columns of predaceous army ants in American and African rain forests, but not in other regions, where the ant swarms are apparently too small to attract followers. Indeed, only two army ant species in the Neotropics, *Labidus praedator* and *Eciton burchelli*, and a few driver ant (*Dorylus* spp.) species in Africa, form swarms that are large enough and dependable enough to regularly attract birds (Gotwald 1995) (see Chapter 7). These army ants raid in huge columns or waves across the forest floor, killing any arthropods and small vertebrates they encounter, and bringing their prey back to their nests to eat. In Neotropical forests, a number of bird species are professional ant followers, rarely feeding away from ant swarms, while others are regular or merely occasional associates. As their family name suggests, some species of antbird are particularly prominent among the ant followers, but certain species of woodcreepers, ground-cuckoos, and tanagers are also "professionals"—that is, they make most of their living following ants. Professional ant followers can cling to vertical perches, such as the slender saplings that are common in the forest understorey, while watching for likely prey fleeing the army ants. The bird species that are most dependent on army ants track the location of several ant colonies and check the temporary ant nests—known as bivouacs—each morning to assess their activity (Swartz 2001). By keeping track of several ant colonies as they go through their cycles of raiding and resting, the birds can be assured of having an active colony to follow each day. The army ants of Africa are followed by flocks of birds unrelated to those in the Neotropics and appear to include fewer professionals. Thrushes, such as the fire-crested alethe (*Alethe diademata*), are particularly prominent at African ant swarms, and several species of large bulbul cling to vertical perches in a manner similar to Neotropical antbirds (Brosset 1990).

## Antbirds

The antbirds of the American tropics are small to medium-sized birds that occupy lowland forests. They belong to two related families of suboscine passerines: a smaller family of ground antbirds (Formicariidae) and a larger family of typical antbirds (Thamnophilidae), with a total of around 250 species in 53 genera. Antbirds are adapted for catching insects in the forest interior; their long, thick, hooked bills with notches on the edge are effective at holding and slicing up small animals. Antbirds often have moderately rounded wings and long feet, an advantage when foraging in the clutter of the forest interior. Many antbirds are members of mixed-species flocks, and 27 species are found in close association with columns of army ants (Willis & Oniki 1978; Chesser 1995). The plumage of antbirds is patterned brown, black, or white, and often has streaked markings. Many species are hard to see in the forest understorey and are more easily recognized by their songs. One of the most striking antbirds is the white-plumed antbird, *Pithys albifrons*, which has an orange-brown body, grey wings, and a dark grey head, with a dramatic white crest just in front of its eyes and a white beard at the base of its bill (Fig. 5.2). The crest is held upright in social interactions, but folds backward in flight. This species is among the commonest professional found at ant swarms throughout much of the Amazon rain forest (Kricher 1997).

**Fig. 5.2** The white-plumed antbird (*Pithys albifrons*) is a specialist feeder on insects and other small animals fleeing from army ant swarms. (Courtesy of Andrew Dennis, taken near Manaus, Brazil.)

**Fig. 5.3** Babblers are one Old World family of little brown birds. Here is a relatively attractive babbler: the golden crowned babbler (*Stachyris dennistowni*) from Luzon, Philippines. (Courtesy of Tim Laman.)

## Babblers and Old World warblers

Most of the little brown birds in the Old World have traditionally been included in either the babblers (Timaliidae), with 200 or so species (Fig. 5.3), or the Old World warblers (Sylviidae), with more than 300 species. Both groups are represented in African and Asian rain forests, but babblers are most diverse in tropical and subtropical Asia and warblers in Africa. Together these two groups form an almost exclusively Old World bird radiation (Sibley & Ahlquist 1990). The 100 or so species of New World warblers or woodwarblers (Parulini) are not related to the Old World warblers, but are nine-primaried oscine relatives of the tanagers (Thraupini). The babblers show a great diversity in size, bill shape, and other characteristics, but differ from the warblers in their generally larger size, high sociability, and the absence of migratory behavior (Cibois 2003). The noisy "babbling" calls between flock members of some species are the origin of the common name for the family and contrast with the pleasant "warbling" song of many warblers.

Recent advances in DNA-based systematics have played havoc with the traditional classification of both the babblers and the Old World warblers (Cibois 2003). All the previous "babblers" of Australia, New Guinea, and Madagascar have been shown to belong to other families, as have several Asian and African species previously included in the Timaliidae. Conversely, both the *Sylvia* warblers of the western Palaearctic and—more surprisingly—the white-eyes (Zosteropidae) of the Old World tropics have been shown to belong within the babblers. These studies have also confirmed that a single North American species, the wren-tit (*Chamaea fasciata*), is a babbler. The newly delimited babblers are still the most diverse bird family in the rain forests of tropical and subtropical Asia, with a secondary center of diversity in West and Central Africa, but are now also represented in the rain forests of Australia, New Guinea, and Madagascar, as well as many Pacific islands, by *Zosterops* white-eyes.

## Forest frugivores

In striking contrast to the drabness and superficial uniformity of most insectivores, many frugivores are big and/or brightly colored, and differences among rain forest regions are obvious to even the casual observer. Indeed, most birds seen by the casual observer are likely to be frugivores. Although insectivores dominate in terms of the numbers of individuals and numbers of species in all rain forests, fruit- and seed-eaters contribute far more to the biomass, because individual birds tend to be bigger. Larger size allows birds to handle larger food items—most fruits are bigger than most insects—and also lets them swallow more food during intensive feeding bouts on fruiting plants, so that time spent exposed to predators while feeding is minimized. Large size also directly reduces the risk of predation by decreasing the number of predators that can attack a particular bird. Lower predation risk allows these birds to make more use of color for signaling to potential mates and rivals although, in many cases, it is only the male that does so. Many of the most spectacularly colored rain forest birds are frugivores, including the resplendent quetzal (*Pharomachrus mocinno*) in Central America, the violet turaco (*Musophaga violacea*) in West Africa, the fairy bluebird (*Irena puella*) in Southeast Asia, and the raggiana bird of paradise (*Paradisaea raggiana*) in New Guinea. When insects become abundant, such as swarming termites after rainfall, even large-bodied frugivores tend to become temporary insectivores, and some of the largest species, such as toucans and hornbills, commonly eat lizards and small birds as well.

Pigeons and parrots take fruits in all rain forest regions, but there are noticeable differences between regions in the other bird groups involved. A Neotropical fig tree with a large crop of ripe figs will attract avian fig-lovers ranging in size from tiny manakins and tanagers to large toucans, and also including guans, trogons, New World barbets, cotingas, tyrant flycatchers, and orioles (Shanahan et al. 2001; Tello 2003). Trumpeters (Psophiidae) feed on any figs that are knocked to the ground. In African rain forests, a similar tree would attract mostly birds from families that are absent from the New World: hornbills, bulbuls, African barbets (Lybiidae), white-eyes, cuckoo-shrikes, and turacos (Musophagidae). Asian rain forests lack turacos and African barbets, and add Asian barbets (Megalaimidae), frugivorous broadbills (Eurylaimidae), leafbirds (Irenidae), and flowerpeckers (Dicaeidae), but the list would be otherwise similar. In New Guinea, where there are no barbets or bulbuls, our fig tree would again draw pigeons, parrots, mynahs, white-eyes, flowerpeckers, and cuckoo-shrikes, plus the single species of hornbill, but also bowerbirds, birds of paradise, and honeyeaters. Cassowaries (Casuariidae) eat fallen figs from the ground. The assemblage in an Australian rain forest would be similar, but less diverse. Madagascan rain forests have relatively few frugivorous birds and our fig tree would attract mostly pigeons, parrots, and bulbuls, as well as the velvet asity (*Philepitta castanea*), a member of an endemic group of suboscines that is related to the broadbills. Some of the rain forest vangas also include some fruit in their diet.

## Turacos

The turacos (Musophagidae) are a very distinctive and exclusively African group of fruit-eating birds, related to the cuckoos, with 23 species in six genera (Turner 1997a). The largest species, the great blue turaco (*Corythaeola cristata*), is 75 cm (30 inches) in length but most of the others are somewhat smaller, around the size of a pigeon. Forest species are often sedentary, remaining in localized areas. Turacos have a long tail, small, rounded wings, and strong long legs which enable them to move readily in the tree canopy. They may have a conspicuous crest on their head and soft, often hair-like feathers. The bright colors of some species rival those of the parrots. One particularly beautiful species is the Livingstone's turaco, *Tauraco livingstonii*, which has delicate apple-green feathers covering its breast and head and a pronounced green crest. A red eye ring and a white eye stripe further enhance its appearance. While the bright colors of many birds result from refraction of light by the structure of the feathers, the blue and red colors of turacos are true pigments.

## Toucans and hornbills

In tropical rain forests from South and Central America, through Africa and Asia, to New Guinea, there are large fruit-eating birds with huge bills, loud calls, and often striking coloration on the bill, head, and neck. Although superficially very similar, these birds belong to two unrelated families, the toucans (Ramphastidae) in South and Central America and the hornbills (Bucerotidae) in the tropics of the Old World (Fig. 5.4). Of the major rain forest areas, only those of Madagascar and Australia are without one or the other family, although New Guinea has only a single species of hornbill.

The long and deep bills of toucans and hornbills have multiple functions in feeding, display, defense, and nesting, but the primary one is probably to extend their reach when perching (Bühler 1997). This is particularly important when feeding on fruits, which are often borne on twigs that are much too thin to support the weight of these heavy birds. Instead, they can grasp a strong support with their feet and use their long bills and long, flexible necks to reach out for fruits that would otherwise be beyond their reach (see Plate 5.2, between pp. 150 and 151). The same bill design is also very useful when reaching into deep nests to pluck out young birds to eat.

These bills look massive and powerful, but appearances are misleading. The bills are actually thin and light. The shape is dictated by design principles. A long bill for firmly holding food items must also be deep to prevent bending. The muscles that close the bill are attached near the base, so only a relatively small force can be exerted at the tip, which is the only place where the upper and lower parts of the bill meet. The bill is used like forceps to pick fruits one at a time, before tossing them skillfully back into the throat. Larger forces can be applied further back from the tip and this open region, which is serrated in many species, is used to crush large items of animal food, such as lizards, snakes, nestlings, and insects, which make up a significant proportion of the diet in all species and dominate it in some hornbills. The basic similarity of the bills in the two groups thus follows directly from simple engineering principles.

(a)

(b)

**Fig. 5.4** Hornbills and toucans are similar-looking birds of the Old and New World. (a) A black hornbill (*Anthracoceros malayanus*) eating fruits in Gunung Palung National Park, Indonesia. (Courtesy of Tim Laman.) (b) This keel-billed toucan (*Ramphastos sulfuratus*) is from the Children's Eternal Rainforest in Costa Rica. This is one of the largest toucans but is only two-thirds the size of the black hornbill. (Courtesy of Dale Morris.)

Despite their many similarities, toucans and hornbills also differ in important ways. Most hornbills are a great deal bigger than even the largest toucans and can eat correspondingly larger food items. Indeed, the biggest hornbills are the largest flying frugivores on Earth. Hornbills also differ from toucans by the presence of a casque, an outgrowth from the top of the already large bill (Woodcock & Kemp 1996). The shape and prominence of the casque varies among species; it is virtually absent in young birds. The function of the casque is still debated, but it is probably involved in both visual communication, through its size, shape, and color, and acoustic communication, through the ability of the large air cavity to amplify calls of appropriate frequencies.

Both hornbills and toucans nest in natural cavities in trees, but the hornbills are unique among birds in that the breeding female of most species seals the entrance to the cavity from the inside, with mud, sticky food, and her own droppings, leaving only a narrow vertical slit (Woodcock & Kemp 1996). Through this, the male feeds her and later her chicks, for 6 or more weeks. Droppings are voided out of the slit and food remains are thrown out, thus maintaining nest hygiene and leaving a useful record, below the nest, of the food brought by the male. The female probably builds such a maternity prison to keep out unwanted visitors, such as snakes and other predators.

## Mistletoe birds

The fruits of most mistletoes—parasitic plants in the families Loranthaceae and Viscaceae that grow on the branches of trees and shrubs—are small enough to be eaten by any frugivore. In tropical Asia, however, mistletoe fruits are largely consumed by tiny flowerpeckers (Dicaeidae), the smallest frugivores in the region (Fig. 5.5). The 40 or so species of flowerpeckers are brightly colored birds with short tails and relatively short beaks. The foraging of mistletoe birds is linked to the biology of the mistletoe plants (Reid 1991). The year-round availability of

**Fig. 5.5** Crimson-breasted flowerpecker (*Prionochilus percussus*) in Gunung Palung National Park, Indonesia. (Courtesy of Tim Laman.)

mistletoe fruits allows these birds to specialize on them, while the unique biology of the mistletoe plants means that the fruits eaten by other frugivorous birds are usually wasted. Unlike most plants, where the seeds can be scattered over the ground, successful dispersal of mistletoe seeds requires that they be attached to a branch of a compatible host species, with a suitable diameter for establishment. All other seeds are wasted. Mistletoe seeds are extremely sticky and it is this stickiness that provides both the means of attachment to a branch and the incentive for a bird to remove them from their beak or cloaca. Flowerpeckers typically defecate strings of sticky seeds, which they wipe off on a branch. The preferred perch diameter of these tiny birds apparently matches the size of branch on which the mistletoe seeds can best germinate and establish a new plant.

There are other mistletoe specialists in other rain forest regions, but the mistletoe–flowerpecker association in tropical Asia seems to be uniquely strong. In New Guinea and Australia, some honeyeaters are important consumers of mistletoe berries, but the Australian mistletoe bird (*Dicaeum hirundinaceum*) is another flowerpecker, the only species that reached the continent. In Africa, some species of tinkerbird (*Pogoniulus*) are specialists, as are several species of *Euphonia* tanager in the Neotropics, as well as the paltry tyrannulet (*Zimmerius vilissimus*) of Central America and several species of cotingas.

## Parrots

Not all fruit-eaters disperse seeds, or even drop them undamaged under the parent plant. Some birds, including most parrots, destroy seeds in the beak and then swallow the pieces. Long-tailed parakeets (*Psittacula longicauda*) are major predators of dipterocarp seeds in Southeast Asia (Curran & Leighton 2000), whereas the huge blue hyacinth macaw (*Anodorhynchus hyacinthinus*) uses its powerful bill to open palm nuts in the Amazonian region. Even the tiny seeds of figs are not safe from parrots, with species such as the miniscule double-eyed fig-parrot (*Cyclopsitta diopthalma*) in Australia and the orange-chinned parakeet (*Brotogeris jugularis*) in Costa Rica able to break them individually in the bill (Shanahan et al. 2001). Although parrots destroy most seeds they encounter, they are also notoriously messy eaters. A significant proportion of the fruits they try to eat are dropped to the ground unharmed. Here terrestrial rodents, browsing mammals, and more benign bird frugivores may eat the fruits and disperse the seeds that they would otherwise not have had access to in the high canopy. In New Guinea, the vulturine parrot (*Psittrichas fulgidus*) appears to specialize on a few strangling fig species (*Ficus*) that produce exceptionally hard-walled figs (Mack & Wright 1998). These parrots may disperse the seeds directly, as well as making the pulp and seeds accessible to smaller and weaker-billed seed-dispersing birds.

The parrots (Psittaciformes) are a widely distributed group of 354 species in around 80 genera (Collar 1997; Juniper & Parr 1998). A varying number of separate parrot families have been proposed in the past, but most authorities now recognize only two: the Psittacidae, with more than 330 species spread over much of the world, and the much smaller Cacatuidae, the 22 species of cockatoos, which range from Australia to the Philippines. Parrots vary widely in size and often are brightly colored. They are easily recognized by a number of features such as the short, thick, and hooked upper bill that can move up and down on a specialized joint (Fig. 5.6). In addition, the upper bill has an enlarged fleshy

(a)

(b)

**Fig. 5.6** Parrots are noisy, colorful birds found in rain forests. (a) Hyacinth macaw (*Anodorhynchus hyacinthinus*) from Brazil. (Courtesy of Tim Laman. (b) Scarlet macaw (*Ara macau*) from South America. (Courtesy of Tim Laman, taken in captivity.)

covering where it joins the head. Parrots often have plump bodies and rounded wings, and are known for their loud, screeching calls.

Parrot diets often overlap with those of another group of muscular-jawed seed destroyers, the squirrels, and it is probably no coincidence that rain forest parrot diversity is highest in the Neotropics (with 50 species in the Amazon basin alone) where there are only seven squirrel species, and New Guinea (with 43 species) where there are no squirrels at all (see Table 3.1). In contrast, African and Asian rain forests have many coexisting species of squirrels and few parrots. In those forests, squirrels vary in size and have a diversity of food preferences and foraging strategies. In forests with both squirrels and parrots, squirrels will aggressively chase parrots out of trees in the squirrel's territory. There are only eight species of parrots in African rain forests and six in mainland Southeast Asia. The one region in East Asia where parrot diversity is somewhat higher is the Philippines, with 12 species, including one cockatoo; it also a region with few squirrels—only two species. Madagascar spoils the pattern, by lacking squirrels but having only three parrots, but this huge island is peculiar in many ways because of its long period of isolation. Not all parrots are seed-eaters: members of the Australasian subfamily Loriinae (loris and lorikeets) are flower-feeders (see below), while the tiny pygmy parrots (*Micropsitta* spp.) in New Guinea and eastern Indonesia forage on tree trunks and branches for lichens, fungi, and small insects.

(c)

Fig. 5.6 (*cont'd*) (c) Rainbow lorikeet (*Trichoglossus haematodus*) from Australia. (Courtesy of Tim Laman, taken in captivity.)

## Pigeons and doves

The pigeons and doves (Columbidae), like the parrots, are a pantropical group of birds in which most species probably destroy more seeds than they disperse. Unlike the parrots, this destruction takes place in a muscular, grit-filled gizzard, rather than the bill, and they feed mostly on relatively small seeds. However, by no means do all pigeons and doves destroy the seeds they swallow. In the rain forests of Asia, New Guinea, Australia, and the Pacific, there is a distinctive radiation of "fruit pigeons", with thin-walled gizzards and short, wide guts, through which even the largest seeds can pass undamaged (Corlett 1998). This group includes the imperial pigeons (*Ducula*) (Fig. 5.7), fruit doves (*Ptilinopus*), and mountain pigeons (*Gymnophaps*), as well as the Australian topknot pigeon (*Lopholaimus antarcticus*). Imperial pigeons, in particular, are among the largest frugivores in the rain forests of this region and can swallow very large fruits (up to 4 cm diameter), which few other birds can handle. This makes the fruit pigeons extremely important seed dispersal agents for plants that bear large fruits with large seeds, such as nutmegs (Myristicaceae) and palms (Palmae). The Madagascan blue pigeon (*Alectroenas madagascariensis*) is a member of the same Indo-Australian radiation as the fruit pigeons (Shapiro et al. 2002) and appears to have a similarly important role in seed dispersal (A. Bollen, personal communication).

Pigeons feed on fruits in the canopy of all rain forests, but the role of species other than fruit pigeons in seed dispersal is unclear. In contrast to the fruit pigeons, these other pigeon species feed mostly on small-seeded fruits, such as figs and, in the Neotropics, *Cecropia*. Seed fate seems to vary, even between bird species in the same genus, from complete destruction to high-quality dispersal, but too few species have been investigated for any clear patterns to emerge (Corlett 1998; Shanahan et al. 2001).

**Fig. 5.7** Pigeons are an important family of fruit-eating birds in rain forests. Here is a green imperial pigeon (*Ducula aenea*) on a fig tree in Palawan, the Philippines. (Courtesy of Tim Laman.)

## Frugivory and extreme courtship behavior

One of the most striking examples of convergence between unrelated bird families in different rain forest regions is the remarkable similarity in both appearance and behavior between the birds of paradise in New Guinea and a number of species of cotingas (Cotingidae) and manakins (Pipridae) in the Neotropics (Johnsgard 1994). The birds of paradise (Paradisaeidae) are probably the best-known bird group in the rain forests of New Guinea (Fig. 5.8a,b). They are related to the crow family (Corvidae), but there is hardly anything crow-like in either their appearance or behavior. The 42 species of birds of paradise vary from starling-like to crow-like in size (Frith & Beehler 1998). What is most astonishing about birds of paradise is the elaboration in males of feathers used in courtship displays. In some species, feathers on the tails, wings, body, and head of males are extraordinarily elongated and often brightly colored or ruffled; presumably, such feathers in some way signal male fitness to the watching female birds and thus desirability of individual males as mates. Male birds often shake or ruffle these feathers while hopping about, with the females watching and evaluating nearby—a display that must be seen to be believed. The King of Saxony bird of paradise, *Pteridophora alberti*, has two exquisite plumes on its head that are several times longer than its body; males impress females by raising these feathers while bouncing on branches. Some species even display while hanging upside down, spreading their long feathers over their head.

These fantastic displays by male birds of paradise are a reflection of an extreme type of mating system in which male and female birds lead entirely separate lives, except when they briefly come together to copulate. To put it simply, males display and females nest (Frith & Beehler 1998). An adult male

**Plate 2.1** Dipterocarp leaves and fruits with two wings (*Dipterocarpus costulatus*) from Malaysia. (Courtesy of K.M. Wong.)

**Plate 2.2** Palms give tropical forests a special look, as shown by this hill sago palm (*Eugeissona utilis*) from Brunei in northwestern Borneo. (Courtesy of K.M. Wong.)

(a)

**Plate 2.3** (a) A tropical slipper orchid (*Paphiopedilum sanderianum*) with long pendant petal tips. This species grows on limestone cliffs in Borneo. (Courtesy of Hans Hazebroek.) (b) Epiphytic orchid (*Bulbophyllum medusae*) from Borneo. (Courtesy of Tim Laman.)

(b)

(c)

(d)

**Plate 2.3** (*cont'd*) (c) Large epiphytic orchid (*Grammatophyllum speciosum*), Borneo. (Courtesy of Tim Laman.) (d) Flower of *Grammatophyllum*. (Courtesy of Tim Laman.)

**Plate 2.4** Tank bromeliad (*Neoregelia pimeliana*) filled with water and in flower. (Courtesy of David Lee, taken in a garden.)

**Plate 2.5** Durians (*Durio* spp.) are unusual fruits of Asian forests, growing both in the wild and in plantations. Fruits vary from the size of a grapefruit to the size of a basketball, and are covered with conical spikes that end in a sharp point. The fruit splits along five suture lines to reveal seeds covered in flesh, variously colored white, yellow, orange, or red, depending on the species. The flesh is often highly aromatic, some would say smelly, with a rich, creamy flavor that is totally unique. Here are a domestic durian on the left and two wild species on right, collected from Borneo. (Courtesy of Tim Laman.)

**Plate 3.1** Spectral tarsier (*Tarsius spectrum*) eating a cockroach in Sulawesi. (Courtesy of Tim Laman.)

**Plate 3.2** Adult male orangutan (*Pongo pygmaeus*), a great ape, eating durian fruit in Gunung Palung National Park, Indonesian Borneo. (Courtesy of Tim Laman.)

**Plate 3.3** Pygmy mouse lemur (*Microcebus rufus*) from Madagascar. (Courtesy of Harald Schuetz.)

**Plate 3.4** Diadem sifaka (*Propithecus diadema diadema*) from Madagascar. (Courtesy of Harald Schuetz.)

**Plate 4.1** Dhole (*Cuon alpinus*), a rain forest relative of dogs, feeding on a carcass at night, from Khao Yai National Park, Thailand. (Courtesy of Tim Laman.)

**Plate 4.2** Malay civet (*Viverra tangalunga*) from Brunei, Borneo. (Courtesy of K.M. Wong.)

**Plate 4.3** Forest elephants (*Loxodonta cyclotis*) at a water hole in the Congo with sitatungas (*Tragelaphus spekei*) in the foreground. (Courtesy of Roger Le Guen.)

**Plate 5.1** Green honeycreeper (*Chlorophanes spiza*), a frugivorous member of the New World tanager family, in Colombia. (Courtesy of Gustavo Londoño.)

**Plate 5.2** Rhinoceros hornbill (*Buceros rhinoceros*), on a fig tree in Gunung Palung National Park, Indonesia. (Courtesy of Tim Laman.)

**Plate 5.3** Purple-throated carib hummingbird (*Eulampis jugularis*) visiting *Heliconia* flowers in Dominica. (Courtesy of Tim Laman.)

**Plate 5.4** Honeyeaters are an important family of flower-visiting birds in New Guinea, Australia, and the South Pacific Islands. This cardinal honeyeater (*Myzomela cardinalis*) from western Samoa is drinking nectar from the red pea-like flowers of a leguminous tree. (Courtesy of Clifford and Dawn Frith.)

**Plate 5.5** Ornate hawk eagle (*Spizaetus ornatus*) from Belize. This small but powerful eagle hunts birds and mammals from a perch inside the forest. (Courtesy of Tim Laman, taken in captivity.)

**Plate 6.1** Tube-nosed bat (*Nyctimene cephalotes*) from Sulewesi. This bat has been captured as part of a research study, and is being fed some banana pulp before being released. (Courtesy of Tigga Kingston.)

**Plate 6.2** Paradise tree snake (*Chrysopelea paradisi*) gliding through the air in Borneo. (Courtesy of Tim Laman.)

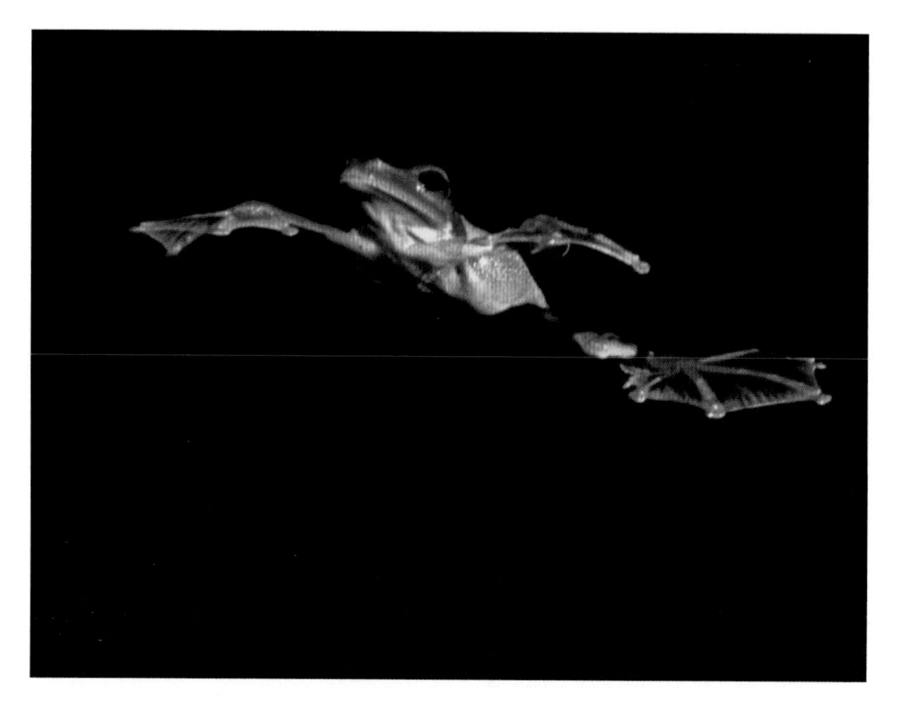

**Plate 6.3** Wallace's flying frog (*Rhacophorus nigropalmatus*), with large webbed feet, in Danum Valley, Sabah, Malaysia. (Courtesy of Tim Laman.)

**Plate 7.1** Male Rajah Brooke's birdwing butterfly (*Trogonoptera brookiana*) from Borneo. (Courtesy of Hans Hazebroek.)

**Plate 7.2** Glass-wing nymphalid butterflies (*Ithomia patilla*) from Costa Rica. (Courtesy of Dale Morris.)

**Plate 7.3** Army ants (*Eciton burchelli*) attacking a scorpion, in Costa Rica. (Courtesy of Dale Morris.)

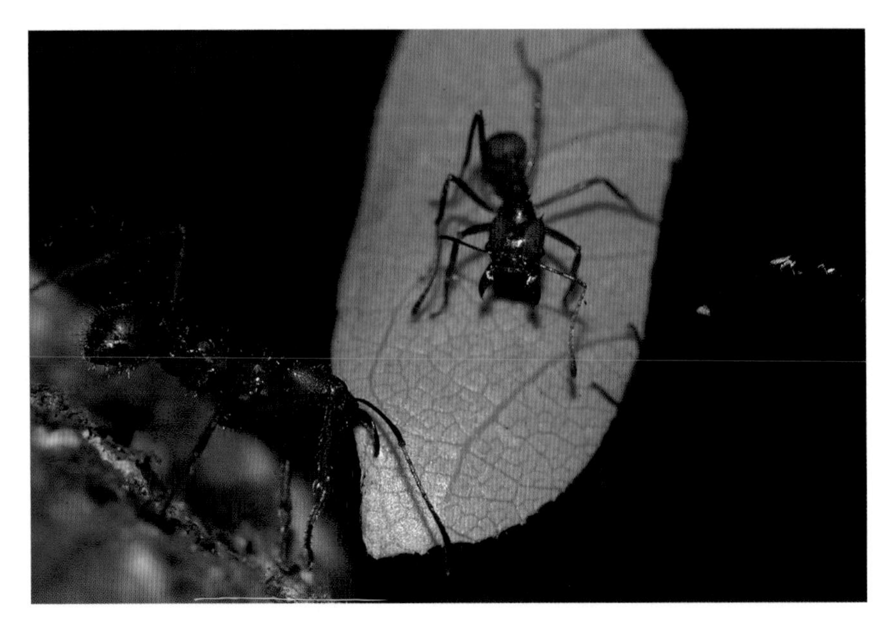

**Plate 7.4** Leaf-cutter ant (*Atta cephalotes*) carrying a leaf fragment, in Costa Rica. The small ant on the leaf is guarding against parasitic flies, which lay eggs on the heads of the ants. (Courtesy of Dale Morris.)

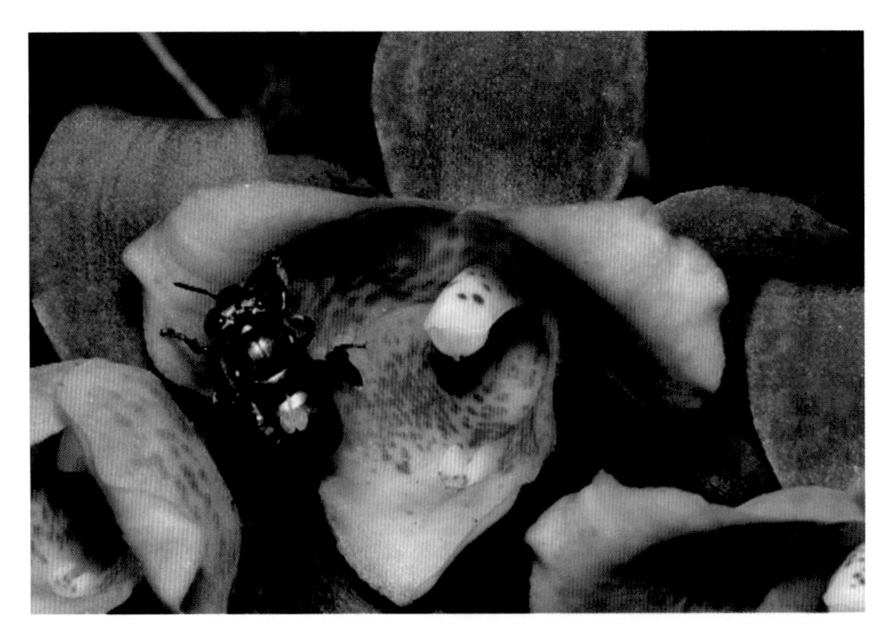

**Plate 7.5** Euglossine bee on a flower of the orchid *Lycaste brevispatha*, from Costa Rica. (Courtesy of Dan Perlman.)

**Plate 8.1** The pattern of deforestation as shown in a Landsat 7 image from 2001 of a region in central Amazonia. The primary highway, paved with World Bank help in 1984, is off the image to the left (west). In this false-color image healthy forest is dark red and the areas of clearing and pasture are light blue. Dark blue areas are those likely recently burned. The eastern side of the river has not been cleared because it has been set aside as a biological reserve and as lands for indigenous people. The image is 69 × 48 km. The large block clearings are typically for pasture. Small holders occupy lots that are typically 500 × 2000 m, arranged along the parallel roads. Over time the clearings in adjacent lots merge. This image was obtained from the Global Land Cover Facility at the University of Maryland; image processing and interpretation by Thomas Stone of the Woods Hole Research Center.

**Plate 8.2** Poaching contributes to the decline of wildlife in protected areas. A sign in Brazil asking motorists to drive carefully to avoid hitting wildlife has been shot up by hunters. (Courtesy of William Laurance.)

**Fig. 5.8** The bird of paradise family and the bowerbird family from New Guinea and Australia have many unusual and conspicuous species. (a) Riflebird (*Ptiloris victoriae*), a bird of paradise, from northeast Queensland, Australia. (Courtesy of Andrew Dennis.) (b) Satin bowerbird (*Ptilonorhynchus violaceus*) in his bower, from Australia. (Courtesy of Dale Morris.) (c) A male Emperor of Germany bird of paradise (*Paradisaea guilielmi*) dramatically displays and calls while perching on a tree trunk, in Papua New Guinea. (Courtesy of Clifford and Dawn Frith.)

spends a large proportion of its time on its display site, which may be solitary or clustered together with other males in a "lek". A female visits one or more of these sites briefly to mate and then she alone is responsible for building the nest and rearing the nestlings. A successful male may mate with many females, but most males will mate with none. What has made the birds of paradise of even more interest to biologists is that different members of the family display the full range of mating systems, from monogamous pairs that forage and raise their young together, to the extreme divergence between sexual roles described above.

The bowerbirds (Ptilonorhynchidae) are a rather distantly related group of birds found in the rain forests of New Guinea and Australia, with a mating system similar to that of the birds of paradise (Frith 1998). Bowerbirds are named for the elaborate courtship sites that the males construct, in which they display to females (Fig. 5.8b) (Johnsgard 1994). The courtship area, or bower, might consist of a cleared circular area on the forest floor, in which a domed tunnel of sticks has been constructed, decorated with colored stones, flowers, snail shells, fruits, and insect parts. When a female is near, a male sings and dances above and on his bower, until the female is ready to mate. In some species, the adult male has spectacular plumage and the bower is simple, while in others the male is dull but the bower is complex. This shows that the bower represents a transfer of sexual attraction from the male's own appearance to the appearance of his bower (Frith 1998).

In the Neotropics, similar extraordinary mating displays are known in the cocks-of-the-rock (*Rupicola* spp.), the red cotingas (*Phoenicircus* spp.), and some manakins (Pipridae) (Ridgely & Tudor 1994; Kricher 1997). As with the birds of paradise, males of these species have spectacular plumage, which they display to the dull-looking females in elaborate courtship displays, either alone or, more often, in leks. Males spend most of their adult lives at the lek. And as with the birds of paradise, the females raise the young alone.

All the bird species with these extreme mating displays are at least partial frugivores and most depend heavily on fruits. This is clearly no coincidence. Fruit is a relatively reliable food source in tropical rain forests so it is possible for a frugivore to meet its daily food requirements in a small proportion of the time available. This gives the males a lot of free time to display and makes it easier for a female to feed both herself and her young. This cannot be the entire story, however, since most frugivores, even in New Guinea and the Neotropics, do not go to these extremes. There is clearly also a very large phylogenetic component, with extreme mating displays confined to a very few evolutionary lineages (Prum 1994). Indeed, once these huge differences between the sexes become established, they may be very hard to reverse, even over evolutionary timescales, because the features that make a male successful in competitive displays are probably incompatible with them contributing to parental care.

Why are these bold courtship displays found in the rain forests of New Guinea and the Neotropics, but not in Southeast Asia, Africa, or Madagascar? It could simply be chance: the combination of factors needed for these displays to evolve may have come together only in bird lineages that did not reach these other rain forests. It could also be a reflection of the reliability of the fruit supply. In particular, the multiyear fruiting cycles in Southeast Asian rain forests may have made it impossible for birds to devote so much of their time to nonfeeding activities. Or it could reflect differences in predation risks. Displaying males are extremely conspicuous and multimale leks in particular must be very attractive to predators. Although most recorded attacks on leks involve raptors (Trail 1987), most displays are close to the ground and thus vulnerable also to mammalian carnivores. Birds of paradise, bowerbirds, cotingas, and manakins evolved and diversified in rain forests that lacked placental carnivores (see Chapter 4), although the cotingas and manakins have had to contend with cats, dogs, mustelids, and procyonids for 2–3 million years since the appearance of the Panama land bridge.

## Birds as seed dispersal agents

In most rain forests, birds are the single most important group of seed dispersal agents. They are particularly important for plants that bear small fruits, while mammals—especially primates—are often more important for large fruits. Birds seem to be most important in New Guinea and Australia, where there are no primates and few mammalian frugivores apart from bats. They may be least important in Madagascan rain forests, where frugivorous birds are few and primates are abundant (see Chapter 3).

Although frugivory might seem to be a simple act, the fate of the seed—and thus the success of seed dispersal—depends critically on how the frugivorous animal selects and processes the fruit, and what it does afterwards. Fruits in each rain forest region are eaten by birds that differ in size and behavior, as well fruit acquisition and processing techniques, from those in other regions (Corlett 2001). The major groups of fruit-eating birds in Asia and Africa usually take fruits from a perch and swallow them whole. Fruit consumption and seed dispersal by these birds is therefore limited principally by their maximum gape width (i.e. the maximum size they can swallow), which is generally larger in bigger birds. Seed dispersing birds in Asian rain forests range in size from 5 g (0.2 oz) flowerpeckers to 3 kg (6.5 lb) hornbills, and the largest fruits that can be swallowed whole range from < 8 to > 30 mm (0.3–1.2 inches) in diameter (Corlett 1998). African rain forests have a similar bird fauna, but a somewhat smaller range of bird sizes.

The New Guinea avifauna includes flowerpeckers and a single hornbill, but also the world's largest frugivorous birds, the flightless cassowaries (*Casuarius*), the biggest of which can swallow whole fruits up to 7 cm in diameter (Stocker & Irvine 1983). Some birds of paradise, another endemic group, have an apparently unique ability to use their feet and bills together to break open the woody capsules of some Myristicaceae and Meliaceae to get at the arillate seeds inside (Beehler & Dumbacher 1996). This behavior may have evolved in response to the absence of primates and squirrels, which open many such fruits elsewhere, and makes the birds of paradise one of the most important groups of vertebrate seed dispersal agents in New Guinea.

Many seeds in Neotropical rain forests are dispersed by endemic bird families with fruit acquisition and processing behaviors that are rarely seen in the Old World. Most fruit-eating suboscines, including manakins, cotingas, and tyrant flycatchers, have wide gapes, take fruits on the wing (rather than from a perch), and swallow them whole (Levey et al. 1994). Tanagers and related groups of nine-primaried oscines have narrow gapes, take fruits while perching, and crush them in the bill, squeezing out all but the smallest seeds, before swallowing the pulp. These "mashers" drop most large seeds under the parent plant, but are important dispersal agents for species with seeds < 2 mm in length, which they cannot separate from the pulp.

What are the consequences of these differences between rain forest regions in the birds that eat fruits and disperse seeds? Unfortunately, the necessary interregional comparisons with standardized methodology have yet to be made. We would predict that fruits targeted at birds would be more often borne on the outside of tree canopies in the Neotropics, where they would be more accessible to those birds that pluck them in flight, while in the Old World they should always be accessible from a perch. This prediction parallels the observed differences in the positions of bird-pollinated flowers, but birds are far more important as seed dispersal agents than as pollinators, so the consequences of bird pollination and bird dispersal should be easy to separate. We might also predict that the small fruits eaten by fruit-mashing tanagers would have smaller seeds than similar-size fruits eaten by similar-size bulbuls. At least for the common pioneer trees, this does seem to be the case, with those in the Neotropics typically having much smaller seeds than those in Africa and Asia, although the greater importance of fruit bats as dispersers of Neotropical pioneers is an alternative explanation for this pattern (Corlett 2001).

## Fruit size and body size

One of the more striking contrasts among regions is the difference in the sizes of both fruits and fruit-eaters between rain forests of the Old and New Worlds. Evidence that Old World fruits are consistently larger than New World fruits comes from a comparison of 1642 species in 236 genera of eight large, flowering plant families that occur in both the New and Old World tropics (Mack 1993): Anacardiaceae, Burseraceae, Lauraceae, Meliaceae, Moraceae, Myristicaceae, Palmae, and Simaroubaceae. In every family, fruit lengths were longer in species from Africa and Southeast Asia than in New World species, with Old World fruit length being twice as long as New World fruit length in species in the families Anacardiaceae and Moraceae. Within the huge fig genus (*Ficus*), Old World species had significantly longer fruits than New World species (average 2.45 cm vs. 1.86 cm). In another comparison, the fruits of the Southeast Asian nutmegs in the genus *Myristica* are significantly longer than the similar Neotropical nutmegs in the genus *Virola* (average 5.9 cm vs. 3.0 cm).

Along with larger fruits, Old World rain forests also have larger frugivores (Cristoffer 1987). This is clearest for the primates (see Chapter 3), terrestrial mammals (see Chapter 4), and fruit bats (see Chapter 6), but is also true for birds when similar groups are compared, such as the toucans and hornbills (see above). The largest flying frugivores are Southeast Asian hornbills and the largest of all avian frugivores are the flightless cassowaries of New Guinea and Australia (see below). Larger frugivores presumably have larger mouths that can process larger fruits more efficiently. Birds have no teeth, so the maximum size of fruit that a bird can process is set by its maximum gape width. Fruit bats have teeth, but are also gape-limited when carrying large fruits away from fruiting trees to process elsewhere.

It is clear that the contrasts in fruit and frugivore sizes must have a common explanation (Snow 1981), but it is far from obvious what this explanation is. Two broad types of explanation for differences between rain forest regions were mentioned in the first chapter—historical and ecological. The long period of isolation experienced by the South American continent during the Tertiary suggests a possible historical explanation. Although large fruits are found in many families of plants in the Old World, the large frugivores are mostly in just a few vertebrate lineages. It is thus conceivable that the smaller size of Neotropical frugivores has resulted largely from the failure of apes and cercopithecine monkeys (see Chapter 3), pteropodid fruit bats (see Chapter 6), and hornbills and fruit pigeons (this chapter) to get there. Smaller frugivores would automatically lead to smaller fruits, since plants that produced large fruits would have less chance of their fruits being eaten and the seeds within them dispersed. There are no comparable sets of fruit data from New Guinea, but the presence there of cassowaries, hornbills, and large pteropodid fruit bats fits with anecdotal evidence that large fruits are common.

Historical explanations are less convincing when it comes to terrestrial frugivores, since South America had its own fauna of megaherbivores until quite recently (see Chapter 4). Indeed, in a classic paper entitled "Neotropical anachronisms: the fruits the gomphotheres ate", ecologists Dan Janzen and Paul Martin argued that a "megafaunal dispersal syndrome" was still recognizable among the fruits produced by Costa Rican trees (Janzen & Martin 1982). They listed a number of tree species in both deciduous forest and rain forest that produce large, indehiscent fruits with well-protected seeds of the type that are dispersed by elephants and other large herbivores in Africa and Asia.

Even for arboreal frugivores and the fruits they consume, a purely historical explanation for the distinctiveness of the Neotropics is not fully convincing. Not only does it require rather a lot of historical accidents to explain the absence of large frugivores, but it also assumes that no indigenous vertebrate groups were capable of evolving large species. What stops toucans or phyllostomid bats from getting as big as hornbills and pteropodid bats? However, ecological explanations for smaller fruits and/or smaller frugivores in the New World are no more convincing. Cristoffer (1987) suggests that small size was "forced" on arboreal frugivores by more fragile vegetation, but offers no evidence for the consistent differences in vegetation structure that this explanation would need. Looking instead for factors favoring large animals or large fruits in the Old World is no easier. The unreliable fruit supply in Southeast Asian dipterocarp forests (see Chapter 2) may favor large frugivores that can cope with a lower quality diet in the long periods between mast years when fruit is unavailable, but this explanation would not apply to the more seasonal rain forests of Africa (Karr 1976a). The height and density of Southeast Asian forests may also favor the production of larger seeds that produce large, shade-tolerant seedlings and require larger fruits to contain them (Foster 1986; Primack 1987). Again, this explanation does not fit well with the greater openness of forest in Africa. Whatever the explanation, or explanations, it is clear that this striking pattern needs further investigation.

## Flower visitors

Any small, brilliantly colored bird that is not a frugivore is likely to be a nectar-feeder. Only in the tropics has the year-round availability of nectar made it possible for birds to specialize on flowers as an energy source, and thus for plants to use them as long-distance dispersers of pollen. Flower-birds have evolved independently in the major rain forest regions, as have bird-flowers, resulting in both some striking parallels and major differences among the regions (Paton & Collins 1989).

Nectar is a great source of easily digestible energy, but it usually contains negligible amounts of protein and lipids. Nectar specialists therefore must also eat invertebrates to balance their diets. Long, thin bills, adapted to obtaining nectar from flowers are probably not ideal tools for catching invertebrates, and nectar specialists seem to be restricted to small insects and spiders, of which they must eat very large numbers. The members of one genus of long-billed Asian sunbirds (*Arachnothera*) are called spiderhunters, because they pick spiders from their webs. Pollen from flowers is also rich in protein, but it is not easy to harvest efficiently, and nectar-feeders seem to vary greatly in the extent to which they can digest it.

## Hummingbirds

The best-known of the avian flower specialists are the 328 species of hummingbirds (Trochilidae), which are the dominant nectar-feeding birds of the New World (Fig. 5.9) (Schuchmann 1999). Their name is derived from the humming sound that their wings make as they rapidly beat the air, up to 70 times a second

(a)

(b)

**Fig. 5.9** Hummingbirds are one of the most distinctive families in the Neotropics, often seen drinking nectar from flowers. (a) A purple-crowned fairy hummingbird (*Heliothrix barroti*) hovers over tubular white flowers in Peru. (b) A white-necked Jacobin humminghbird (*Florisuga mellivora*) feeding on an orange legume flower in Peru. (Both courtesy of Luis A. Mazariegos.)

in small species, creating a sound like a small high-speed fan. Most hummingbirds are totally aerial in foraging, moving and darting rapidly in flight and hovering upright when removing nectar from flowers. The brilliant, iridescent plumage of many species makes them a conspicuous feature of rain forests from the lowlands up to the alpine zone in the Andes, with as many as 20 species in any one location. Their small legs and feet are suited just for perching on twigs, and they never walk on the ground.

Hummingbirds are attracted to flowers with plentiful, relatively dilute nectar. In the most specialized flowers, the nectar is located at the base of a long corolla tube, often red in color, with a narrow opening, and there is no landing platform for visitors that cannot hover. The birds extract nectar by rapid licking with their long, grooved tongues, which are forked near the tip. Hummingbirds may defend clumps of flowering plants against members of their own species, as well as chasing away insects and birds of other, larger species. Like the manakins and birds of paradise described above, male and female hummingbirds lead separate lives, with only the females involved in nest building and rearing the young. However, although males must engage in mating displays in leks or individually, the high energy demand of the hummingbird life style does not permit the extravagant courtship displays seen in some frugivores.

The bills of different hummingbird species vary in length and degree of curvature, with the longest and most curved bills matching the shape of the flowers they visit. Thus, the swordbilled hummingbird (*Ensifera ensifera*), with a 10 cm (4 inch) bill—longer than the rest of its body—visits flowers with a similarly long corolla tube, such as the passionflower, *Passiflora mixta*. Perhaps the most extreme case of correspondence between bill and flower shape is found between the white-tipped sicklebill (*Eutoxeres aquila*), which has an extraordinary scythe-shaped bill matching exactly the shape of the flowers on which it depends. The latter include those of a few species of *Heliconia* and, at higher elevations, lobeliad species in the genus *Centropogon*, which have equally curved corollas (Stein 1992). Sicklebills are heavy for hummingbirds (weight around 12 g, 0.5 oz), so hovering is energetically expensive. Consequently, these birds perch while feeding. An even more curious example comes from the rain forests of St Lucia in the West Indies, where the purple-throated carib (*Eulampis jugularis*) is the sole pollinator of two species of *Heliconia* (see Plate 5.3, between pp. 150 and 151) (Temeles et al. 2000). One *Heliconia* species has relatively shorter and less curved flowers than the other and is visited mostly by the male hummingbirds, which have shorter and less curved bills than the females. The other *Heliconia* species, with longer and more curved flowers, is pollinated by the females, which have correspondingly longer and more curved bills. This pattern of shorter bills for males than females is true in other hummingbird species as well.

Hummingbird species in an area form communities with complex ecological interactions, involving competition for nectar resources, coevolution with plants, and predator avoidance (Feinsinger 1976; Feinsinger & Colwell 1978; Bleiweiss et al. 1997; Bleiweiss 1998). These communities have been intensively investigated for the insights they can provide on community structure. Phylogenetic analyses have revealed that in one geographical region these communities do not represent a single evolutionary radiation; rather, they have formed as a result of a complex mixture of evolutionary radiation, migration, and colonization of species in combination with local species extinctions. In particular, some of today's lowland rain forest hummingbirds, notably the brilliant hummingbirds and bee hummingbirds, apparently evolved from Andean ancestors and later colonized the lowlands. Among the hummingbird groups that evolved in the lowlands, there is a basic phylogenetic division between the relatively dull-colored hermit hummingbirds (Phaethornithinae) that live in the forest interior and the much more diverse "typical" hummingbirds (Trochilinae) that visit flowers in the forest canopy and forest edge.

## Sunbirds and honeyeaters

Sunbirds (Nectariniidae) are the most important group of flower-visiting birds in Africa and tropical Asia (Cheke et al. 2001). The greatest concentration of species is in Africa, where communities of sunbirds can be almost as diverse as those of hummingbirds in the Neotropical lowlands. Many species also occur in India and Southeast Asia. Like hummingbirds, they are small and often brightly colored, and sometimes their feathers have a brilliant metallic sheen. The sexes often differ greatly in appearance. In many species, the bill is curved and elongate in a manner reminiscent of hummingbirds, although the bills of these sunbirds are curved down whereas hummingbird bills are sometimes curved upwards. The tongue is tubular for most of its length, then split in half at the tip, as an adaptation for licking up nectar. In contrast to hummingbirds, sunbirds have strong legs and sharp claws for climbing around inside of foliage. They usually take nectar while perching, so plants adapted to encourage their visits must provide some form of perch (Fig. 5.10). When perches are unavailable, sunbirds can hover for a few seconds, though not with the grace and stability of hummingbirds.

**Fig. 5.10** In the Old World, sunbirds are common flower visitors, which perch while feeding. Here is a somewhat dull, female, brown-throated sunbird (*Anthreptes malacensis*) in a garden setting in Malaysia. (Courtesy of Morten Strange.)

Two species of sunbird reached New Guinea, and one of them also occurs in northeast Australia. However, the major flower visitors in both Australia and New Guinea are the honeyeaters (Meliphagidae) (see Plate 5.4, between pp. 150 and 151), which are found as far west as Bali and eastward—until their recent extinction there—to Hawaii (Lindsey 1998). Honeyeater species vary greatly in their dependence on nectar, with many species having omnivorous diets, including fruits and insects, in which nectar plays only a minor role. Although the bills of many species are adapted for nectar feeding, being elongated and curved, they are rarely as specialized as those of sunbirds or hummingbirds. Honeyeater tongues are elongate, channeled, can extend beyond the bill, and have a brush-like tip for taking up nectar. Most honeyeaters are larger than other nectar-feeding birds and most are a dull green, brown, or grey, although some striking exceptions exist. The Sulawesi myzomela (*Myzomela chloroptera*), a brilliantly colored honeyeater, on the island of Sulawesi, which lies midway between New Guinea and mainland Southeast Asia, is amazingly similar in both appearance and behavior to the unrelated crimson sunbird (*Aethopyga siparaja*), with which it coexists.

### Other nectar-feeders

In addition to the three main groups of nectar-feeding birds, each rain forest region also supports other groups of more or less specialized flower visitors. In the Neotropics, these include a variety of flowerpiercers, honeycreepers, tanagers, and orioles, which are all nine-primaried oscines (see Plate 5.1, between pp. 150 and 151). Some species of flowerpiercers (*Diglossa*) are the most nectar-dependent of these birds but, as their name suggests, their bills are specially adapted for piercing the long corolla tubes of hummingbird-pollinated flowers and stealing the nectar without transferring any pollen. However, flowerpiercers visit flowers with short corollas in the normal way and can then be important pollinators.

White-eyes are regular flower visitors from Africa to Australia, and many other Old World birds, including leafbirds, bulbuls, and flowerpeckers, take substantial amounts of nectar (Corlett 2004). In Madagascar, the two species of sunbird asity (*Neodrepanis*) resemble sunbirds in size, diet, and their down-curved bills and long, tubular tongues, but are in fact suboscine relatives of the broadbills (Lambert & Woodcock 1996). In Australia and New Guinea, a specialized group of parrots, the brush-tongued loris and lorikeets, rival the honeyeaters in their dependence on flowers with easily accessible nectar, although they are very destructive feeders, often damaging the flowers (Brown & Hopkins 1996). These small parrots can apparently harvest and digest pollen more efficiently than the slender-billed nectar specialists (Gartrell & Jones 2001).

The Hawaiian Islands have been home to an incredible evolutionary radiation of more than 50 bird species in the Hawaiian honeycreeper subfamily (Drepaniidae), derived from a single colonizing species of seed-eating finch (*Carpodacus* sp., Fringillidae). DNA studies suggest that this colonization event may have occurred only 2–5 million years ago (Fleischer & McIntosh 2001), yet the honeycreepers have radiated into a huge range of shapes, sizes, colors, habits, and habitats. This radiation included many nectar-feeders, such as the orange-red I'iwi (*Vestiaria coccinea*), which evolved a long, curved bill to feed from the similarly curved flowers of the endemic lobelias. Other honeycreepers, such as the scarlet 'Apanane (*Himatione sanguinea*), had short bills and fed on the more accessible nectar of the 'Oh'ia lehua (*Metrosideros polymorpha*). A smaller radiation of honeyeaters in Hawaii produced six species, all of which are now extinct, as are many of the honeycreepers. Introduced Japanese white-eyes (*Zosterops japonica*) are the now the commonest visitors to flowers with easily accessible nectar.

### Birds as pollinators

Flower-birds and bird-flowers are found in all rain forests, but their independent origin in each region has resulted in major differences in the role of birds as pollinators. Hummingbirds have the greatest dependence on floral nectar and have produced the most spectacular examples of coevolution between plants and birds. However, a large majority of the plants pollinated by hummingbirds are herbaceous: either forest floor herbs, such as the many species of *Heliconia*, or canopy epiphytes, particularly in the families Bromeliaceae and Gesneriaceae. Hummingbirds also pollinate shrubs and lianas, but very few species of tree. Hummingbirds are probably poor pollinators for trees because territorial species defend such clumped nectar resources against other birds, which must greatly reduce cross-pollination (Schuchmann 1999). At the other extreme, the honeyeaters are mostly more or less omnivorous, with a much lower dependence on nectar but, unlike hummingbirds, honeyeaters visit and probably pollinate many rain forest trees (Brown & Hopkins 1996). These opportunistic nectar-feeders are often aggressive to other birds while feeding, but do not defend one tree throughout its flowering period. The sunbirds seem to be intermediate in their degree of specialization, with most species eating more insects than hummingbirds do, as well as small fruits. Sunbirds pollinate mostly large herbs, including gingers (Zingiberaceae) and bananas (Musaceae), and mistletoes (Loranthaceae), but also a range of shrubs, climbers, and trees (Corlett 2004). Overall, pollination by birds seems to be most important in the Neotropics and New Guinea, and least important in Southeast Asia.

Despite the large differences in size, morphology, and behavior between the three major groups of flower-visiting birds, bird-pollinated ornamental plants from one region attract the flower-visiting birds of other regions when they are planted there. This suggests a universality to the bird-pollination syndrome, despite the very different birds involved. Large, red, tubular or two-lipped flowers, with copious quantities of dilute nectar and no scent are recognized as "bird-flowers" by hummingbirds, sunbirds, and honeyeaters alike.

## Ground-dwellers

Ground-living birds are often large and usually drab in color, although there are some striking exceptions. Invertebrates, seeds, and fallen fruits are the main foods available at ground level, and many of the large ground-feeding birds probably eat all three, although in widely varying proportions. However, the diets of most of these shy and elusive birds are inadequately known.

### Cassowaries, elephant birds, and tinamous

By far the largest ground-living birds in tropical rain forests and, indeed, the largest of all forest birds, are the four species of flightless cassowaries (Casuariidae) of New Guinea and Australia, which feed mostly on fruit (Fig. 5.11a) (Davies 2002). The double-wattled (or southern) cassowary (*Casuarius casuarius*), one of the few bird species shared between the rain forests of New Guinea and Australia, can attain a height of 1.8 m (6 feet) and a weight of 65 kg (145 lb), making it the second largest bird after the ostrich. Even the smallest cassowary species, the dwarf cassowary (*C. bennetti*), can measure more than 1 m in height. The feathers of these species are reduced to a coat of quills, which probably protects the body from scratches in the dense vegetation. Cassowaries are also noted for their yellowish casque, a horny crown on the head, which gives them a royal appearance, a bright, often purplish coloration of the head and neck, and bright red wattles. Cassowaries are well able to defend themselves, and all four species have been known to kill people with a kick from their powerful clawed feet. Recent studies of captive cassowaries have shown that their booming calls include a very low-frequency component that may be important for communication over long distances in dense rain forest (Mack & Jones 2003). These birds can swallow large fruits whole (up to 7 cm diameter in the case of the double-wattled cassowary in northeast Queensland) and even the biggest seeds pass undamaged through their guts, making them very important seed dispersal agents in these forests, which lack large mammalian frugivores (Stocker & Irvine 1983). The elephant birds (*Aepyornis* spp.) of Madagascar were even larger than cassowaries—the biggest species may have attained a height of 3 m—and they may have played a similar role in the dispersal of large seeds before their extinction some time within the last 2000 years (Davies 2002).

No other rain forest region has anything remotely resembling a cassowary or elephant bird, but the most characteristic ground-dwelling birds of Neotropical forests, the tinamous (Tinamidae) are members of an equally ancient bird family (Davies 2002). Cassowaries, elephant birds, and tinamous, along with the nonforest ostriches, rheas, and emus, and the kiwis and extinct moas of New Zealand, make up one of the three major lineages of birds, the paleognaths or "old jaw" birds. The present distribution of these birds strongly suggests a common origin on the southern supercontinent of Gondwana (Cracraft 2001). Tinamous, however, are much smaller than cassowaries, looking somewhat like a partridge, with their plump bodies and rounded wings (Fig. 5.11c). Tinamous can fly, but cannot sustain flight for more than a short distance and prefer to walk or run. They often hide from danger, rather than taking flight. They are shy, unspectacular birds, which like many understorey birds are more likely to be heard than seen. Their loud, whistling calls at dusk are among the strongest and most pleasing in the Neotropical forest. Tinamous eat a lot of fallen fruit, along with seeds and invertebrates, but, unlike cassowaries, apparently grind up most seeds in the gizzard.

### Pheasants and their relatives

Most tropical forests also support ground-dwelling species from the order Galliformes, which includes the familiar pheasants, partridges, grouse, and quail

(a)

(b)

**Fig. 5.11** Birds are sometimes important ground feeders. (a) The double-wattled cassowary (*Casuarius casuarius*) is a large ground feeder in New Guinea and Australia. (Courtesy of Jessie Cohen, taken at the US National Zoo.) (b) Crowned pigeon (*Goura victoria*) from Yapen Island in New Guinea. (Courtesy of Roger Le Guen.)

of temperate regions. A variety of evidence suggests that the galliforms, along with the related anseriforms (ducks, geese, and swans), make up a second major bird lineage, the Galloanserae, which also appears to have had a Gondwanic origin (Cracraft 2001). The order is divided into a variable number of families by different authors.

Although most members of this order are terrestrial, the major representatives in Neotropical forests, in the family Cracidae, are more arboreal (del Hoyo 1994). This is particularly true of the guans and chachalacas, which feed at all levels in the forest, but the larger curassows generally feed on the ground. Curassows, like other cracids, appear to be mostly vegetarian, and eat considerable amounts of fallen fruits. They have a powerful gizzard that can crush even large, hard seeds, so they are probably not effective seed dispersal agents for the seeds of most species they consume (Yumoto 1999).

African rain forest Galliformes include the endemic guineafowls (Numididae), the francolins (*Francolinus*), and the Congo peafowl (*Afropavo congensis*), but the true pheasants in the family Phasianidae are largely confined to Asia (Madge & McGowan 2002). Some rain forest pheasants are exceptions to the rule that ground-feeding birds are drab. Males of several species, such as the aptly named

(c)

(d)

**Fig. 5.11** *(cont'd)* (c) Great tinamou (*Tinamus major*) from Costa Rica. (Courtesy of Gustavo Londoño.) (d) Male great argus pheasant (*Argusianus argus*) from Malaysia. (Courtesy of Morten Strange.)

peacock pheasants (*Polyplectron*), have spectacular tail feathers with iridescent markings. The tails are kept folded, however, except when the males display them deliberately during sexual displays. Fruit, both fallen and plucked directly from low-growing plants, seems to be an important part of the diet of many rain forest pheasants, and the larger species, such as the great argus (*Argusianus argus*), can swallow large fruits whole (Fig. 5.11d). Very few frugivores occupy the understorey of Asian rain forests, so pheasants are potentially important seed dispersal agents (Corlett 1998).

Forest pheasants are confined to mainland Southeast Asia and the main islands on the Sunda Shelf. Another galliform family, the megapodes (Megapodidae), has an almost exactly complimentary distribution, being found on the eastern Indonesian Islands, New Guinea, and Australia. This distribution may not necessarily have resulted from competitive exclusion between the two groups, however, since the limits of the megapode range also coincide with the absence of some of the most important mammalian predators, such as civets and cats. Megapodes may be particularly vulnerable to ground predators because of the extraordinary ways in which they incubate their eggs: in giant compost heaps warmed by the heat of decomposition or in burrows excavated in soil that is heated by the sun or volcanic activity (Jones et al. 1995). Megapodes provide no parental care at all for their chicks, which have to dig themselves out from up to 150 cm (60 inches) underground and then learn to find food and avoid predators without any parental guidance. All six New Guinea species are mound-builders, piling damp leaf litter and other materials to a height of 1 m or more and covering several square meters. The mounds of another megapode, the Australian brush-turkey (*Alectura lathami*), can weigh from 2 to 4 tons. In contrast, the maleo (*Macrocephalon maleo*) of Sulawesi does not construct mounds, but lays its eggs either at inland geothermal sites or on sun-warmed beaches. Bizarre breeding habits aside, megapodes resemble the other rain forest Galliformes in most other respects: they are dull-colored, medium to large birds, with a plump body and relatively small head.

In all rain forest regions, species of Galliformes are preferentially hunted because they are mostly ground-dwelling and relatively large (McGowan & Garson 2002). As a result, their densities have often been greatly reduced and many species are threatened with extinction.

## More pigeons and doves

Although there are many species in the canopy (see above), the pigeons and doves (Columbidae) are also widely represented on the forest floor (Gibbs et al. 2001). Ground pigeons are particularly diverse in New Guinea, which is home to the biggest of all living pigeon species, the almost turkey-sized crowned pigeons in the genus *Goura*. The elaborate fan of feathers that forms a crest on the head is displayed during courtship bows (see Fig. 5.11b). New Guinea rain forests also support the magnificent pheasant pigeon (*Otidiphaps nobilis*), which not only looks like a pheasant but moves like one. Most ground-feeding pigeons are probably seed predators rather than dispersers, but the types and proportions of seeds damaged vary greatly between species (Corlett 1998). The emerald dove (*Chalcophaps indica*), named for its emerald-green wings, is found from tropical Asia to northern Australia; it feeds on fallen fruits and seeds with some small seeds surviving passage through its gut. In contrast, the Nicobar pigeon (*Caloenas nicobarica*), which is confined to small tropical islands in the Oriental and Australian regions, has a thick-walled stomach lined with horny plates that are used, together with swallowed stones, to grind up even large and very hard seeds.

DNA extracted from museum specimens of the extinct dodo (*Raphus cucullatus*) of Mauritius and the related (and equally extinct) solitaire (*Pezophaps solitaria*) of Rodrigues has been used to show that these giant, flightless ground pigeons were most closely related to the Nicobar pigeon, with this group in turn close to the crowned pigeons of New Guinea (Shapiro et al. 2002). We can only speculate on the past role of these birds in the rain forests of Mauritius and Rodrigues.

### Other ground-dwelling families

In addition to these widespread families, other ground-living families are characteristic of particular regions. Pittas (Pittidae) are medium-sized, largely insectivorous suboscines of the rain forest floor from Africa to Australia, but with a center of diversity in Southeast Asia (Lambert & Woodcock 1996). Their often bright colors are confined, in most species, to the underparts, with the upper parts generally having cryptic patterns, so pittas are difficult birds to spot in deep shade. Rain forests in New Guinea and Australia also support the endemic insectivorous logrunners and chowchilla (Orthonychidae) (Garnett 1998). Madagascan forests have the endemic thrush-sized, rail-like mesites (Mesitornithidae) and ground-rollers (Brachypteraciidae) (Morris & Hawkins 1998). New World rain forests have trumpeters (Psophiidae): long-legged, chicken-sized birds that are named for their loud calls. Trumpeters defend large group territories, in which they feed on fallen fruits and invertebrates.

## Woodpeckers

Woodpeckers (Picidae) are the most specialized of the many forest birds that forage on the trunks and larger branches of rain forest trees (Fig. 5.12a) (Winkler & Christie 2002). Woodpeckers pick insects and other invertebrates off the trunk surface and probe into holes with the aid of an extensible and often sticky tongue. Woodpeckers can also dig into the bark and wood with their powerful bills, which gives them access to a source of food—wood-boring insects—that is not available to any other bird. In addition, many woodpeckers consume considerable amounts of fruit when it is available. They also use their bills to excavate holes for roosting and breeding. Woodpeckers range in size from tiny piculets (species of *Picumnus* and *Sasia*), 10 cm (4 inches) or less in length, to giants like the crimson-crested woodpecker (*Campephilus melanoleucos*) in South America and the great slaty woodpecker (*Mulleripicus pulverulentus*) in tropical Asia, which is 50 cm long.

The woodpeckers are most diverse and conspicuous in Southeast Asia. Here, up to 16 species can coexist and several species may participate in the same mixed-species flock, which is not usually observed elsewhere (Styring & Ickes 2003). In South American rain forests, up to a dozen species of woodpeckers can coexist within in an area, along with a similar number of bark-feeding woodcreepers (*Dendrocolaptidae*). In contrast, only a few woodpecker species, none very large, are found in African rain forests and the family has not reached Madagascar, New Guinea, or Australia. In those areas, other birds pick insects off tree trunks and branches: the sickle-billed vanga (*Falculea palliata*) in Madagascar

(a)

(b)

**Fig. 5.12** Birds that can make tree holes. (a) Spot-breasted woodpecker (*Chrysoptilus punctigula*) from Colombia. (Courtesy of Gustavo Londoño.) (b) Red-crowned barbet (*Megalaima rafflesii*) from Borneo. (Courtesy of Tim Laman.)

and three species of riflebird (*Ptiloris*) in New Guinea and Australia. None of these birds, however, is able to excavate for wood-boring insects or to make its own nest holes. Rain forests in Madagascar and New Guinea also have specialized mammals—the prosimian aye-aye (*Daubentonia madagascariensis*) in Madagascar and the marsupial striped possums (*Dactylopsila*) in New Guinea—

with morphological adaptations, including chisel-like incisors and an elongated middle finger, for extracting wood-boring insect larvae (see Chapter 3).

Holes excavated by woodpeckers are often taken over by other species of birds, such as toucans and starlings, as well as by mammals. Natural tree holes are relatively rare, particularly in young forests, and only a few other bird species can excavate their own, most notably the barbets (Fig. 5.12b) and the trogons. Thus the diversity and abundance of woodpeckers might be expected to have a strong influence on the availability of nesting and roosting sites for cavity-using non-excavators. This possibility does not seem to have been investigated in the tropics, but the diversity of hole-nesting birds and mammals in New Guinea, where both woodpeckers and barbets are absent, suggests that the relationship, if any, is not simple.

## Birds of prey

Birds of prey, also known as raptors, are diverse and abundant in tropical rain forests, with a huge range of sizes and diets (Fig. 5.13; see Plate 5.5, between pp. 150 and 151) (Ferguson-Lees & Christie 2001). The threat these birds pose to other animals in the forest is reflected in a variety of antipredator adaptations, including cryptic colors, vigilant behavior, and mixed-species associations, which occur in both insectivorous birds (see above) and even some forest primates. For many forest animals, avoiding being eaten by a bird of prey is at least as important an occupation as finding food to eat. Antipredator adaptations make

**Fig. 5.13** Philippine eagle (*Pithecophaga jefferyi*) in the Philippines. (Courtesy of Tim Laman, taken in captivity.)

prey very difficult to detect, so the great majority of forest raptors spend most of the day sitting motionless on a perch, watching for any movement. For the human visitor, the frustrating result of that behavior is that neither predator nor prey is easily seen, and the forest often appears completely devoid of vertebrate life. Hiding is not the only possible defense, however. Some species of pitohui (*Pitohui*) in New Guinea have evolved defensive toxins that are present in their skin and feathers and have a striking brick red and jet black color scheme that may serve as a warning to birds of prey (Dumbacher & Fleischer 2001).

Rain forest raptors belong to two related families: the Accipitridae, which includes the hawks and eagles, and the Falconidae, which includes the falcons and Neotropical caracaras (Ferguson-Lees & Christie 2001). The New World vultures, although traditionally included with the other raptors, are now considered to be a separate group distantly related to the storks. Eagles large enough to threaten the biggest arboreal animals in the forest are found in most rain forest areas. In the Neotropics, the formidable harpy eagle (*Harpia harpyja*) pursues monkeys through the canopy with surprising speed and agility, and will also capture sloths, opossums, and other mammals (Brown & Amadon 1968). The crowned eagle (*Harpyhaliaetus coronatus*) plays a similar role in African rain forests, where it is a major predator on adult monkeys, including the largest species, as well as taking a range of other animals, such as civets, duikers, and hornbills (Struhsaker & Leakey 1990). There have even been reports of human remains in their nests.

In Asia, the changeable hawk eagle (*Spizaetus cirrhatus*) and the Asian black eagle (*Ictinaetus malayensis*) are big enough to threaten large arboreal mammals, but the former takes mostly terrestrial prey and the latter seems to be a specialist feeder on the contents of birds nests (Ferguson-Lees & Christie 2001). The critically endangered great Philippine (or monkey-eating) eagle (*Pithecophaga jefferyi*) may have been a primate specialist, but in its few remaining strongholds, where monkeys are scarce, it also eats colugos, civets, and squirrels, as well as large birds and reptiles (Fig. 5.13). Primates do not occur in New Guinea and Australia, but the largest eagles present in rain forest areas, the New Guinea eagle (*Harpyopsis novaeguineae*) and wedge-tailed eagle (*Aquila audax*), respectively, are capable of killing the biggest arboreal marsupials, including possums and tree kangaroos (Brown & Amadon 1968). The New Guinea eagle also preys on terrestrial mammals, such as forest wallabies, and there is one report of a small child being taken (Ferguson-Lees & Christie 2001).

In contrast, the island of Madagascar does not have birds of prey that are big enough to threaten adults of the largest primate species. The largest forest raptor in Madagascar, Henst's goshawk (*Accipiter henstii*), takes sleeping individuals of only the smaller nocturnal lemurs, but even the larger, diurnal lemurs give alarm calls in the presence of birds of prey (Wright 1998). It has been suggested that these calls have been retained from a time when a larger eagle, as big as the crowned eagle of Africa, was present, but they may also reflect a continuing threat to infants and juveniles from smaller raptors.

Most major rain forest areas also have a specialized raptor preying on snakes and other reptiles that live in the forest canopy: serpent-eagles in Asia, Africa, and Madagascar, and several species of hawks and the laughing falcon (*Herpetotheres cachinnans*) in the Neotropics (Thiollay 1985). Other species in all rain forests specialize on small birds, arboreal mammals, or tree frogs and lizards, but all raptors are opportunists, prepared to take any suitable sized prey if given a chance.

At the other end of the size scale, the smallest birds of prey, such as the tiny falconets, feed mostly on insects. In the Neotropics, however, the thrush-sized tiny hawk (*Accipiter superciliosus*) appears to specialize on hummingbirds (Stiles 1978), the similar-sized bat falcon (*Falco rufigularis*) on bats captured at dawn and dusk, and the tiny pearl kite (*Gampsonyx swainsonii*) on lizards. Honey buzzards (*Pernis, Henicopernis*) are specialist predators on wasp nests in Asia and Africa, while the red-throated caracara (*Daptrius americanus*) fills the same niche in the New World. Even the most aggressive wasps reportedly stay away from this latter species, suggesting that it has some form of chemical repellent (Thiollay 1991). In the Neotropics, but not in other regions, gregarious aerial insectivores, such as the swallow-tailed kite (*Elanoides forficatus*) and plumbeous kite (*Ictinia plumbea*), hunt in flight above the forest canopy. Another feeding habit confined to this region is shown by the hook-billed kite (*Chondrohierax uncinatus*), which specializes on arboreal snails.

## Scavengers

In most tropical habitats, vultures are among the most important scavengers. The birds that have evolved to fill this niche in the Old and New Worlds are unrelated, yet, seen together in a zoo aviary, are almost indistinguishable. They are all large birds, with powerful hooked bills, bare heads and necks, and massive wings used for their energy-saving, soaring flight (Houston 1994). When it comes to tropical forests, however, the differences between the two lineages are immediately apparent: the New World vultures are the dominant scavengers in Neotropical forests, while Old World vultures are entirely absent from forested regions. The reason for this is straightforward: Old World vultures rely on their acute eyesight to detect carcasses from the air, which is useless when a dense tree canopy intervenes, while at least one genus of their New World counterparts (*Cathartes*) makes use of its well-developed sense of smell. How this fundamental difference in scavenging communities arose is not clear. Fossil evidence indicates that both groups of vultures were present at both Old and New World sites in the past (Houston 1994), suggesting that forest vultures may have been lost from Africa and Asia, rather than simply failing to evolve there. Today, the major scavengers on carcasses in Old World forests are mammals or opportunistic birds of prey, such as the crowned eagle.

Dying animals provide a large potential food supply for scavenging birds, but it is necessary to find a corpse before it is consumed by mammals and various groups of insects, or rendered inedible by bacteria. Neotropical turkey vultures (*Cathartes aura*) cannot easily find newly dead animals, but meat decays quickly in the tropics and they are very efficient at locating day-old carcasses. They fly low over the canopy, gaining lift from the updrafts on the windward side of emergent trees, and descending to the ground only when the smell of food is detected. The yellow-headed vultures (*Cathartes* spp.) feed in the same way as the turkey vulture. The king vulture (*Sarcoramphus papa*) appears to lack this acute sense of smell, so it flies high above the canopy, depending on the *Cathartes* vultures to locate carcasses and then following them to join the feast (Houston 1994).

## Night birds

As a general rule, the day belongs to the birds in the rain forest while the night belongs to the mammals. But just as two major groups of mammals, the primates and squirrels, have become fully adapted to a diurnal life style, the owls and nightjars have successfully invaded the night, with the owls becoming adapted to larger, vertebrate prey, while the nightjars and related families specialize on insects (Fig. 5.14).

Owls (Strigiformes) are the nocturnal counterparts of the diurnal birds of prey, although they are not closely related. There are two families of owls: the barn-owls (Tytonidae, 16 species) and typical owls (Strigidae, *c.* 190 species). Barn-owls (in two genera, *Tyto* and *Phodilus*) are found in rain forest only in Madagascar, Southeast Asia, New Guinea, and Australia, and on some small tropical islands (Bruce 1999). One species, the greater sooty owl (*Tyto tenebricosa*), is a rain forest specialist in New Guinea and Australia, taking arboreal marsupials up to the size of a 900 g (32 oz) ringtail possum, as well as terrestrial mammals and birds.

Little is known about the ecology of most of the typical owls of tropical rain forests, but the great range in size suggests that they are ecologically diverse, despite their similar overall appearances (Konig et al. 1999). The smallest are the pygmy owls or owlets, with some species a mere 15 cm (6 inches) in length. Old and New World species are currently placed in the same genus (*Glaucidium*), but DNA evidence suggests that they are not closely related. They feed on insects and small vertebrates, and some species, such as the collared owlet (*Glaucidium brodiei*) of tropical Asia, kill birds as big as themselves. Some small owls, including the Amazonian pygmy owl (*Glaucidium hardyi*), are partly diurnal, and when seen during the daytime are mobbed by small birds. The larger owls are more diverse and can kill larger prey. In Madagascar, owls are important predators of small lemurs. The biggest rain forest owl in Africa is Shelley's eagle owl (*Bubo shelleyi*), which weighs more than 1 kg and can capture large, nocturnal, flying squirrels. The closely related forest eagle owl (*Bubo nipalensis*) in mainland Asia can kill birds as large as a peafowl while, in the Neotropics, the spectacled owl (*Pulsatrix perspicillata*) takes birds and mammals up to the size of an opossum. Unrelated owl species that specialize on fish are found in African and Asian rain forests.

The nightjars (Caprimulgidae) are a cosmopolitan and relatively uniform group, while four related nocturnal families have more restricted distributions: the oilbird (Steatornithidae) and potoos (Nyctibiidae) in the Neotropics, the frogmouths (Podargidae) from tropical Asia through to Australia, and the owlet-nightjars (Aegothelidae) in New Guinea and Australia (Holyoak 2001). While some species spend the day hidden from predators in tree holes or dense vegetation, most roost on branches (potoos and frogmouths) or on the ground (most nightjars), and rely on the highly cryptic (camouflage) colors and variegated patterns of their plumages to avoid detection (Fig. 5.14b).

The nightjars are adapted for capturing insects in flight. They have small bills but a very large gape and are able to open their mouths both vertically and horizontally. Nightjars hunt visually, and their large eyes have a reflecting tapetum behind the retina, which increases the efficiency of light gathering in low light intensities, as well as making their eyes shine like a cat's in the beam of a flashlight. The potoos have larger, hooked bills and also hunt insects in flight or from a perch. The aptly named frogmouths have large heads and massive bills,

(a)

(b)

**Fig. 5.14** Night birds. (a) Buffy fish owl (*Ketupa ketupu*). (Courtesy of Tim Laman, taken in Sabah, Malaysia.) (b) Jungle nightjar (*Caprimulgus indicus*) camouflaged on the ground, in India. (Courtesy of N.A. Aravind.)

particularly in the genus *Podargus* of Australia and New Guinea. The Papuan frogmouth (*P. papuensis*) is as big as a large owl and hunts from low perches by scanning the ground for insects and small vertebrates. The owlet-nightjars have forward-facing eyes and broad, flat bills, which give them an owl-like appearance. They feed by catching insects in flight or on the ground.

The final member of this group of night birds is the bizarre oilbird (*Steatornis caripensis*), a single Neotropical species in its own family, which is the only known nocturnal frugivorous bird (Thomas 1999). Oilbirds roost and breed gregariously in caves, where they use a crude form of echolocation to navigate in the dark. When foraging, however, they depend on their highly sensitive vision and, perhaps, their sense of smell. Oilbirds are totally frugivorous and even rear their young on an exclusive diet of low-protein fruit pulp, which leads to a very slow growth rate. They feed largely on the lipid-rich fruits of species in the laurel family (Lauraceae) and palms, which they typically pluck off in flight and swallow whole. The combination of their gregarious habitats and exclusively fruit diet can make them locally important as seed dispersers.

## Migration

In the relative constancy of the tropical rain forest environment, we would expect migration to be unnecessary—and indeed, most rain forest bird species are more or less resident year round. Staying put in a permanent territory with predictable food resources, known predators, and hiding places has obvious advantages. However, a minority of species—but billions of individual birds—migrate each year between the tropics and the temperate region. Most migrant species are apparently of tropical origin; their members fly north every summer to breed, taking advantage of the annual superabundance of insects and, possibly, lower predation risks (Karr 1980). Fewer tropical species fly south to temperate latitudes, which cover a much smaller area of land in the southern hemisphere. Long-distance migrants are found in all rain forest regions but they are least important in Africa, where most migrants winter in the extensive grassland and savanna environments, rather than in the forest, and are most important in Southeast Asia (Fig. 5.15). Indeed, the migrant Siberian blue robin (*Luscinia cyane*) is one of the most common understorey birds in winter in the Malay Peninsula (Wells 1990). In the Neotropics, the abundance and diversity of wintering migrants (mostly woodwarblers, tyrant flycatchers, vireos, and thrushes) in rain forests increases northwards from Amazonia, where numbers are few. In Mexico, Central America, and the West Indies, migrants can be a seasonally important component of the avifauna (DeGraaf & Rappole 1995). In the summer breeding season, these Neotropical migrants dominate some forest bird communities in the eastern and northern United States.

Migrations also occur within the tropics on various scales. It is hypothesized that the long-distance migrations described above evolved from these more local movements. The best-known cases occur in Central America, where a number of species have been shown to make more or less regular altitudinal migrations. Most are fruit- or nectar-feeders and may be forced to move by large seasonal fluctuations in the availability of these foods. In the highlands of Costa Rica, Mexico, and Guatemala, for example, the frugivorous resplendent quetzals make seasonal elevational movements that apparently track the fruiting periods of the

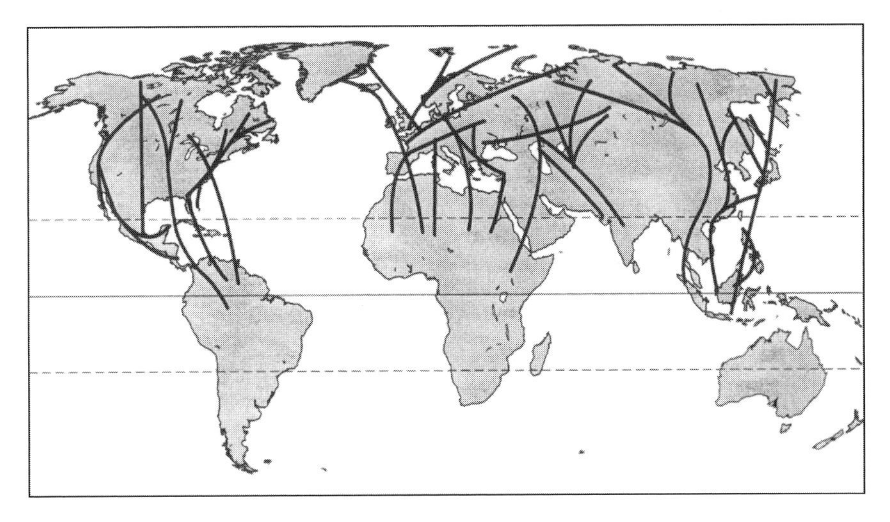

**Fig. 5.15** Major bird flyways. Note that African flyways are north of the rain forests and the Neotropical flyways begin mainly north of the Amazon basin. New Guinea, Australia, and Madagascar are not part of the flyways. Of the five major rain forest regions, only Southeast Asian rain forests represent an end-point for a major flyway. (From Brown & Lomolino 1998; after McClure 1974 and Baker 1978.)

Lauraceae trees on which they feed (Powell & Bjork 1995). Such movements have obvious consequences for the design of protected area systems, which must encompass the full range of habitats used by such birds (Powell & Bjork 2004).

## Comparison of bird communities across continents

Ornithologists have often speculated about the structure of bird communities in different parts of the world. They have been particularly interested in learning if the species in a community are organized in the same way. For example, do birds in different communities have the same range of feeding behaviors? Are the proportions of feeding specialists and generalists similar? The problem with making such comparisons is that field biologists in different parts of the world have typically gathered and analyzed their data in different ways.

### A pioneering study

In order to compare bird communities in different rain forests accurately, it is necessary to use the same methods in each place. So far, the most comprehensive attempt to compare rain forest bird communities across continents was undertaken by ornithologist David L. Pearson during the 1970s (Pearson 1977). Using the same methods, he studied bird communities at six sites: three in western Amazonia, in Ecuador, Bolivia, and Peru, and one each at Kutai in Indonesian Borneo, Maprik in Papua New Guinea, and at Makokou in Gabon in Central Africa. At each site 200–700 h were spent observing birds in their community and recording foraging heights and foraging techniques used, such as gleaning an insect from a plant surface, sallying to catch an insect on the wing, snatching prey from a surface, pecking and probing at bark to get at hidden prey, and eating fruit. The study was conducted during daylight hours, and did not include raptorial birds, such as hawks, or nighttime species, such as nightjars. The observations were made along a circular path 2.5–3 km (1.5–2 miles) long. By observing birds within 25 m (80 feet) on each side of the path, a total of around 15 ha (35 acres) was covered for each forest.

While this study is unique in comparing data on bird communities in six countries using the same methods, it also has several major problems. Only one 15 ha plot was examined at each site, and just one plot was used to represent the entire African and Asian continents and the island of New Guinea. No replicates were used to determine if these plots were typical or abnormal. The observations at each site were made over a period of several months in only 1 year. It is unknown if the period of observation was typical for the year, or if the year itself was typical. Possibly bird abundances and even the numbers of species may have differed at other times of the year or in other years.

Vegetation varied among plots, and was expressed as the mean number of leaves present at various heights in the canopy; that is, the number and heights of leaves a thin pole would touch if a person extended it straight up from where the person was standing (Fig. 5.16). The Borneo plot, dominated by dipterocarp

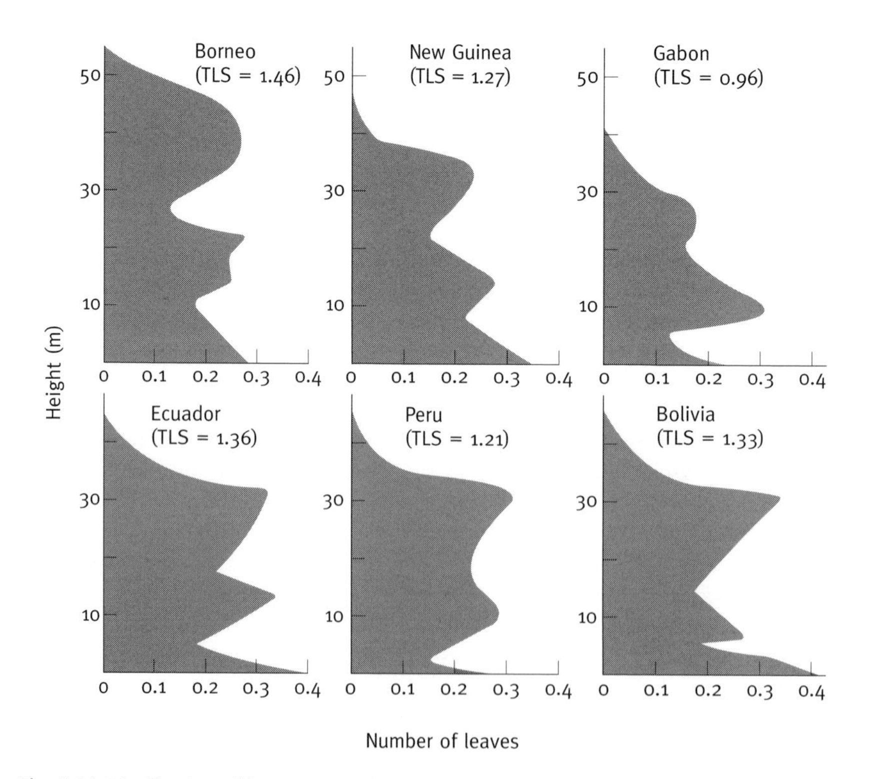


**Fig. 5.16** Distribution of leaves at six forest sites. The average number of sightings of leaves was determined in seven height strata: 0–2 m, 2–6 m, 6–14 m, 14–26 m, 26–33 m, 33–40 m, and greater than 40 m. The total leaf sightings (TLS) is also given; this is the average number of leaves that would be touched by an imaginary pole stretching from the ground to the top of the canopy. (From Pearson 1977.)

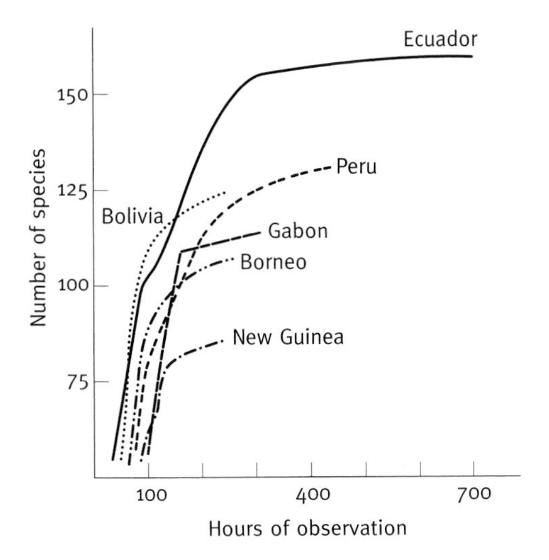


**Fig. 5.17** Cumulative number of bird species found on six field sites after a specified number of hours of observation. (From Pearson 1977.)

trees, had the tallest trees at over 50 m in height, and a high percentage of its foliage occurred in the canopy at heights of 30–50 m. The greatest contrast was with the more open Gabon forest, with trees primarily below 40 m, relatively few canopy leaves, and most of the foliage below 20 m. The New Guinea and Amazonian forests were intermediate in both height and leaf distribution.

## Key results

Despite the limitations, Pearson's study provided many valuable insights, some of which highlight the regional differences discussed earlier in the chapter. Each of the plots contained a surprisingly high percentage, 53–70%, of the forest birds known to occur in the area. The 15 ha plot in New Guinea, for example, contained 70% of the forest bird species known from the area. The overall richness of birds varied widely among plots, with more than 125 species for each of the three Amazonian plots, and fewer than 90 species for the New Guinea plot (Fig. 5.17).

In analyzing the patterns of foraging in the six forests, the most striking observation is how similar foraging techniques are across forests. The most common foraging technique is gleaning, followed by fruit eating, and then sallying (i.e. taking short flights from a perch to capture insects). Snatching prey and probing/pecking are methods of intermediate frequency, with flower visiting and army ant following less common. Some differences among regions were explained by the presence or absence of particular bird families. The foraging technique that varied most among forests was probing/pecking. At Amazonian sites, where probing/pecking is common, the technique is employed by woodpeckers (Picidae) and woodcreepers (Dendrocolaptidae). Woodcreepers are absent from the Borneo and Gabon plots where probing/pecking was less common and was exhibited primarily by woodpeckers. Both woodcreepers and woodpeckers are absent from New Guinea; at this site, probing/pecking was used at a very low rate only by a tiny parrot (*Micropsitta pusio*).

The tendency of bird species to leave or arrive in the plots temporarily was greatest for fruit-eating birds, such as pigeons, toucans, and parrots, that tracked the maturation of fruit crops. It was least for insect-eating birds, except for those species that made long-distance migrations to other continents.

The technique of following army ant swarms occurred in the South American and West African plots, which had specialized antbirds and huge army ant swarms. This technique was not observed in plots in Borneo and New Guinea, which lacked specialized antbirds and large swarms of foraging army ants.

Certain patterns of bird foraging behavior can be related to the co-occurrence of other groups of animals, most notably mammals. For example, the presence of fruit-eating behavior is highest in New Guinea, where there are no primates or squirrels to compete for fruit, and only a relatively low density of cuscuses and tree kangaroos. Around 50% of the bird species in New Guinea eat some fruit, with 19% of the species observed feeding on fruit more than 90% of the time. In contrast, only around 25% of bird species ate fruits at the Bolivia, Borneo, and Gabon sites. The plot in Bolivia only had one primate species and a few fruit-eating rodents but had a high level of fruit-eating behavior. The presence of fruit-eating birds was lowest in Gabon, where there are also eight species of fruit-eating squirrels, six species of fruit-eating ruminants, and 13 species of primates.

There is some evidence for substitution of one group of birds by other groups on different continents. In the New World, woodcreepers are common and intensive probers on bark along with woodpeckers. In the Old World plots, small squirrels take over this ecological role, with many primates also foraging along trunks in Gabon. In Borneo, gliding lizards (*Draco*) also probe for insects. In New Guinea, where there are no woodpeckers, woodcreepers, or squirrels, a large flightless tiger beetle (*Tricondyla aptera*) was observed to probe in bark crevices along the trunk, foraging just like a woodcreeper elsewhere.

## Conclusions and future research directions

The study by Pearson (1977) illuminated many significant differences in bird communities between the six forests. The differences may have been due to the fact that the forests have different families of birds. An alternative possibility is that the species in the bird communities were responding to competition with members of the mammal communities, principally the abundant squirrels of the Old World and the varying abundance of primates. Also worth considering is the role of the dipterocarp family and multiyear fruiting cycles in reducing the abundance of animal prey, such as insects, in average years in Southeast Asian forests. Furthermore, the forest height and climates of each of the forests differ, with a pronounced dry season and lower forest in Gabon and everwet conditions and taller forest in Borneo. Lastly, hunting by people had depressed the populations of large animals in Ecuador and New Guinea to levels below the other sites, affecting the community structure of the entire forest.

What can be said is that Pearson's groundbreaking study described major differences among six widely separated rain forests. It is up to a new generation of tropical ecologists to expand his approach to include more sites over a period of several years, and to include in the study key groups of mammals such as primates and squirrels. The relative abundance of basic building blocks of the food chain, such as new leaves, flowers, fruit, and insects, also need to be tracked.

Such a cross-continental approach could help to reveal the relative importance of evolution, biogeography, and current environmental conditions in explaining differences in bird communities. These studies could also provide insight into, for example, the structure of mixed-species flocks in different rain forests rather than, as tends to happen at present, simply extrapolating Neotropical studies to the rest of the world.

Such cross-continental studies also need to be combined with investigations of the human impact on bird communities. In many areas of the world, large birds have been intensively hunted by traditional people for food, and some brightly colored birds have been hunted for their feathers. With the arrival of guns, the intensity of this collection has increased. Large frugivorous birds, such as hornbills or cassowaries, are the only dispersal agents for some of the largest fruits, so their loss may have a long-term impact on the plant community. Bird populations have also been decimated by the introduction of diseases and predators, particularly on tropical islands such as Hawaii and Guam (Rodda et al. 1992; van Riper et al. 2002). So although we need investigations of bird communities, we also need to be aware that the bird communities we see today may not represent the original structure of the community, but may rather reflect species that are most resistant to the ever-increasing human impact.

Moving beyond observations to experiments on their ecological roles may be particularly difficult with forest birds. Experimental manipulations of rain forest bird communities are likely to be both difficult and controversial, while comparisons between intact bird communities (if they exist) and those impacted by hunters are usually confounded by simultaneous impacts on other components of the rain forest community, such as mammals or plants. The exclusion of all birds from branches or small trees with netting is possible (Van Bael et al. 2003), but selective exclusion is probably not practical. One possible approach would be to study the impact of reintroducing key bird species to forest areas from which they have been eliminated by hunting. Such studies would obviously require the cooperation (or exclusion) of hunters, but could have major educational and conservation benefits, as well as helping to understand the role of particular bird species in the rain forest community.

## Further reading

Bleiweiss R. (1998) Origin of hummingbird faunas. *Biological Journal of the Linnean Society* **65**, 77–97.

Corlett R.T. (1998) Frugivory and seed dispersal by vertebrates in the Oriental (Indomalayan) region. *Biological Reviews* **73**, 413–448.

Corlett R.T. (2004) Flower visitors and pollination in the Oriental (Indomalayan) region. *Biological Reviews* **79**, 497–532.

Karr J.R. (1989) Birds. In *Tropical Rain Forest Ecosystems* (eds, Lieth H. & Werger M.J.A.). Elsevier Scientific Publishing, Amsterdam, Netherlands, pp. 401–416.

Karr J.R. (1990) Birds of tropical rainforest: comparative biogeography and ecology. In *Biogeography and Ecology of Forest Bird Communities* (ed, Keast A.). SPB Academic Publishing, The Hague, Netherlands, pp. 215–228.

Pearson D.L. (1977) A pantropical comparison of bird community structure in six lowland forest sites. *Condor* **79**, 232–244.

Woodcock M. & Kemp A.C. (1996) *The Hornbills: Bucerotiformes (Bird Families of the World)*. Oxford University Press, Oxford.

# Fruit Bats and Gliding Animals in the Tree Canopy

Birds are the most conspicuous group of animals in the forest canopy, but there are also other animals flying or gliding through the forest. These animals are remarkably different in each forest region, with less evidence for the convergence in form and function that we have seen in rain forest bird communities. Some of these differences appear to be related to patterns of food availability—a key concern for highly mobile animals—while others may reflect differences in forest structure. Some may be simply the results of biogeographical accidents. Bats are one such group of highly mobile animals, replacing the birds as the dominant nighttime vertebrates in the air above and inside the forest canopy. All rain forests have bats and on all but the most isolated of oceanic islands there are species that eat fruits and disperse their seeds. Another group is the gliding vertebrates, which are principally found in Asian forests. These animals use their gliding ability to move widely through the forest without having to descend to the ground. Gliding animals have different diets, but they all solve the challenge of getting enough to eat by moving among many trees rather than staying mostly in one place.

## Fruit- and nectar-feeding bats

Bats are the most species-rich group of mammals in tropical forests, and they are surprisingly well studied (Emmons 1995). Many biologists find them fascinating because of their unique adaptations for flight and foraging at night. Although the ancestors of all bats were probably insectivores or omnivores (Simmons & Conway 2003), modern bats have a huge range of diets: from fish and small terrestrial vertebrates, to blood, insects, nectar and pollen, fruit, and leaves. Here, we consider only the fruit- and nectar-eating bats. Confined to the tropics and subtropics, these plant-visiting bats make up about a third of the world's 1100 or so bat species. There are two main reasons for this focus. Firstly, these bats play an important role in all rain forests as pollinators and seed dispersal agents. In island rain forests, in particular, they are keystone species on which the long-term survival of many plant species depends (Fujita & Tuttle 1991). Secondly, these bats provide one of the clearest examples of convergent evolution, where unrelated groups of organisms have evolved to play similar ecological roles in the rain forests of the Old and New Worlds.

Fruit bats are abundant in all rain forests, except on the most remote Pacific islands, but their conspicuousness varies between regions. In the Old World tropics, before the sun starts to set, the first detachment of fruit bats appears on the horizon. High overhead, hundreds fly fast and purposefully over the village and head toward the forest. Their huge size recalls a more prehistoric scene; with wings that stretch up to 1.8 m (6 feet) across, silhouetted against the darkening sky, they look more like pterodactyls than mammals. This nightly ritual as the fruit bats leave their daytime arboreal roosts prompts not so much as an upward glance by the villagers. The large trees that were black with roosting bats during the day are soon vacated. Although the combined effects of hunting and habitat loss are making such stirring sights far less common, this evening scene is still the most popular image of fruit bats in the Old World tropics and can still be seen from Africa and Madagascar, to India, across Southeast Asia to Indonesia, New Guinea, Australia, and out into the Pacific. Yet such imagery is as alien to the New World tropics as it would be to the temperate zones. New World fruit bats are considerably smaller in size, do not roost in huge tree colonies, and belong to an entirely different group of bats. Although fruit bats in the Old and New World are united by their consumption of fruits and nectar, they are evolutionarily and ecologically worlds apart.

In the Old World, all the plant-visiting bats belong to the family Pteropodidae, the only family in the suborder Megachiroptera (the "megabats") (Fig. 6.1). All pteropodids are herbivorous, and most depend on fruit and a variable amount of

(a)

(b)

**Fig. 6.1** Some Old World fruit bats. (a) Rodrigues Island fruit bat (*Pteropus rodricensis*). (Courtesy of Tom Kunz.) (b) Female flying fox with pup (*Pteropus pumilus*) in the Philippines. (Courtesy of Tom Kunz.)

nectar and young leaves for their diet. The nectar specialists have traditionally been placed in a separate subfamily, Macroglossinae, but recent molecular studies have shown that their long narrow muzzles, reduced teeth, and protrusible, brush-tipped tongues are the result of convergent adaptations to nectar feeding that have arisen independently several times among the megabats (Alvarez et al. 1999). Pteropodids range in size from around 15 g to 1.5 kg (0.5 oz to 3.3 lb).

The fruit bats of the Neotropics, in contrast, are all members of the predominately insectivorous suborder Microchiroptera (the "microbats"), in the family Phyllostomidae. This family has the greatest diversity of diets of any mammalian family. In addition to frugivores, the Phyllostomidae also includes species feeding on insects, blood, small vertebrates, nectar, or omnivorous mixtures of these items. The diets of most phyllostomids are poorly known, but feeding strategies seem to be flexible in many species, with the proportion of fruit and nectar varying seasonally. Fruit-eating phyllostomids range in size from 5 to 100 g.

Although there is still some debate as to the evolutionary relationships of the bat suborders, it is clear that the frugivorous habit evolved independently in these two families (Van Den Bussche & Hoofer 2004). How have selection pressures differed or converged to shape the faunas of the two regions?

### Patterns in species diversity

The overall species diversity of the pteropodids and phyllostomids is very similar. The pteropodids include around 185 species in 42 genera, of which 20% are African and the rest are distributed from Asia to New Guinea, Australia, and the western Pacific. All but the dozen or so nectar specialists are predominantly frugivorous. There are around 160 species of phyllostomids distributed within 55 genera. Most phyllostomid species eat at least some fruit, but only around half are predominantly frugivorous.

Although the global diversity of pteropodid frugivores is higher than that of phyllostomid frugivores, the story is reversed at the local level. The number of coexisting frugivorous bat species is up to three times greater in Neotropical rain forests than it is in the Paleotropics. Twenty-five species of fruit-eating phyllostomids have been captured in Manu National Park, Peru, and 20 species on Barro Colorado Island, Panama. In contrast, the most diverse sites in Africa, Asia, and New Guinea rarely support more than a dozen species of pteropodids (Findley 1993; Fleming 1993; Simmons & Voss 1998). How can we reconcile the greater local species diversity of frugivorous phyllostomids with the greater global diversity of frugivorous pteropodids? The explanation seems to be in part biogeographical and in part ecological.

The greater global diversity of pteropodid fruit bats partly reflects their wider global distribution, throughout tropical and subtropical Africa, Asia, New Guinea, Australia, and the western Pacific, while phyllostomids are confined to the Americas. More important, however, is the fact that the distribution of Old World fruit bats is not continuous. Not only was there a largely separate radiation of fruit bats in Africa, but also around 62% of pteropodid species are restricted to islands (Fleming 1993). The corresponding figure for island phyllostomids is only 12%. Islands have two contrasting influences on diversity. On the one hand, they promote rapid speciation; fruit bats flying out over the ocean, perhaps blown off course by a storm, land in an unoccupied island and establish a new population. This population becomes isolated on the island, and founder effects, different selection pressures, or simple genetic drift ultimately result in divergence from the source population. However, while island biogeography promotes this kind of speciation on geographically isolated islands, it also sets limits to local diversity as a consequence of the species–area effect and distance to source populations. Species richness on islands typically increases with island area and with proximity to the mainland or other islands (MacArthur & Wilson 1967), and this classic tenet of island biogeography theory has been demonstrated for Old World fruit bats (Altringham 1996). What results is a high geographical turnover in moving from island to island; that is, individual islands have their own species, and the species on each island are taxonomically distinct but ecologically similar. The result is that islands have a high percentage of endemic species (species found in no other place), but a limited absolute number of local species. The many Old World islands thus explain the apparent paradox of high global diversity despite the low local diversity.

### Species diversity and food resources

Island biogeography is just part of the answer, and the question remains—why are fruit bat communities in large tracts of rain forest in Africa, Asia, and New Guinea still so much less diverse than their Neotropical counterparts? One theory attributes it to differences in the diversity and distribution of food resources in time and space (Fleming et al. 1987). This would influence foraging ecology, morphology, and community structure, which in turn affect the local diversity of plant-visiting bats. One major difference between rain forest regions is the greater diversity of fleshy-fruited shrub species in Neotropical rain forests, in comparison with rain forests elsewhere (Gentry 1982). These understorey shrubs (such as the many species of *Piper*) provide food for an entire guild of frugivorous bats in the Neotropics in a way unmatched in Asia and Africa. The relative paucity of both understorey frugivorous birds and bats in Old World rain forests attests to the impact this difference has had on vertebrate communities (Wong 1986; Francis 1994). Similarly, plants in the Neotropical forest canopy supply a variety of flowers and fruits adapted for bats, while the density of such plants seems to be lower in Old World rain forests, although quantitative data are lacking. Even in disturbed areas, many of the pioneer trees and shrubs in the Neotropics, such as the numerous species of *Cecropia*, produce fruits that are consumed by bats, while the fruits of most Old World pioneers are consumed by birds. Within the Old World tropics, the rain forests of Southeast Asia provide a particularly unpromising environment for fruit-dependent bats. Fruit production in the forest canopy is dominated by the wind-dispersed fruit of dipterocarp species, and resource scarcity is compounded by the multiyear seasonality of so many non-dipterocarp species, as described in Chapters 1 and 2.

## Feeding habits

Annual consistency in the availability of fruits is thought to have promoted dietary specialization in the Neotropical bat fauna (Giannini & Kalko 2004). In contrast, the more generalized feeding habits of Old World fruit bats may be a

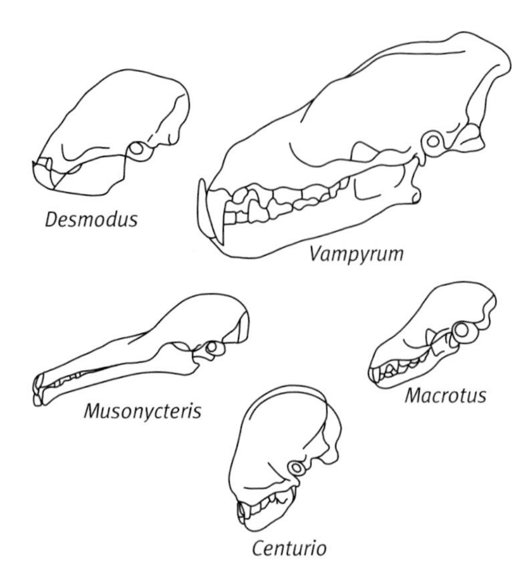


**Fig. 6.2** A diversity of skulls from New World phyllostomid bats, illustrating adaptations for feeding on different food types. *Musonycteris* is the most extreme nectarivore, probing into flowers with its long snout. *Centurio* is the most extreme frugivore and probably eats overripe fruit. *Vampyrum* is a large carnivore, eating birds and bats rather than insects. *Desmodus* is a blood-feeder (sanguinivore), while *Macrotus* is a more generalized feeder, eating mostly insects and fruits, and sometimes a bit of nectar. (Courtesy of Patricia W. Freeman.)

response to patchiness of food in space and time in Old World forests (Fleming 1993) (see Chapter 2). In general, New World fruit bats have short, wide faces and a nonprotrusible tongue; sharp serrated teeth are used to cut fruits into small bites. Within this general pattern, New World phyllostomids exhibit a broad range of skull and jaw shapes—suggesting that they have become specialists on particular types of fruit or other food items (Fig. 6.2) (Fleming 1993). Differences in teeth reflect the diversification in diet (Freeman 1998, 2000). Indeed, tooth diversity in the New World Stenodermatinae subfamily, which contains about half of the genera in the family, is greater than that of all other Microchiroptera combined (Freeman 2000). Such specialization may reduce food overlap and permit more species to coexist in any one locality. Each species may specialize on different sizes and textures of fruits, with larger species taking larger fruits than smaller species (Dumont 2003). The variable nonfruit—usually insect—component in the diet of many frugivorous phyllostomids (Herrera et al. 2002) may also have permitted greater specialization in the fruit component of the diet, since the bats can fill gaps in the supply of their preferred fruits—or in their nutritional value—by consuming more invertebrates.

In contrast, the Old World pteropodids generally show only limited departure from the ancestral skull shape, although there is some variation in tooth and skull morphology (Dumont 2004). Typically, the jaws of pteropodids are long and allow bigger bites with a wider gape than phyllostomids; the cheek teeth are relatively blunt. The bats rely on their protrusible tongue to manipulate food, and in most species only the juice, edible pulp, and small seeds are actually swallowed. The food is chewed and manipulated so that the juice can be extracted by pressing the tongue against the rigid palate on the roof of the mouth. The fibrous pulpy remains and any large seeds are than spat out as reject pellets ("spats") from the side of the mouth. This spitting habit of Old World fruit bats is the origin of a story from India in which God gave Moses two chances to add to creation. Having crafted a very nice squirrel in his first attempt, Moses then went on to form a winged rodent, what we would today call a fruit bat. Just as the creature had been imbued with life and was taking flight, Moses cried out in great consternation, "'My Lord and my God, do not let it fly, for I have forgotten to give it an anus'. 'Too late', replied God, and so the bat flies without an anus" (Fitzgerald 1923, cited in Marshall 1983). Of course this bat has an anus, but the frequency at which it spits out rejected food suggests otherwise. This fruit-processing behavior is not confined to Old World pteropodids, nor is it universal among them. New World stenodermatine fruit bats feeding on figs and other fibrous fruits also spit out less digestible materials, while the pteropodid *Syconycteris*, which consumes only the softest fruits, does not produce spats (Dumont 2003).

The tongue-feeding frugivory of the Old World fruit bats limits the degree of specialization that differences in tooth morphology might otherwise permit, as seen in New World frugivores. As a consequence, there is much greater dietary overlap among Old World fruit bats, and they appear to be rather diverse in their choice of food, in many cases being sequential specialists—switching preferred food plants as they become available through the season. Furthermore, many fruiting plant species in the Old World are visited by a number of different species of bats (Marshall 1983, 1985). In contrast, greater predictability and diversity of fruit sources in New World forests has apparently allowed fruit bats to be relatively sedentary, with less emphasis on energetically efficient, long-distance flight that is required to exploit the patchily dispersed food base of the Old World forests. The flexible nonfruit component in the diet of many phyllostomids may also encourage sedentary behaviour, since the bats can change their diet during fruit shortages, rather than travel further.

Within this general pattern, there are some Old World fruit bats that have more specialized morphology and feeding behaviors (Dumont 2003). Species that feed on hard fruits have relatively strong jaws with larger muscles, and species that specialize in nectar feeding have narrower snouts than fruit-eating bats (Cogan 2001). A few examples can be used to illustrate this diversity. The tube-nosed fruit bats (subfamily Nyctimeninae), which are named for their prominent tubular nostrils and are most diverse in New Guinea, have compact, broad, rounded skulls (see Plate 6.1, between pp. 150 and 151) similar to those of the Neotropical fruit bats in the genus *Artibeus* (see Fig. 6.4). The monkey-faced *Pteralopex* species of the Solomon Islands and Fiji have powerful jaws and robust, multicusped teeth with which they can break open nuts and eat the kernels inside. The harpy fruit bats (*Harpyionycteris*) of the Philippines and Sulawesi have multicusped teeth that are inclined strongly forward, perhaps as an aid in plucking and grasping fruit.

## Flying behavior

Flight efficiency is related to the shape of the wing. Long, narrow wings give a high aspect ratio and insure energetic efficiency over long distances. Flight speed depends on the relationship between body mass and wing area. A relatively heavy body with a small wing area results in high wing loadings and fast flight. Fleming (1993) demonstrated that the pteropodids have higher aspect ratio values and wing loading than their presumed ancestors. Thus, pteropodids tend to be fast, efficient fliers, and many species travel considerable distances in a night. In the extreme case, the large flying fox *Pteropus vampyrus* from Southeast Asia has a wingspan of 1.7 m (5.5 feet) and can make nightly foraging trips of 30–50 km (18–30 miles) (Lim 1966). Moreover, a few species in diverse genera of Asian and African pteropodids also have populations that migrate seasonally in those parts of their range where there are distinct wet and dry seasons (e.g. *Eidolon* spp., straw-colored fruit bats; *Epomophorus* spp., epauleted fruit bats; *Pteropus* spp., flying foxes; and *Rousettus* spp., rousette fruit bats (Marshall 1983)). These are broad generalizations, and there are also some small, forest-dwelling pteropodid species with small home ranges that feed almost exclusively in the local forest (Norberg & Rayner 1987).

In the Neotropics, the greater diversity of foods available to the phyllostomids, in combination with the greater availability of fruits in the forest interior, has had two consequences. First, there is less need for long-distance capabilities—the furthest that these bats are known to travel in a night is 10 km (Fleming 1993). At the same time, these bats need to be maneuverable in order to forage within the cluttered environment of the forest interior for fruits and floral nectar (Dumont 2003). So the primary selection pressure on the wing is for maneuverability, which relates inversely to both aspect ratio and wing loading. Thus what we expect to see are shorter, broader wings (lower aspect ratio) and lower wing loading (smaller body size). And this is indeed what is found in these bats. Not only are phyllostomids much smaller bats (5–120 g, 0.2–4 oz) than most of their fruit-eating paleotropical counterparts (up to 1500 g, 3.3 lb), but there has been a trend towards lower aspect ratio relative to the presumed ancestral form (Fleming 1993).

## Foraging behavior

A dispersed food base and highly efficient flight has promoted colonial roosting and flock foraging in many Old World fruit bats. Nearly two-thirds of pteropodid species are communal roosters, and colonies in some species can number several thousands of bats. These colonies are predominantly tree roosting and sometimes are found in urban areas if they are protected from hunting. One conspicuous colony is found on the campus of the Indian Institute of Science in the heart of Bangalore in southwest India. In addition, at least 27 species use caves (Marshall 1983).

Some of these cave-roosting bats, members of the genus *Rousettus*, are the only Old World fruit bats known to echolocate. They do so by tongue-clicking, a broad-band, low-frequency method that is useful for orientation within dark caves and is possibly also used for foraging. However, the great majority of pteropodid bats locate fruits and flowers by vision and smell. Old World fruit bats have large eyes and a correspondingly large visual cortex (Altringham & Fenton 2003). In many Old World plant species that target bats, the flowers or fruits are produced in a way so they can be readily detected and used by bats. The flowers and fruits often extend beyond the leaves on branches or are produced on the trunk, and they are often light in color to contrast with the foliage (Dumont 2003). Many Old World species begin to forage before sunset, both to take advantage of the light during foraging and because of the long commuting time to food sources. They often land, at least briefly, on the plants where they are feeding (Fig. 6.3a). For large species, such as *Pteropus*, this may compensate for lack of maneuverability in flight, since a bat can land on the outer branches of a tree and then climb around inside the canopy to find the fruits. Communal

(a)

(b)

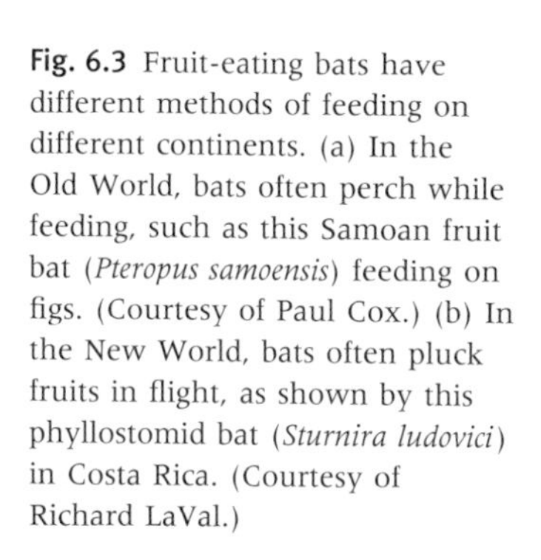

**Fig. 6.3** Fruit-eating bats have different methods of feeding on different continents. (a) In the Old World, bats often perch while feeding, such as this Samoan fruit bat (*Pteropus samoensis*) feeding on figs. (Courtesy of Paul Cox.) (b) In the New World, bats often pluck fruits in flight, as shown by this phyllostomid bat (*Sturnira ludovici*) in Costa Rica. (Courtesy of Richard LaVal.)

roosting may allow bats to share information about the locations of short-lived but superabundant feeding areas. This behavior also makes it possible for individuals to forage together as a flock, further increasing the awareness of potential foods as well as the potential dangers of predators such as hawks (Marshall 1983; Fleming 1993). Single-species flocks are most common, but mixed-species flocks—that is flocks with two or more species of bats—are known to visit bat-flowers in West Africa (Marshall 1983).

In Neotropical bats, foraging within the forest is further facilitated by the use of echolocation for orientation. Vision and olfaction are also used during foraging, as in Old World fruit bats, but at least some Neotropical bats can, if necessary, forage in complete darkness using echolocation alone. Many Neotropical species begin foraging only 1 or 2 h after sunset—considerably later than most pteropodids. Most species take fruits or nectar in flight—often hovering briefly—rather than landing on the plant (Fig. 6.3b). These bats can visit plants along a regular route or trapline, in contrast to most of the larger Old World bats that tend to concentrate at large food sources.

Real differences exist between the bat faunas in the Americas and the Old World, and these have clearly impacted on how bat communities are organized on a local scale. In the most obvious way, we can describe the communities of wide-ranging generalist Old World pteropodids versus the more localized and specialized New World phyllostomids (Fleming 1986; Heideman & Heaney 1989). However, this is something of an oversimplification; when we look within communities in each region of the world, we start to see parallels in foraging ecology that are again shaped by the distribution of food resources (Dumont 2003). Many of the small Old World fruit bats of the forest interior are, like the phyllostomids, localized in their distribution, although in Southeast Asia they are forced to move in response to the flowering and fruiting cycles of the dipterocarp forest. In both cases, guilds of bats are responding to the abundance of fruit and flowers in the tree canopy. In all rain forests, tree fruit crops of canopy trees tend to be large but widely dispersed, necessitating relatively long commuting flights. To some extent it is a question of temporal and spatial scales that distinguish the two regions. In the Neotropics, where fruiting is more regular and predictable, a large canopy frugivore such as the common fruit bat *Artibeus jamaicensis* might be considered to have a large home range at 3 $km^2$ (Bonaccorso 1979), whereas the Asian nectar-feeding dawn bat *Eonycteris spelaea* might travel over 30 km to a food source (Start & Marshall 1976).

As well as the conspicuous Old World fruit bat that forages over large areas in unobstructed air spaces, and roosts in large colonies, McKenzie et al. (1995) also described a guild of Old World bats that fly in the partially cluttered air spaces beneath tree canopies, roost in small colonies, and forage locally. Although not all are confined to the understorey, there are clear parallels here with the guild of understorey frugivores described for the Neotropics and typified by the short-tailed fruit bat *Carollia perspicillata* (Bonaccorso 1979; Fleming 1988). Both groups feed on small-seeded fruits of understorey trees or shrubs with extended fruiting periods but smaller daily fruit crops. Asynchronous fruiting or sequential specialization by the bat provides a steady (but often small) supply of food throughout the year. Home ranges tend to be restricted as a consequence, though Old World colonies may move their roost sites to be closer to the fruit sources (McKenzie et al. 1995). The spotted-winged fruit bat (*Balionycteris maculata*) of Malaysia is one of the smallest pteropodids (13–15 g) and is typical of this group.

It feeds off small figs and other fruits in the forest understorey, roosts in small harem groups in hollowed out ant nests or epiphytic ferns, and forages within a 1 km radius of the roost (Hodgkison et al. 2003).

## Bats as pollinators and seed dispersal agents

Bats are effective but expensive pollinators. They can carry large pollen loads for long distances between widely separated plants, but these flights must be fueled by large volumes of nectar. Not only is the reward expensive, but the flowers themselves must be larger and more robust than flowers targeted at pollinating insects, such as bees. The greater importance of bat pollination in the Neotropics than elsewhere must, at least in part, reflect the lower costs of attracting hovering visits from tiny (6–25 g, 0.2–0.9 oz) phyllostomid nectar specialists than accommodating the generally larger and less maneuverable pteropodids. Many Neotropical epiphytes, herbs, and climbers produce bat-pollinated flowers that are smaller and more delicate than the flowers of the trees and shrubs pollinated by Old World bats (Tschapka & Dressler 2002). Interestingly, many of the plants producing flowers targeted at Neotropical nectar specialists belong to endemic or near-endemic families, such as the Bromeliaceae (the pineapple family), Cactaceae (the cactus family), and Marcgraviaceae (a family without a common name), suggesting a long period of coevolution.

Even in the Neotropics, however, there are larger bat-flowers, often produced by trees, which are targeted at the many larger phyllostomids (some > 100 g) that include some nectar in a more omnivorous diet (Tschapka & Dressler 2002). These flowers look (at least, to human eyes) very like those visited by Old World pteropodids and belong to many of the same families, including the Bignoniaceae, Bombacaceae, Leguminosae, and Lythraceae. The readiness with which pteropodids visit the flowers of Neotropical bat-plants grown as ornamentals in the Old World, and with which phyllostomids visit Old World bat-plants planted in the Neotropics (Marshall 1985), suggests a universality to the bat-pollination syndrome which is surprising in view of the very different sensory systems of the two bat families.

Bats are often very wasteful seed dispersal agents, depositing most seeds from the fruits they eat under their day roosts or in temporary nighttime "feeding roosts" near the fruiting plants. Seeds that are small enough to be swallowed, however, are scattered widely as bats defecate in flight, in contrast to birds, which usually defecate from perches. The size threshold for swallowing seems to be around 2–3 mm diameter for most pteropodids (Corlett 1998), but is more variable among phyllostomids, some of which swallow and defecate much larger seeds. Wide dispersal across open areas is particularly important for the pioneer shrubs and trees that invade natural or human-made clearings, so it is not surprising that many of the most important Neotropical pioneers are dispersed primarily by bats, including the many species in the genera *Cecropia*, *Muntingia*, *Piper*, *Solanum*, and *Vismia*. What is surprising is that bats are generally much less important than birds as dispersers of woody pioneers in the Old World, although there are a number of important exceptions. Where Neotropical pioneers, such as *Cecropia peltata*, *Piper aduncum*, and *Muntingia calabura*, have been introduced to the Old World, they are rapidly adopted by the local bats, making the paucity of native bat-dispersed pioneers even harder to explain.

Bats have traditionally been viewed as benign seed dispersers, which drop or defecate seeds without damaging them. It is becoming apparent, however, that this is not necessarily true for all species of bats and all seeds. *Pteralopex*, a pteropodid from the Solomon Islands with particularly large and complex teeth, cracks the nuts of *Canarium* species (Flannery 1995b), while a frugivorous phyllostomid, *Carollia perspicillata*, has been reported to destroy the seeds of *Anacardium* (Bonaccorso 1979). Most surprisingly, two species of *Chiroderma* in Brazil have been shown to destroy the tiny seeds of figs by separating them in the mouth and then audibly crunching them up (Nogueira & Peracchi 2003). As these examples make clear, there is a long way still to go before we have a complete understanding of the complex relationships between bats and plants.

## Fruit bat conservation

The great diversity and abundance of fruit bats in the tropics means that they are a key part of rain forest biodiversity. Moreover, their role as pollinators and seed dispersers means that they are often not only of economic importance, but are also important to the ecological health of these forests (Charles-Dominique 1986; Cox et al. 1991; Fujita & Tuttle 1991). Given the differences between the Old and New World bat faunas, are there differences in the risks or conservation measures needed in the two regions? In both cases, habitat loss is a major issue (Kunz & Racey 1998; Pierson & Racey 1998; Racey & Entwistle 2003). In the New World, recent work suggests that the canopy frugivores are particularly at risk because they depend upon continuous tracts of primary forest to provide both roost sites and canopy fruits (Ochoa 2000). The increase in early successional tree and shrub species in habitats disturbed or fragmented by selective logging and small-scale shifting cultivation may actually favor the feeding behavior of the small Neotropical understorey frugivores, but even these species may require primary forest for roosting. It has been suggested that the relative abundance of canopy frugivores versus small frugivores at a feeding site may act as an indicator of forest disturbance and the health of the bat community (Schulze et al. 2000).

In the Old World, island endemic species are a high priority for conservation. Endemic island species are a major conservation concern in almost all groups of animals, and bats are no exception (Mickleburgh et al. 1992; Rainey 1998). In many parts of the Indo-Pacific and Asia, the common threats of habitat destruction and disturbance have been compounded by the human consumption of bats (Rainey 1998). Large body size and colonial roosting make many pteropodid species an attractive food item. International trade in fruit bats in Guam and the North Mariana Islands has spread the threat to populations across the region. The absence of hunting pressure combined with dietary flexibility, commuting ability, and tendency to migrate seasonally may cushion the impact of forest fragmentation and degradation to some extent on some of the large, open-space megabats. Several of the smaller species, such as the short-nosed fruit bat (*Cynopterus brachyotis*) are actually more common in secondary forest and gardens, as can be seen in some species in the New World as well. However, species like the black-capped fruit bat (*Chironax melanocephalus*) and the tailless fruit bats (*Megaerops* spp.) are rarely reported outside of undisturbed forest and may be particularly at risk from habitat loss.

## Box 6.1 Two contrasting fruit bats

Many of the contrasts between Old World and New World bats can be illustrated by a comparison of the following two species. With its dark fur and wings and distinctive pale facial stripes running above the eye from the nose-leaf to the ear, the common fruit bat *Artibeus jamaicensis* is a striking animal (Fig. 6.4a). Sharp white teeth set against a black tongue give a fierce appearance. These 50–60 g (1.8–2.1 oz) bats are large by New World standards and are thickset and muscular. The common fruit bat is a typical New World canopy frugivore and feeds mainly on fruit, especially figs. Not only is it often one of the most common bat species in Neotropical forests, but it is also one of the most important seed dispersers in early successional forest. As is typical for most New World fruit bats, these bats roost in small groups relatively near to the feeding grounds. Adult males try to attract several females into a harem by defending a secure roost site in a tiny tree hollow or cave. Like other New and Old World species, this species may even create tents by biting through the midrib of a palm frond so that the leaflets collapse down to provide a hidden roost space (Kunz & McCracken 1996). Less successful bachelor males roost together in the foliage and in leaf tents. On Barro Colorado Island in Panama, bats usually traveled less than 3 km (1.8 miles) per night (Handley et al. 1991), although in Mexico individuals might go as far as 10 km (Morrison 1978).

This New World bat provides a series of contrasts with its Old World counterpart. With a wingspan of 1.7 m and weighing in at 1.5 kg, the large (or Malayan) flying fox *Pteropus vampyrus* lives up to its name and is the world's biggest bat. The blackish-brown body is offset by a beautiful mantle of orange-brown fur around the back of the head and neck, and the species is found from southern Burma and Thailand to the Philippines and western Indonesia. The bats roost in large established colonies on the open branches of trees, especially in areas of mangrove and coastal palm swamp. In the 1920s, colonies with as many as 150,000 squawking, fidgeting, flapping bats were recorded, but hunting and habitat loss have reduced roost sizes to a few hundred bats in most places (Heaney & Heideman 1987), and the Philippine subspecies is likely to become extinct within 20 years. Their diverse diet includes fruit, flowers, nectar, pollen, and leaves. In captivity, they can eat half their body weight in food in a day; to meet these dietary requirements in the wild, they may fly long distances (30–50 km) in a night (Lim 1966; Kunz & Jones 2000), and they often show territorial behavior toward other individuals at the same tree. Although they can cause damage to some orchard crops, they are important pollinators of several ecologically and economically important trees in the region.

## Gliding vertebrates

Despite the striking differences between the Old and New World communities, fruit bats are found in all the major rain forest areas. In contrast, if you see animals gliding through the rain forest during the day that are not birds, you are probably in Southeast Asia. This region has a unique abundance and diversity of gliding animals not seen elsewhere. The curious puzzle of gliding animals in Southeast Asia is very revealing of continental differences in rain forest ecology. Southeast Asian forests are inhabited by over 60 species of gliding animals, included in 16 genera, representing at least six independent evolutionary origins of this remarkable trait and suggesting a common ecological pressure (Emmons & Gentry 1983; Dudley & DeVries 1990; Laman 2000). These gliding animals include flying squirrels, flying lemurs (colugos), flying *Draco* lizards, flying geckos, flying frogs, and most incredibly, flying snakes.

The 40 of so species of flying lizards (*Draco* spp.) are 12–25 cm (5–10 inches) long and can soar from tree to tree by expanding their ribs to make a brightly

(a)

Fig. 6.4 Two contrasting bats. (a) The common fruit bat (*Artibeus jamaicensis*) of Neotropical forests. (Courtesy of Jessie Cohen, taken in captivity.) (b) An example of the large fruit bats of the Old World is the golden-crowned fruit bat (*Acerodon jubatus*) from the Philippines. At more than 1.2 kg, this is among the heaviest of all bats. (Courtesy of Tim Laman.)

(b)

colored fan (Fig. 6.5). This fan, along with an expandable pouch on the chin of the males, known as a dewlap, is part of the territorial display of these species. These lizards are highly maneuverable and social, with males gliding around a single tree to find a better spot to make a territorial display. Males can rocket

(a)

(b)

**Fig. 6.5** Gliding lizards (*Draco cornutus*) from Danum Valley, Sabah, Malaysia. (a) Lizard gliding between trees. (Courtesy of Tim Laman.) (b) Close up of a lizard with its ribs spread out to form a gliding surface. (Courtesy of Tim Laman.)

into the air to chase another male away. Even though these species are almost exclusively arboreal, feeding on ants, termites, and other insects found on the tree, they can descend to the ground to lay their eggs. These lizards are by no means rare, and as many as seven species can coexist in one area of forest. The flying geckos (*Ptychozoon* spp., also known as parachuting geckos) have webs of skin along their tail, body, and webbed feet. These geckos are completely arboreal, feeding on insects and laying eggs on the tree bark. When pursued, they jump off and glide to the ground or a lower perch.

The five species of flying snakes in the genus *Chrysopelea* splay out their ribs to catch the air after leaping from a branch and undulate their bodies as if they were swimming (see Plate 6.2, between pp. 150 and 151) (Socha 2002). The body width approximately doubles while gliding and the ventral surface becomes slightly concave. The lateral undulations are thought to contribute to lift, but the precise mechanism is not clear. These snakes can maneuver in flight to avoid obstacles and attain a glide ratio (the ratio of horizontal distance to height lost) of 3.7, which is the same as *Draco* lizards. They are exclusively arboreal and feed on lizards and other small vertebrates.

The most mobile gliding vertebrates are the giant flying squirrels of the genus *Petaurista*, which glide up to 450 m (1500 feet) between trees. They are nocturnal and arboreal, feeding on young leaves, fruit, nuts, twigs, and insects. When fully extended for gliding, these animals are 1 m long. Cartilage extends from their wrists to expand the skin flaps even further. Giant flying squirrels can forage over several kilometers during a night to visit widely scattered food sources. Several smaller species of flying squirrel coexist with these giants, with a total of 15 species in Borneo alone.

There are also two strange species in their own order, the Dermoptera (meaning "wings of skin"). Known misleadingly as flying lemurs—which they are not—or, better, as colugos, the two species are *Galeopterus variegatus*, a widespread species of Southeast Asia (Fig. 6.6), and the more restricted *Cynocephalus volans*, found in the southern Philippines. Colugos are found from the lowlands to the mountains, in primary and secondary forest, and even in rubber gardens and coconut plantations. Molecular studies have shown that the colugos belong to a group of mammals that includes the primates, tree shrews, and rodents, but their precise relationships within this group are still unclear. Weighing between 1 and 1.75 kg (2.2–3.8 lb), colugos have a large extensible skin membrane that stretches along the neck and sides of the body to the fingers and toes as well as going to the tail. This allows the animals to glide up to 130 m between trees in search of young leaves to eat. Over the course of an evening, an animal might cover a distance of 1 or 2 km. In flight, the nocturnal animals stretch out their membranes to form a dark square kite, with the head in the middle of the leading edge. Their superiority as gliders is demonstrated by the ability of colugo mothers to carry their infants while foraging. When resting from flight, colugos hang upside down on branches wrapped in their membranes, resembling shy lemurs enfolded in large fur blankets. Their relative clumsiness on the ground, as well as their immediate tendency to climb the nearest tree, bears testament to their tree-dwelling nature.

The animals described here are considered gliders because they all have structures that increase their surface area, allowing them to be better airfoils, but they lack the ability to self-propel. The squirrels, flying lemurs, lizards, and

**Fig. 6.6** Colugo (*Galeopterus variegatus*) gliding through the night air in Bako National Park, Sarawak, Malaysia. (Courtesy of Tim Laman.)

geckos are true gliders that can alter direction in flight, brake to slow down, and make a heads-up landing on a tree trunk. The snakes can glide and maneuver as well as the other gliders, but they land on foliage and branches rather than tree trunks. The flying frogs of the genus *Rhacophorus* are the weakest gliders, using additional air surfaces provided by their long webbed fingers and toes to slow descent and prevent tumbling (see Plate 6.3, between pp. 150 and 151). Even so, flying frogs can alter the angle of their feet and hands, changing direction in mid-flight, and even glide from tree to tree using adhesive pads to stick on the landing site. These frogs spend their lives in treetops feeding on insects, except when they descend to the ground to mate and lay eggs in pools of water, often the wallows of wild pigs.

This abundance of gliding animals in Southeast Asia contrasts with other rain forests. The outlying Asian rain forests of the Western Ghats in India have two species of flying squirrels, a *Chrysopelea* flying snake, a *Draco* flying lizard, and a *Rhacophorus* gliding frog. The rain forests of New Guinea have only one widespread glider, the tiny marsupial sugar glider (*Petaurus breviceps*). Africa has six species of gliding rodents, the squirrel-like anomalures (Kingdon 1997), and one reported species of flying lizard. Madagascan rain forests lack gliders altogether, and only a few species of gliding hylid frogs with well-developed webbed feet have been reported from tropical America. This paucity of gliding ability in South American animals is particularly surprising, considering that this continent has some squirrels and a great abundance of lizards, snakes, frogs, and rodents.

### Why are gliding vertebrates abundant only in Southeast Asian forests?

What could be the explanation for the great diversity of gliding animals in Southeast Asia and their almost complete absence from South America? The repeated evolution of gliding species in Southeast Asia suggests that there has been some constant ecological pressure driving this process over the millions of years needed to produce such adaptations. Gliding has generally been viewed as either a means of escaping predators, by allowing animals to move between trees without descending to the ground, or as an energetically efficient way of traveling long distances between scattered food resources. But what is special about Southeast Asian rain forests?

Scientists have proposed various theories to explain the diversity of gliding animals in Southeast Asia (Emmons & Gentry 1983; Dudley & DeVries 1990; Laman 2000). The first theory might be called the tall trees hypothesis. The forests of Southeast Asia are taller than forests elsewhere (see Chapter 2), due to domination by tall trees in the dipterocarp family and also, possibly, lower wind speeds, so there is a more advantageous situation for gliding between trees. Taller trees perhaps allow for longer glides and the opportunity to build up speed in a dive before gliding. This argument has several flaws, however. First, gliding animals are found throughout the Southeast Asian region, even in relatively short-stature forests found in the northern range of the rain forest in China, Vietnam, and Thailand. Some gliders also thrive in low secondary forests, plantations, and even city parks. Clearly, gliding animals do not require tall trees for their activities. In addition, many gliding animals begin their glides from the middle of tree trunks, not even ascending to the tops of the trees to take off. Within the tree canopy, different gliding species live at different heights, and begin their glides at different heights, again demonstrating that the tall height of dipterocarp trees is not critical for their glides.

A second theory, which we might call the broken forest hypothesis, speculates that the tree canopy in Southeast Asian forests has fewer lianas (woody vines) connecting tree crowns than Neotropical and African forests. As a result, animals must risk descending to the ground or glide to move between trees. In addition, the tree canopy is presumed to be more uneven in height in Asian forests due to the presence of scattered, emergent dipterocarp trees with lower trees between them, again favoring gliding animals. Yet ecologists who work in different regions of the world observe tremendous local variation in tree height, canopy structure, and abundance of lianas, depending on the site conditions of soil, climate, slope elevation, and local disturbance. One can find many locations in Southeast Asia where there are abundant lianas and numerous connections between trees, and similarly many Amazonian forests with few lianas. The fact that many primates, civets, and nonflying squirrels move constantly between tree canopies in Asian forests without gliding certainly suggests that forest structure is not the main driving force behind the abundance of gliding animals.

A final theory, that we could call the food desert theory, suggests that it is the presence of the dipterocarp family itself that is driving the evolution of gliding species. The gliding animals consist of two main feeding groups: leaf-eaters and carnivores that eat small prey such as insects and small vertebrates. How would such animals be affected by living in a dipterocarp forest? For leaf-eating animals, the dipterocarp forest is a leaf desert. Dipterocarp trees often comprise 50% or more of the total number of canopy trees in a forest and over 95% of the large trees. Despite their abundance, dipterocarp leaves are unavailable to most vertebrate plant-eaters because of the high concentrations of toxic chemicals in their leaves. Flying squirrels, flying lemurs, and other flying animals all avoid eating dipterocarp leaves; they are similar to the shipwrecked sailors marooned on a raft, floating on an ocean and yet dying of thirst. These vertebrate leaf-eaters must travel widely through the forest, bypassing the dipterocarp trees, to find the leaves they need to eat. And gliding is a more efficient manner of traveling between trees than descending to the ground and walking, or jumping between trees.

In a similar manner, carnivorous animals, such as lizards and geckos, may need to search more widely for food due to the lower abundance of insects and other prey. The low insect abundance in dipterocarp trees is caused by the irregular flowering and fruiting cycles at 2–7-year intervals, causing a scarcity of the flowers, fruits, seeds, and seedlings that are the starting point of so many food chains (see Chapter 2). With fewer insects, there are in turn fewer spiders and centipedes. There also appears to be a lower density of small vertebrates in the tree canopy and on the forest floor in Southeast Asian forests than in rain forests elsewhere (Inger 1980; Allmon 1991). The lower abundance of invertebrate and small vertebrate prey in dipterocarp forests forces animals such as lizards and geckos to move between tree crowns in search of food, with gliding being the most efficient means. Gliding would also facilitate finding breeding partners, a critical consideration when a population is at a relatively low density.

## Conclusions and future research directions

In earlier chapters, examples of differences among rain forests were developed using plants, mammals, and birds. In this chapter, we have focused on animals that live in and above the forest canopy. After primates, some of the most striking differences among rain forest areas are found in the frugivorous bats. Old World fruit bats are larger on average than New World fruit bats, and tend to land while feeding on fruits, in contrast to smaller New World bats that usually take fruits on the wing. Old World bats detect fruits by their smell, shape, and color, while New World bats supplement these senses by echolocation. Old World bats tend to be sequential specialists, eating a wide range of fruit species over the year and supplementing their diet with young leaves, while New World bats often specialize on a core group of fruit species throughout the year, but may also consume a varying amount of invertebrates. There are similar differences between the Old and New World nectar-feeders.

Despite the many differences on the bat side of the relationship, there is plenty of evidence that Old World bats can locate, eat, and disperse the seeds of New World bat-fruits grown in Old World gardens and locate and pollinate New World bat-flowers. Similarly, New World bats recognize Old World bat-fruits and flowers planted in the New World. A striking consequence of the convergence in fruit characteristics has been the "escape" and spread of several Neotropical pioneer trees and shrubs into disturbed habitats in the Old World. However, despite the large amount of anecdotal evidence, there has been no systematic study of bat–plant relationships across the evolutionary divide. Are fruit and nectar sources really interchangeable? Are all fruits and flowers targeted at one group of bats recognized by the other group, or is it just a few conspicuous examples that have given this impression? Do Old and New World bats show similar behavior when visiting the same fruit or flower species? Are they equally effective as seed dispersal agents or pollinators? We predict that seed dispersal, which does not require a precise match between fruit and frugivore, will be relatively insensitive to the origin of the bat involved. In contrast, pollination, which often does require an accurate fit between bat and flower, is less likely to be successful with the "wrong" bat. We must also emphasize that studies involving exotic plants—particularly potentially invasive pioneers—should be based on the many existing plantings and should never involve the introduction of new plant species to an area.

Another important topic in bat biology is the impact of their extinction on plant communities. In all rain forests, bats are the principal pollinators or seed dispersers of many tree, shrub, and vine species, and in the Neotropics they also pollinate many herbs and epiphytes. Predicting the impact of bat extinctions is difficult, since the degree of dependence by a particular plant on a particular bat is rarely known. Bats are more important as pollinators and seed dispersers in the Neotropics than elsewhere, but the greater diversity of the bat communities there may limit the impact of a single extinction. In the Old World, it is the biggest bats that are most vulnerable from hunting and habitat loss and these are also the species most likely to have an irreplaceable role in seed dispersal, because they consume larger fruits and fly longer distances. Plant-feeding bats are particularly important on islands because they have reached many that are too remote or too small to support any other mammals. When a bat species goes extinct on an island, due to hunting or disturbance of its roosting trees by logging, there may be no other species to take its place pollinating flowers and dispersing seeds. Island bats may be similarly vulnerable to the loss of key flower or fruit resources. Most island bats are pteropodids, but phyllostomids play an important role on some Caribbean islands. Although many observers have warned of the potential impact of island bat extinctions, there has been little systematic study of their importance and it is vital that such studies start before it is too late to find intact bat communities for comparison.

In the Neotropics, bats are particularly significant in the recovery of forest in cleared and cultivated areas because of their tendency to defecate seeds of pioneer trees while flying, whereas birds usually defecate from perches. There is much less evidence for a major role of fruit bats in early forest succession in the Old World, but this may simply reflect the absence of comparative studies. If the relative importance of birds and bats in rain forest recovery differs between the Old World and New, this could have significant implications for the way in which such degraded landscapes are managed. The relative importance of nocturnal bats and diurnal birds in dispersing seeds into degraded and open habitats can be assessed very easily by comparing the "seed rain" into seed traps during the day and during the night.

The abundance of gliding animals in Asian rain forests and their rarity in rain forests elsewhere provides a strong argument that dipterocarp forests have unique properties that have influenced the evolution of its animal community over millions of years. The most likely explanation is that the multiyear periods between episodes of mass flowering and fruiting in dipterocarp forests, and the low diversity of edible leaf material, forces animals to forage more widely via gliding. The shortage of plant food may, in turn, lead to a low density of insects, other canopy invertebrates, and small vertebrates for canopy-dwelling reptiles, frogs, and snakes to eat. The same factors must also influence the ecology of other, nongliding, animal species in these forests, as has already been suggested for the primates (see Chapter 3).

The weakness with this line of reasoning is that comparative studies of the abundance of insects, lizards, frogs, snakes, and other animals in the rain forest canopy have not been undertaken at enough sites to confirm this impression. Do Southeast Asian forests really have a lower density of insects and insect-eating vertebrates than do comparable forests elsewhere? Such studies could yield a wealth of new insights. Particularly valuable would be studies comparing gliding animals in Southeast Asia with the ecologically most similar species in African and Neotropical rain forests. How would a Bornean flying snake differ in population density, physiology, foraging behavior, reproductive behavior, demography (especially injury and mortality from falls), and dispersal from a similar-sized snake species, as closely related as possible and eating the same type of food, that lives in the Congo River basin or Amazonia? Repeating the same studies for frogs, reptiles, and squirrels would also be extremely valuable. Organizing such cross-continental comparative studies should be a priority for tropical ecology.

In this and past chapters, we have discussed the relationships between plants and vertebrate groups. In the next chapter, we will devote our attention solely to insects, the most species-rich group in the tropical rain forest.

## Further reading

Dudley R. & DeVries P. (1990) Tropical rain forest structure and the geographical distribution of gliding vertebrates. *Biotropica* **22**, 432–434.

Emmons L.H. (1995) Mammals of rain forest canopies. In *Forest Canopies* (eds, Lowman M.D. & Nadkarni N.M.). Academic Press, San Diego, CA, pp. 199–223.

Fleming T.H. (1993) Plant-visiting bats. *American Scientist* **81**, 160–167.

Fleming T.H., Breitwisch R. & Whitesides G.H. (1987) Patterns of tropical vertebrate frugivore diversity. *Annual Review of Ecology and Systematics* **18**, 91–109.

Kunz T.H. & Fenton M.B. (eds) (2003) *Bat Ecology*. University of Chicago Press, Chicago, IL.

Kunz T.H. & Racey P.A. (eds) (1998) *Bat Biology and Conservation*. Smithsonian Institution, Washington, DC.

Laman T. (2000) Wild gliders: the creatures of Borneo's rain forest go airborne. *National Geographic* **Oct**, 68–85.

Nowak R.M. (1994) *Walker's Bats of the World*. Johns Hopkins University Press, Baltimore, MD.

Chapter 7

# Insects: Diverse, Abundant, and Ecologically Important

Plants, birds, and mammals are the best-known and best-studied components of tropical rain forests, but they make up only a tiny fraction of the total number of species—probably less than 1%. Invertebrates are the dominant animals of the rain forest, contributing most of the species and the overwhelming majority of individuals (Primack 2002). Of all invertebrates, the insects contribute most of the biomass and probably most of the species, although it must be admitted that we know very little about the rain forest diversity of such species-rich, noninsect groups as mites and nematodes.

In contrast to the plants, birds, and mammals, where most species have already been described and named by scientists, rain forest insects are much less known. Certain groups of large, popular insects, such as butterflies, dragonflies, large bees, and long-horn beetles are reasonably well known, but smaller and less colorful insects, including many beetle families and moths, are poorly known. A million or so insects have been given scientific names, but most of these are from the temperate regions. Yet the majority of insect species live in the tropics, and most live in the rain forest. If estimates of 5–10 million insect species are correct (Ødegaard 2000), then only 10–20% have been described so far. Sort through a leaf litter sample for tiny insects in any tropical rain forest and you are likely to catch at least one species new to science! Of the species that do have names, the name is usually all that is known. The ecology of tropical insects is largely unknown.

Some general patterns are apparent, however. First, insect faunas appear to be broadly similar in all the major rain forest areas that have been sampled, with the same insect orders and even families found in most rain forest areas. Both the relative abundance of different insect groups and the relative importance of different feeding guilds seem to be more or less the same. These similarities contrast with the major differences between the vertebrate faunas of the major rain forest regions discussed in previous chapters. Second, Neotropical forests seem to have more species than those in other regions, at least in the canopy. Third, there are conspicuous and ecologically important exceptions to the first two generalizations! We aim here to illustrate both the general patterns and some of the major exceptions.

Although many rain forest insect groups are very poorly known, there are some significant exceptions. Butterflies are well collected in some rain forests, and it is likely that more than 90% of the total species have been described, making them the best group to illustrate the general patterns of diversity in rain forest insects. It would be premature, however, to assume that all other insect groups show the same patterns, as described later. The social insects—ants, bees, wasps, and termites—are also relatively well studied and illustrate the overwhelming importance of insects in the ecology of tropical rain forests better than any other insect groups, as well as providing striking exceptions to the rule of relative uniformity among rain forest insect faunas.

## Butterflies

Butterflies are both the best known and the best loved group of insects, and most species live in tropical rain forests. A mere 500 ha (1250 acres) of lowland rain forest at Garza Cocha in Ecuador has more butterfly species—676—than the whole of North America (DeVries 2001). Species richness is lower in the Old World, with a small Neotropical country like Costa Rica having more species (1044) than either peninsular Malaysia (777) or New Guinea (785). The Neotropics, with 5341 species, has roughly double the species of all of Africa (2729 species). Even so, one can still see more species in a day spent walking in an Asian or African rain forest than in a lifetime in more temperate climates. Tropical rain forests also support the biggest and most brightly colored butterflies as well as an incredible variety of smaller and less conspicuous species.

The insects we know as butterflies are only a fraction of the species in the huge order Lepidoptera, most of which are known as moths. The division between butterflies and moths is fairly arbitrary, and the precise boundaries differ between authors. The skippers (Hesperiidae) have often been included as "honorary butterflies" despite some moth-like characteristics because, like the true butterflies, they are active during the day, in contrast to moths, which are typically active at night. More recently, the Neotropical Hedyloidea have been identified as the closest relatives of the butterflies and are sometimes called "nocturnal butterflies" (Yack & Fullard 2000). The true butterflies, considered here, are most commonly divided into four families (Papilionidae, Pieridae, Lycaenidae, and Nymphalidae), although some authors recognize more. The relative contribution of each family to the total butterfly diversity is very similar in Asia, Africa, and the Neotropics, although several important subfamilies are largely or entirely restricted to one region (DeVries 2001). The Lycaenidae and Nymphalidae together account for around 80% of the total species, with just the lycaenids accounting for 50% of the total (Table 7.1). The Madagascan butterfly fauna, in contrast, is strikingly deficient in lycaenids (only 17% of the species) and disproportionately rich in nymphalids (67%), largely in the subfamily Satyrinae, which includes the species known as satyrs and wood nymphs. New Guinea's butterfly fauna is also somewhat different, with a higher proportion of pierids than other rain forest areas, largely in the genus *Delias*, the jezebels (Parsons 1999).

**Table 7.1** Variation in the contribution of each butterfly family to species richness among different faunas. The percentage of total faunal richness by each family is given in parentheses.

| Fauna | Papilionidae (swallowtails) | Pieridae (whites) | Nymphalidae (nymphs) | Lycaenidae (blues) | Total |
|---|---|---|---|---|---|
| All of Africa | 80 (2.9) | 145 (5.3) | 1107 (40.6) | 1397 (51.2) | 2729 |
| Zaire | 48 (3.7) | 100 (7.6) | 607 (46.5) | 551 (42.2) | 1306 |
| Kenya | 27 (3.7) | 87 (12.1) | 335 (46.5) | 271 (37.6) | 720 |
| Southern Africa | 17 (2.3) | 54 (7.2) | 265 (35.6) | 409 (54.9) | 745 |
| Madagascar | 13 (4.9) | 28 (10.7) | 175 (66.8) | 46 (17.5) | 262 |
| Australia | 18 (6.5) | 35 (12.6) | 85 (30.6) | 140 (50.3) | 278 |
| New Guinea | 41 (5.2) | 146 (18.6) | 222 (28.3) | 376 (47.9) | 785 |
| Malaysia | 44 (5.8) | 44 (5.8) | 273 (35.9) | 400 (52.5) | 761 |
| Costa Rica | 42 (4.0) | 71 (6.8) | 438 (41.9) | 493 (47.2) | 1044 |

From DeVries (2001).

## Papilionidae: the swallowtails

The Papilionidae are commonly known as swallowtails, because in many species —but by no means all—there is a long "tail" extending from the rear of each hindwing. The family is distributed worldwide, though most species are tropical (Fig. 7.1). Although they make up a relatively small proportion of the total butterfly diversity, they are mostly large and colorful butterflies that are very popular with collectors. Papilionid diversity is greatest in the New World, but the largest and most spectacular species are found in the Old World.

The largest of all butterflies are the magnificent birdwings (*Troides* and *Ornithoptera*), which occur from Sri Lanka through Southeast Asia to New Guinea and Australia (see Plate 7.1, between pp. 150 and 151). The world's largest butterfly is Queens Alexandra's birdwing (*Ornithoptera alexandrae*), in which the relatively dull-colored black and yellow females can attain wing spans of up to 28 cm (11 inches) (Parsons 1999). These butterflies usually fly high above the forest canopy; many early specimens are peppered with holes as a result of being brought down with a shot gun! The males are smaller but more colorful, with a bright yellow body and iridescent blue or green markings on the wings. Birdwing caterpillars feed on poisonous *Aristolochia* vines, incorporating the poison into their bodies to make both the caterpillar and the adult distasteful to potential predators.

This family also provides Africa's largest butterflies, the African giant swallowtail (*Papilio antimachus*), with a wingspan of up to 23 cm, and the slightly smaller giant blue swallowtail (*P. zalmoxis*), both of which fly high in the canopy of rain forests throughout Central and West Africa. The males of both species can be collected by their attraction to human urine, but the females are rarely seen and the caterpillars are unknown. The males of the giant blue swallowtail have a distinctive noniridescent blue color, which is produced by the selective scattering of shorter wavelengths by structures in the wing scales (Huxley 1976). In contrast to the iridescent blue of most other blue butterflies, this color does not depend on the angle from which it is viewed. The appearance of *P. zalmoxis* in flight has been likened to fragments of blue sky.

(a)

(b)

**Fig. 7.1** Some papilionid butterflies. (a) Blue papilionid butterfly on large yellow flowers, from Queensland. (Courtesy of William Laurance.) (b) Two swallowtail butterflies mating, from Kenya. (Courtesy of Harald Schuetz.)

## Pieridae: the whites

The family Pieridae consists of small to medium-sized butterflies, most of which are white, yellow, or orange leading to their common names, the whites and sulphurs: indeed, the yellow European species are probably the origin of the word "butter-fly". The cabbage butterflies are well-known temperate zone members of the group. Some species, such as the jezebels (*Delias*) of Southeast Asia and New Guinea, have black or red patterns on a pale background. Throughout the tropics, pierids are a major component of the butterfly aggregations on exposed riverbanks and puddles along muddy rain forest tracks. These aggregations may contain several species of butterfly, but they are usually arrayed in groups of the same species. This behavior, called "mud-puddling", occurs almost exclusively with males. It is thought that they are obtaining essential minerals, particularly sodium, which are then transferred to the females during mating (Beck et al. 1999). This is necessary because land plants have a very low sodium content.

## Lycaenidae: the blues

The Lycaenidae account for almost half of all butterfly species, although their typically small size makes them less conspicuous than the other families. They show an incredible diversity of shapes and colors. Many species are blue on the upper side and have common names such as blues, hairstreaks, coppers, and metalmarks. Many have one to three pairs of "tails" on the hindwings which, in extreme cases, such as the plane (*Bindahara phocides*) of tropical Asia, can be up to twice as long as the body. These tails are often associated with dark spots on the hindwings, and it has been suggested that the combination resembles eyes and antennae. This apparently deceives predators into striking at the false "head" and thus allows the butterfly to escape with only minor damage (Robbins 1981).

The lycaenids are even more diverse in terms of their life histories than in their appearances. Adults may feed on nectar, fruits, or the carcasses of animals, while the caterpillars range from herbivores to strict carnivores, feeding on insects, such as aphids or ant larvae. Many lycaenids have complex associations with ants (Braby 2000). In some cases these relationships are mutually beneficial; the caterpillars secrete a fluid rich in sugar and amino acids from their "honey glands" that attracts ants to act as bodyguards against predators and parasites. In other cases, however, the benefits are all in the butterfly's favor. One of the most extreme examples of such a relationship is the moth butterfly (*Liphyra brassolis*), which ranges from India through Southeast Asia to New Guinea and northern Australia. This species is among the largest of the lycaenids, with a wingspan of 8 cm (3 inches), and is easily mistaken for a moth. The caterpillars live in the arboreal nests of the weaver ants, *Oecophylla smaragdina*, and feed on the ant larvae (Braby 2000). The moth then pupates in the nest, and the emerging adult is covered with sticky scales that clog the jaws, legs, and antennae of any ants that attack it on its way out of the nest. The female moth lays her eggs on the trunk and branches of trees where *Oecophylla* nest.

Among the 10 subfamilies of the Lycaenidae, the Riodininae are sometimes treated as a separate family, Riodinidae. This is a largely Neotropical group, although there is also a scattering of species throughout the Old World tropics

(DeVries 1997). The common name, metalmarks, refers to the metallic appearance of raised areas on the wings of some species. This subfamily includes some of the most beautiful of all butterflies, but many species are rarely seen inhabitants of the forest canopy.

## Nymphalidae: the nymphs

For diversity of adult form, no other butterfly family matches the Nymphalidae, commonly known as the nymphs (Fig. 7.2; see Plate 7.2, between pp. 150 and 151). They are referred to also as "brush-footed" butterflies because the front pair of legs are greatly reduced in size and often hairy or brush-like. They range in size from small to very large and include some of the most conspicuous species as well as some of the best-camouflaged species of butterflies. Although the family is represented in all rain forests, several of the 10–12 subfamilies and most of the genera have more restricted distributions. These differences, coupled with the conspicuousness of many nymphalids, make the butterfly faunas of the different rain forest regions distinctive to even the casual observer.

The largest nymphalids and probably the largest Neotropical butterflies are the owl butterflies (*Caligo* spp.), named for the big eye-shaped markings on the underside of each hindwing (Fig. 7.2b). These giant butterflies fly mainly at dawn and dusk, when they can easily be mistaken for bats. Owl butterflies are also notable for their mating behavior, which resembles that of some rain forest birds, such as the birds of paradise and the manakins (see Chapter 5). The male butterflies aggregate in "leks" at display sites along the forest edge, where they compete for perches while the females visit to choose a mate (Srygley & Penz 1999).

The most spectacular nymphalids are the morphos (*Morpho* spp.), which are also confined to the Neotropics (Fig. 7.2c). The wings of most male morphos display a metallic blue of an intensity unmatched by any other butterflies. As with many butterfly colors, this blue is not created by pigments, but by optical effects in the microscopic layered structure of the scales. Such structural colors do not fade like pigments, and dead specimens retain the brilliance of the living butterfly, a characteristic exploited to the morphos' cost by the makers of butterfly jewelry. The blue of morpho wings has continued to intrigue physicists because the color changes very little with viewing direction, which may be important in mutual recognition. Recent studies suggest that both interference and diffraction effects are involved (Kinoshita et al. 2002). In morphos, the undersides of the wings are brown, so the resting butterfly is almost invisible against tree bark. The alternate flashes of blue and brown in flight, as the wings open and close, may also make them hard for a predator to follow.

The males of many nymphalid species are territorial, flying from their perch to investigate passing butterflies. Male morphos, for instance, can often be attracted by waving a blue cloth of the appropriate hue. Another group of Neotropical nymphalids, the cracker butterflies (*Hamadryas* spp.), produce short bursts of loud cracking noises as they defend their territories, probably by buckling a stiff part of the wing (Yack et al. 2000). To a human observer these sounds can be audible 50 m (160 feet) away, but they are usually only produced when the butterflies are very close to each other (< 10 cm), suggesting that their own hearing is not as sensitive. These are the only butterflies known to produce

(a)

(b)

**Fig. 7.2** Some Neotropical nymphalid butterflies. (a) Glass-wing nymphalid butterfly (*Ithomia patilla*) from Costa Rica. These butterflies feed on bird droppings to obtain extra nitrogen for egg production. (Courtesy of Dale Morris.) (b) Owl butterfly (*Caligo* sp.) from Costa Rica. (Courtesy of Tim Laman.)

sound for communication, but a species of *Heliconius* (also Nymphalidae) has recently been shown to both produce and react to audible wing clicks (Hay-Roe & Mankin 2004).

Many nymphalid butterflies escape from predators by camouflage. In the

(c)

(d)

**Fig. 7.2** (*cont'd*) (c) Morpho butterfly (*Morpho achilles*) from Colombia. (Courtesy of Gustavo Londoño.) (d) Heliconia butterfly (*Heliconius hecale zuleika*) from Costa Rica. (Courtesy of Ethan Scott.)

Indian leaf butterfly (*Kallima inachus*), the uppersides of the wings are brightly colored, but the undersides, which are all that is visible in the resting butterfly, are an almost perfect imitation of a dead leaf in shape, color, and pattern. There is even a "midrib" down the middle of each wing. The jungle queen (*Stichophthalma louisa*) in the montane rain forests of Vietnam takes camouflage even further, with both sides of the wing resembling dead leaves. When flying in a slow, zigzag course it can easily be mistaken for a falling leaf (Novotny et al. 1991).

In striking contrast, many other nymphalids have conspicuous colors and patterns. The tree nymphs (*Idea* spp.) of Asian rain forests, for example, are large and slow-flying, with grayish-white wings, appearing like a winged seed or piece of paper blowing in the wind, and making them easy prey for a hungry flycatcher. As with the birdwing butterflies mentioned above, however, these conspicuous nymphalids are often poisonous, or at least distasteful, as a result of both chemicals acquired from the plants on which their larvae feed and additional toxins consumed or synthesized by the adults. Such defenses are effective only if vertebrate predators learn to associate the unpleasant experience with the striking appearance of the butterfly. Predators cannot be expected to learn to avoid lots of different patterns, so unrelated unpalatable butterflies have often evolved very similar mimicry patterns. Experience with one member of the group teaches the predator to avoid all similar-patterned butterflies in future (Kapan 2001).

Palatable butterflies can also gain protection by mimicking unpalatable species, which has lead to the evolution of complex multispecies "mimicry rings", including both unpalatable and palatable species, with the latter usually in a minority for obvious reasons (Mallet & Gilbert 1995). Day-flying moths may also join these mimicry rings. A particularly striking example is the "tiger-striped" mimicry ring in Neotropical forests, consisting of butterflies and moths from several families, all of which are striped in yellow, orange, black, and brown. Up to five mimicry rings with different color patterns have been described from one Neotropical forest. Similar rings occur in rain forests elsewhere, mostly centered on unpalatable nymphalids. In some cases, the mimicry involves wing shape and flight behavior as well as pattern, although size seems to be unimportant.

Another group of nymphalids, the largely Neotropical heliconines, also feed on poisonous passionflower plants. They advertize their poisonous contents through warning patterns on both surfaces of their characteristically long and narrow wings (Fig. 7.2d). Female heliconines have an apparently unique method of feeding on pollen; they first accumulate it at the base of the proboscis while visiting flowers, and then dip the pollen into the nectar (Gilbert 1972). Amino acids from the pollen leach into the nectar and are ingested by the butterfly. The extra nitrogen acquired in this fashion increases egg production and permits an adult life span of several months, which is very long for butterflies. The nitrogen is also used to synthesize cyanide, making the butterfly even more toxic to birds. The lure of extra nitrogen for making eggs also explains the behavior of some species in a related group of Neotropical nymphalids, the Ithomiinae, in which the females feed on the droppings of the antbirds (Gotwald 1995). Up to a dozen may follow one swarm. These ant butterflies avoid becoming additional prey for the antbirds by accumulating toxins from their food plants, again advertizing their presence by warning coloration.

## Ants

Ants and termites, along with some species of wasps and bees, are distinguished from most other insects in that they form long-lived colonies with overlapping generations and have a division of labor between fertile queens and sterile workers (Hölldobler & Wilson 1990). In some ways the whole colony acts as a single "superorganism", with an ecological impact at least as great as an individual vertebrate. Less than 2% of known insect species are social, but they account for a large fraction of insect abundance. In the rain forest of central Amazonia, it has been estimated that social insects (mostly ants and termites) account for around 80% of the total insect biomass and a third of the total animal biomass (Fittkau & Klinge 1973). The dominance of social insects in tropical rain forests would probably appear even greater if their role in ecological processes such as biomass harvesting, mineral nutrient cycling, and soil excavation were measured.

The most important group of social insects, in terms of numbers, biomass, and ecological impact, is undoubtedly the ants. Ants are found in all layers and microhabitats of the forest and make use of a huge range of food resources. In comparison with termites, ants are a conspicuous part of the rain forest fauna because many species forage actively on the ground and on plant surfaces. The highly visible ant species are only a small fraction of the total, however; most species are small and secretive. Some species forage on the forest floor, while others live in the forest canopy above. Although certain species, such as Neotropical leaf-cutting ants (in the genera *Atta* and *Acromyrmex*) and the Malaysian giant ant (*Camponotus gigas*), nest in the ground and forage mostly in the canopy, many more species spend their entire lives in one stratum, forcing scientists interested in sampling the entire ant fauna to use a variety of techniques for their research. In the few cases where this has been done, the total number of ant species found is staggering: the current record is 524 ant species in just 6 ha (15 acres) of lowland rain forest in Sabah, Malaysia (Brühl et al. 1998). This compares with about 700 species in the whole of the United States and Canada, and around 10,000 species in the world.

Ants are a widely dispersed family of insects, facilitated by the ability, in most species, of a single, winged female to start a new colony. Only Hawaii and other highly isolated Pacific islands lack native ants (Wilson 1996). As a result of this wide dispersal, there are many similarities between the ant faunas of the major rain forest areas. There are similar numbers of species in tropical Asia, Africa, and the Neotropics and three of the largest ant genera—the *Camponotus, Pheidole,* and *Crematogaster*—are found in all rain forests, except in the remote Pacific. There are, however, many exceptions to this basic uniformity and several of these are of great ecological significance.

### Army ants

Most of the ant diversity on the rain forest floor is made up of small and inconspicuous species living in small colonies (with fewer than 150 workers) in decaying twigs and other short-lived nest sites, and foraging 1 m or so from their nests (Byrne 1994). Many of these ants are thought to be primarily predators on tiny litter invertebrates or scavengers, but some also harvest seeds from bird droppings and may have a large impact on the regeneration of small-seeded plant species.

At the other extreme, there are ground-dwelling ants that form huge colonies and forage over vast areas. The most spectacular of these species are the swarm-raiding army ants of Africa and the Neotropics. Until recently, it was thought that the army ant syndrome had originated independently in the New and Old World species: one of the classic examples of convergent evolution. Recent molecular studies, however, have shown that all the true army ants had a common ancestor in the mid-Cretaceous, around 105 million years ago (Brady 2003). Both this date and the modern distribution of the army ants are consistent with an origin on Gondwana, before its final break up (see Chapter 1). The similarities between the major lineages can now be seen not as convergence, but as a striking example of long-term "evolutionary stasis", with a suite of shared characteristics retained despite the long separation.

True army ants in the Old and New Worlds share a number of important characteristics (Gotwald 1995). The first of these is the group raiding and retrieval of animal prey that gives rise to their common name. This collective foraging gives them access to a far wider range of prey, in particular large invertebrates and small vertebrates, than is available to a solitary forager (see Plate 7.3, between pp. 150 and 151). The second, which is probably a consequence of the first, is that the colonies are nomadic, moving on when the supply of prey is exhausted. The third is that the queen ants are highly modified in comparison with all other ant species. Army ant colonies can be enormous but they include only a single fertile queen. Army ant queens must therefore produce extraordinary numbers of eggs throughout their lifetime: estimated to be as many as 3–4 million per month in the African species, *Dorylus wilverthi* (Gotwald 1995). Not surprisingly, these queens do not look much like ordinary ants. They are much larger than their workers—*Dorylus* queens can exceed 50 mm (2 inches) in length and are among the largest ants known—and have a greatly enlarged abdomen. In most ants, new colonies are founded by a single, winged female, but in army ants, the queens are wingless and new colonies are produced by division, or "budding" off of parts of the colony, as in the honeybees. This explains both the existence of separate radiations in the Old and New World, and the absence of army ants from Madagascar, since dispersal across water is unlikely.

(a)

**Fig. 7.3** (a) Army ant (*Eciton* sp.) soldier from La Selva, Costa Rica. (Courtesy of Tim Laman.)

(b)

(c)

**Fig. 7.3** *(cont'd)* (b) Driver ants (*Dorylus wilverthi*) from Tanzania. (Courtesy of Dale Morris.) (c) African driver ant soldier (*Dorylus* sp.) with workers below, from Kibale Forest, Uganda. (Courtesy of Tim Laman.)

The most dramatic development of the army ant syndrome is found in a few species of *Eciton* in the Neotropics and *Dorylus* (often known as the "driver ants") in Africa, with colonies consisting of hundreds of thousands or millions of workers, which forage on the surface in huge swarms (Fig. 7.3). In *Eciton*, a swarm of ants moves forward on a broad front, sweeping up everything in their path or driving more mobile species before them. In *Dorylus*, the snake-like feeder columns expand at the front into a broad crown. Although the threat these swarms pose to large vertebrates has often been exaggerated, they will consume any prey they can overcome: mostly invertebrates, often other social insects, but also the occasional small and helpless vertebrate, such as bird nestlings. Prey items too large for an individual ant to carry back to the nest—which can be 200 m (650 feet) from the raid front—are cut up or transported by teams of ants and slung beneath their bodies. The Neotropical *Eciton burchelli* has a specialist porter caste of large workers, which shuttle at high speed between the nest and the swarm front, while African *Dorylus wilverthi* relies on teams of smaller, slower workers for the same task (Fig. 7.3b) (Franks et al. 2001). Much of the more mobile prey escapes the ants, only to be snapped up by birds and other vertebrates that follow these surface raids for this very reason (see Chapter 5).

Although army ants, including species of *Dorylus* and *Aenictus*, occur in Southeast Asian rain forests, these form much smaller and much less conspicuous colonies. *Dorylus laevigatus*, for instance, is found throughout Southeast Asia, but lives a largely subterranean existence, except for small, short-lived raids on the surface at night (Berghoff et al. 2003). One colony contained an estimated 325,000 workers, compared with millions in some African species. There is no Asian equivalent of the spectacular and devastating swarm raids of Africa and the Neotropics, and there are no records of antbird followers. Perhaps Asian rain forests do not provide the high density of insects needed to support these huge colonies? Even in Africa and the Neotropics, the majority of army ant species do not move on the surface, but conduct their raids entirely underground or beneath the litter layer. These species perhaps specialize on ants or termites as prey. Such ants are found in all the major rain forest regions, except Madagascar, but have not attracted the same attention as the more conspicuous surface-raiding species, although they may be of equal ecological importance.

In addition to the "true" army ants, in the subfamilies Aenictinae (Africa to Australia), Dorylinae (Africa and Asia), and Ecitoninae (America), elements of army ant behavior are seen in species scattered in other subfamilies. Some Asian species of *Leptogenys* (Ponerinae) are generalist, group-raiding, nomadic predators that could reasonably be termed army ants (Gotwald 1995). Their colonies are smaller than those of the true army ants, however, and the differences between the queen and her workers are much less pronounced. Two Asian species of *Pheidologeton* (Myrmicinae) are also swarm-raiders with massive colonies, but have winged queens and do not change their nest sites as frequently as the true army ants. This may, in part, be a result of their extremely broad diet that includes insects, earthworms, snails, spiders, and other invertebrates, and the corpses of vertebrates, as well as a variety of seeds and large fruits (Moffett 1987). These ants have a very complex division of labor and a more than 500-fold range in dry weight between the largest and smallest workers. Both *Leptogenys* and *Pheidologeton* have been reported to drive insects and small vertebrates ahead of their swarms, but there are no reports of birds or other vertebrates foraging at the swarm front.

## Leaf-cutter ants

In contrast to the generally wide distribution of ant groups, leaf-cutter ants are unique, conspicuous, ecologically important, and found only in the New World.

Leaf-cutter ants are the fungus-growing ants in the tribe Attini, known as attines; specifically, the 50 or so species in the genera *Acromyrmex* and *Atta*. One of the most characteristic sights in Neotropical rain forests is the columns of leaf-cutter ants carrying pieces of leaves or flowers back to their underground nests (Fig. 7.4; see Plate 7.4, between pp. 150 and 151). Each ant holds its leaf section above its

**Fig. 7.4** (a) Leaf-cutter ants carrying leaf fragments on a log. (Courtesy of Rob Bierregaard.) (b) Nest of leaf-cutter ants with rotting leaves covering the surface, from La Selva, Costa Rica. (Courtesy of Tim Laman.)

head, giving rise to the nickname of umbrella or parasol ants. Certain early estimates suggested that they might be responsible for consuming up to 12–17% of total leaf production in tropical forests (Cherrett 1989), but it now appears that such high percentages are confined to disturbed areas. In mature rain forests, leaf-cutter ants consume a relatively small proportion of the total leaf production, although they have a significant impact on their preferred plant species, which may lose up to 40% of their leaves (Wirth et al. 2003). They also have an impact through their harvesting of flowers.

No other ants are completely herbivorous, presumably because the ant digestive system, unlike that of termites, cannot cope with a diet that is so high in cellulose and so low in protein and sugars. Herbivory for the attines is made possible by a unique and obligatory symbiosis with fungi. Molecular evidence suggests that this relationship evolved only once, more than 50 million years ago, when a member of the fungus family Lepiotaceae (Agaricales, Basidiomycotina) was domesticated by an attine ancestor (Currie et al. 2003). Several primitive genera of attines form small colonies of a few hundred workers that cultivate their fungus on the droppings of herbivorous insects and other dead plant matter. The two genera of leaf-cutters, in contrast, form enormous colonies with, in some species, several million workers. These workers cut fresh leaves, flowers, and fruits with their mandibles, traveling more than 100 m (325 feet)—and several hours—from their nest and climbing up to 30 m into the canopy. Back at the nest, this fresh material, along with fallen flowers, leaves, and stipules gathered from the forest floor, is taken into underground chambers, where other ants process it into a soft pulp on which the fungus is planted. These chambers are also occupied by a diverse array of animals that parasitize the ant colony or just use the chamber as a place to live.

The ants keep their fungus garden free of microbial "weeds" by both physical and chemical means, but recent research has discovered that a specialist garden parasite, the ascomycete fungus *Escovopsis*, can rapidly overgrow even ant-tended gardens (Currie et al. 2003). This fungus is kept in check by a previously unrecognized third partner in the ant garden symbiosis, a filamentous bacterium (actinomycete) in the genus *Streptomyces*, which produces antibiotics that specifically target *Escovopsis*. This bacterium is cultured by the ants on specialized body surfaces and has been found in all the attines so far examined. When a young queen leaves the colony to found a new one, she carries the garden fungus in her mouth and the "weed-killing" bacterium growing on a specialized region of her cuticle. Phylogenetic analyses suggest that *Escovopsis* originated in the early stages of fungus cultivation by ants (Currie et al. 2003). One very interesting question is how the leaf-cutter ants' antibiotics have remained effective for so long, given that antibiotic resistance can arise so rapidly in human pathogens (Beattie & Hughes 2002).

The ants feed on protein- and sugar-rich knobs (gongylidia) that bud off from the fungus. For the ant larvae, the fungus is the only food, but leaf-cutter workers also drink sap directly from the cut edges of the leaves that they harvest. The fungus not only acts as an external digestive system, giving the ants access to cellulose-rich plant material (although it is not clear if the fungus actually digests the cellulose) (Abril & Bucher 2002), but also overcomes the plant's chemical defenses. Leaf-cutters avoid many plant species, presumably because they are unsuitable for fungus cultivation, but they still collect a greater variety of plants than is known for any other herbivore.

## Canopy ants

It is a long walk from a nest at ground level up into the forest canopy; only relatively large and fast-moving species like the leaf-cutter ants do this regularly. In the Neotropics, the 2.5 cm (1 inch) long "bullet ants" (*Paraponera clavata*) nest at the base of trees and ascend the trunk to forage in the canopy, with an average round trip of over 100 m. This widespread ant is one of the most noxious to humans, with both a painful bite and a wasp-like sting that can cause a person's hand to swell for days. The Malaysian giant ant, *Camponotus gigas*, is even larger (up to 3 cm) but is largely nocturnal, is far less aggressive, and has no sting. Like the bullet ant, this species nests underground but forages mostly in the canopy, with each colony occupying a huge three-dimensional territory (Pfeiffer & Linsenmair 2001).

Most canopy ants, in contrast, spend their whole lives there, with many species nesting in tree hollows or accumulations of leaf litter. The weaver ants, *Oecophylla*, common predatory ants from Africa to tropical Asia and Australia, make nests by joining leaves together with silk produced by their larvae, which the ants use like living shuttles (Fig. 7.5) (Hölldobler & Wilson 1994). Other ant species construct nests from "carton", a cardboard-like mixture of soil, chewed-up wood pulp, and ant secretions. As many as 61 ant species have been collected by "fogging" a single subcanopy tree in Sabah, Borneo with insecticide (Floren & Linsenmair 2000). Most of these species were rare, however, and there were only one to three species common in each tree, with 2–12 additional somewhat common species. In tropical plantations and disturbed secondary forests, several aggressive and abundant ant species establish nonoverlapping territories, so the canopy is divided into an "ant mosaic". Each dominant species has a characteristic set of subdominant species, and both dominants and subdominants influence

Fig. 7.5 Weaver ant (*Oecophylla smaragdina*) nest from Malaysia. (Courtesy of David Lee.)

the non-ant fauna within the territory. Ant mosaics also occur in the canopy of at least some undisturbed rain forests, but investigations into their biology and significance are only just beginning (Dejean & Corbara 2003).

Even the most basic questions about most species of canopy ants cannot yet be answered—such as, what do they eat? Many species are known to be predators on other insects, but it is not obvious how the most abundant invertebrates of the forest canopy, in terms of both number and biomass, can live as carnivores on less abundant prey. It has therefore been suggested that the principal food of the commonest species is plant carbohydrates, including nectar from extrafloral nectaries (see below) and "honeydew"—the sugary sap excreted by plant-sucking homopteran insects, such as scale insects, aphids, and treehoppers. Even the most predatory of species may derive much of their energy from these sources. In Australian rain forests, for example, all colonies of the weaver ant *Oecophylla smaragdina* tend homopterans (as well as some lycaenid caterpillars, see above), often constructing sheltering "pavilions" out of leaves woven together with larval silk in the same way they make their nests (Blüthgen & Fiedler 2002).

Studies of the nitrogen isotope ratios of canopy ants have shown that, unlike *Oecophylla*, many of the most abundant species get very little of their nitrogen from predation (Davidson et al. 2003; Hunt 2003). In many cases, the isotope ratio is similar to that of herbivorous insects, suggesting that most of the nitrogen in the diet comes from plant sources. Apart from the plant exudates and honeydew mentioned above, these plant sources could include pollen and fungi. The most abundant species of canopy ants actively farm homopteran insects, protecting them from predators and parasites. Plant-sucking insects are undersampled by the insecticidal "fogging" used to collect canopy insects, but the abundance of their associated ants suggest that the ant–homopteran partnership may be responsible for more consumption of plant biomass than all other invertebrates and vertebrates together (Davidson et al. 2003). In a sense, these canopy ants and their plant-sucking homopterans may be the major "herbivores" in the rain forest.

### Ant gardens and ant-epiphytes

Ant gardens are one of the most striking differences between New World and Old World rain forests. In the Neotropics, the carton nests of canopy ants are often associated with a limited number of epiphytic plant species, forming so-called ant gardens. These are not simply chance associations. The plants are there because the ants have "planted" the seeds of certain species in the nest carton (Benzing 1990). The seed coats of some species contain essential oils, often a variant of methyl salicylate, known commonly as winter green oil. These essential oils are similar to ant pheromones, which attract the attention of the ant and presumably influence the ants to pick up the seeds and bring them to the nests. Seeds of several of the ant-plant species in the genera *Codonanthe* and *Anthurium* even have the shape and color of ant pupae, which may further promote their entry into the ant colony. As well as seed dispersal, the plants benefit from the nutrient-rich carton, which often contains vertebrate feces, and they probably also benefit by the protection the ants give them from both insects and vertebrates that might eat them. The ants, in turn, benefit from the supply of fruits and elaiosomes (oily structures on many ant-dispersed seeds), as well as the structural support that the plant roots give the nest. The plants include bromeliads, figs, cacti, aroids (Araceae), and gesneriads. Despite the thousands of species of epiphyte species growing in the Amazon, there is a small core group of a dozen or so species from eight different plant families that are consistently found in ant nest gardens throughout the Amazon, suggesting that this is a coevolved reciprocal relationship between plants and ants. A variety of unrelated ant species are involved, but many ant gardens are occupied by two coexisting species, the larger and fiercely aggressive *Camponotus femoratus* and the smaller *Crematogaster limata*. The exact nature of the symbiotic relationship between these ant species—termed parabiosis—is unclear, but they share foraging trails and nest in different chambers of the same garden.

Ant gardens are much less common in Old World rain forests. Instead, in Asia and New Guinea, a number of epiphytes produce ant houses in which the plant itself forms some type of hollow structure that is occupied by ants (Benzing 1990). Rather than a few insect species forming gardens in which plant species from many families grow, ant houses are formed by relatively few genera of plants, but are occupied by a diverse group of canopy-dwelling ants, with *Crematogaster* species being most common. The ant-house plants form structures that have clearly evolved to be occupied by ants and do not have any other obvious function to the plant. Dramatic examples are found in the genera *Myrmecodia* and *Hydnophytum* (Rubiaceae), which range from Southeast Asia to New Guinea, Australia, and Fiji. In these genera, the plant stems form a tuber, which is honeycombed with tunnels leading from the surface to chambers inside. Ants occupy these chambers to raise their young, but also store droppings, debris, and dead ants in refuse chambers. Ants derive obvious benefit from the relationship by having a strong, secure nest; the benefit to the plant comes from the ability to absorb essential mineral nutrients from the decaying animal material in the refuse chambers and the protection afforded by the ants, which readily attack insects and other animals encountered on the plants. Species of *Dischidia*, in the milkweed family, form very different ant houses. *Dischidia* plants grow as epiphytic vines with thick purplish leaves that press against tree stems like overturned bowls. Ants establish homes in the hollow space formed between the leaf and the tree stem. These ants are typically fierce and protect the plant in exchange for a place to live. In Southeast Asia, ant houses are extremely common in heath forests on nutrient-poor sandy soils and in regenerating forests. One survey of a 25 × 9 m (80 × 30 feet) site recorded over 500 ant-house plants (Benzing 1990). Ants occupied virtually every one of these plants. Such ant houses and associated species are present in the New World as well, including the bromeliad species *Tillandsia bulbosa*, but their abundance and importance appears to be far less than in Asia and New Guinea.

### More plant-ants and ant-plants

The impact of leaf-cutter ants on the plants they harvest for their fungus gardens is entirely negative, but many ant species are involved in relationships with plants in which both partners benefit (Beattie & Hughes 2002). The ant gardens and ant-epiphytes mentioned above are examples of this. Many other rain forest plants have extrafloral nectaries—that is, nectar-producing structures that are not inside flowers—that are visited by ants. The proportion of trees and shrubs reported to have such nectaries varies greatly between rain forest sites, from as low as 12% (Malaysia) to as high as 40% (Cameroon), but differences in definitions, methodology, and sample sizes mean that these different percentages should be interpreted with caution (Blüthgen & Reifenrath 2003). The ants in turn defend the plant against herbivorous insects. The nectaries are sited so as to attract ant defenders to the most vulnerable parts of the plant, such as young, expanding leaves, softer stems, and flower buds. Some plants also produce solid "food bodies" for the same purpose. These food bodies are protein-, lipid-, or carbohydrate-rich nodules produced on the edges of young leaves that are eagerly collected by ants.

These relationships of nectaries and food bodies are relatively unspecialized and opportunistic: they involve many species of plants and many species of ants. There are also a much smaller number of more specialized relationships, which in some cases are obligatory for both partners. Although the details of these relationships vary tremendously, the common feature is that the plant provides a nesting space for the ants, as well as food in exchange for protection. Two of the most conspicuous and best-studied examples have evolved independently in unrelated but ecologically similar tree genera: *Macaranga* (Euphorbiaceae) in Southeast Asia and *Cecropia* (Cecropiaceae) in the Neotropics (Fig. 7.6). Both genera consist of small, fast-growing, large-leaved pioneer trees that grow naturally in tree-fall gaps and often become very abundant along logging roads, as well as in abandoned fields and other human-made clearings.

The 280 or so species of *Macaranga* range from West Africa through tropical Asia to Fiji. Many species have casual relationships in which ants are attracted to extrafloral nectaries and food bodies, but 24 species from the wetter parts of Southeast Asia have developed obligatory relationships, usually with ant species in the genus *Crematogaster* (Fiala et al. 1999). In these *Macaranga* species, the ants nest in hollow stems and feed on food bodies produced by the host plant, as well as honeydew from sap-sucking scale insects that they keep inside the stems. Several *Macaranga* species have a bluish, waxy, surface covering over the stems that is very slippery for most insects but presents no problems to their specific ant partners (Federle et al. 1997).

The costs to the plant can be high: in *Macaranga triloba*, food body production accounts for 5% of the daily biomass production (Heil et al. 1997). In return, the plant receives protection from herbivorous insects and some pathogenic fungi. The ants also bite off any part of a foreign plant that comes into contact with their host, thus preventing overgrowth by climbers, which are very common in the well-lit habitats that most *Macaranga* species require. Ant-free individuals of these *Macaranga* species do not survive long.

Like the Asian *Macaranga* species, many of the more than 100 species of *Cecropia* in Central and South America have hollow stem internodes and produce food bodies on specialized structures at the base of the leaf stalk. The ants in this case are usually members of the Neotropical genus *Azteca*, and they forage entirely on their host tree. Formerly, scientists thought that these ants derived all their nutrition from the food bodies, supplemented in some cases by honeydew produced by aphids. However, recent studies with carbon isotopes have suggested that the food bodies are used mostly as food for the larvae and that the worker ants derive most of their nutrition from consumption of insects found on their hosts (Sagers et al. 2000).

**Fig. 7.6** (a) Ants on the shoot of a *Macaranga capensis* plant, from the East Usambara Mountains, Tanzania. This species is under investigation for its ant–plant relationships. (Courtesy of Norbert Cordeiro.) (b) *Cecropia* plant in flower, from Panama. (Courtesy of Roger Le Guen.)

The *Azteca* ants provide the same services to their *Cecropia* hosts in the Neotropics as their *Crematogaster* counterparts do for *Macaranga* in the Old World tropics. However, *Azteca* species vary considerably in how actively they defend their trees—some species are virtually useless in this role. The isotope study mentioned above also showed that *Cecropia* plants derive most of their nitrogen from debris left by the ants in abandoned internodes, and it is possible that this, rather than defense, is often the major benefit to the host tree. Plant nutrition has previously been shown to be the basis of the relationship between ant-epiphytes and the ants that occupy the chambers which these plants provide for them. Shortage of nutrients is an obvious problem for epiphytes living high up in the forest canopy, because they are not rooted in the soil. Nitrogen in particular is also likely to be a limiting factor for fast-growing pioneer trees, such as *Cecropia*.

## Termites

At first sight, termites seem very similar to ants, and the two are sometimes confused by observers. Both are small social insects with distinct castes, nonflying workers, and nests of varying complexity. In most other ways, however, these two groups of insects are very different. Ants are related to bees and wasps, and some species have stings. Their immature stages are helpless larvae and pupae, and the entire colony is female, except for the relatively brief period when the short-lived males are produced to mate with the young queens (Wilson 1972). In contrast, termites are related to cockroaches and praying mantises. The immature termites progressively resemble the adults and there is no pupal stage. The workers are both male and female. Termite colonies are founded by a long-lived "king" as well as a queen. Most ants are carnivores, although many other types of tissues are consumed, while the diet of termites is dominated by cellulose.

Termites are very abundant in tropical rain forests but much less conspicuous than ants. A few species, such as the Southeast Asian processional termites, *Hospitalitermes*, travel in ant-like columns in the open air, carrying their resources like leaf-cutter ants. In contrast, most termites remain concealed in their nests and construct long-lived covered trails when they must cross open areas. The nests of many species, however, are much more conspicuous than their inhabitants (Fig. 7.7a). Some termites build nests on the soil surface, such as the mushroom-shaped nests of *Cubitermes* in Africa, while others, including many wood-feeding *Nasutitermes*, construct large, ball-shaped nests attached to trees at various heights. Many species, however, make nests inside dead wood or below the ground, which are invisible from the outside.

Estimates of total termite abundance at rain forest sites mostly exceed a thousand individuals per square meter, with a total biomass of at least 2000 kg per hectare (Martius 1994). To put this in perspective, this is greater than the total biomass of primates at most rain forest sites. Termites are the dominant decomposers in lowland tropical rain forests. They consume up to a third of the annual litter fall and, although they digest only part of this, their activities break up woody litter into fragments that are more easily attacked by other decomposers such as beetles, fungi, and worms. The activities of soil-feeding termites also have large, but little understood, effects on soil structure and chemistry. The nests tend to concentrate nutrients, and nest material is used as a fertilizer for crops in several parts of the tropics. Bacteria in termite guts also fix nitrogen and

(a)

Fig. 7.7 (a) Termite mound in Africa. (Courtesy of Olivier Langrand, Conservation International.) (b) Nozzle-headed nasute termite soldiers (white bodies) guarding an open passageway in Costa Rica. The workers are larger and darker in color. (Courtesy of Tim Laman.)

(b)

produce methane, a potential greenhouse gas. The termites are also prey for a huge range of more or less specialized predators, from army ants to Neotropical anteaters (Myrmecophagidae), African and Asian pangolins (*Manis* spp.), and the bizarre short-nosed echidna (*Tachyglossus aculeatus*) of New Guinea and Australia, an egg-laying monotreme related to the platypus.

(c)

**Fig. 7.7** *(cont'd)* (c) Close-up of a nasute termite (*Nasutitermes takasagoensis*) soldier, from Japan. (Courtesy of Kenji Matsuura.)

## Termite diversity

There are approximately 2650 known termite species, most of which are confined to the tropics and subtropics (Abe & Higashi 2001). Fifty or more species can coexist in one rain forest site. The richest termite faunas are in West Africa, followed by South America, Southeast Asia, New Guinea, Madagascar, and Australia (Davies et al. 2003). This pattern is different from most other groups of rain forest organisms, for which the Neotropics is usually the richest area, perhaps because termites are less sensitive to drying. Although the termites originated in the Mesozoic and several genera, such as *Nasutitermes*, are pantropical, there are large differences in the composition of termite faunas in the various rain forest regions. Termite queens cannot fly far on their own, although some species may be transported long distances after being picked up by air currents. In general, the biggest differences between regions are in the soil-nesting groups. More than a century after it was sterilized by a volcanic eruption, the island of Krakatau, between Java and Sumatra, still has no soil-nesting termites (Gathorne-Hardy et al. 2002). Soil-feeding species are particularly poor at crossing water gaps and the low diversity of soil-feeding termites in Southeast Asia, Madagascar, and Australia seems to reflect the isolation of these regions from the major radiations of soil-feeder diversity in Africa and South America (Davies et al. 2003). By contrast, termite groups that feed inside dead wood are very widespread, since they can be "rafted" across ocean gaps in floating wood.

Rain forest termites belong to three (of a global total of seven) families. The dry-wood termites (Kalotermitidae) and damp-wood termites (Rhinotermitidae or Termopsidae, depending on the classification) feed almost exclusively on wood, which they digest with the aid of symbiotic flagellate protozoa and bacteria. There can be 20 or more species of flagellates in the hindgut of each termite species, constituting up to a third of the total body weight (Radek 1999). In a very real sense, the termite we see is a termite–protozoan–bacteria partnership. The dry-wood termites are pantropical but seem to be rare everywhere in rain forests. However, since they are largely confined to dead branches and trunks in the canopy, they have probably been undersampled. The damp-wood termites feed mostly on moist and partly decomposed wood on the forest floor and appear to be more common. Neither group of these so-called "lower termites" is seen by most visitors to the rain forest because they feed on the wood within which they nest.

Members of the third termite family, the Termitidae, are known as "higher termites" because of their more complex social organization involving a true sterile worker caste. They lack the flagellate protozoans on which the lower termites depend for cellulose digestion, relying instead on their own enzymes and a diverse community of bacterial symbionts in their guts. The Termitidae dominate in all tropical forests but the four subfamilies have very different distributions. Soldiers of the largest subfamily, the Nasutitermitinae (the "nasute" or long-nosed termites), can be recognized by a long projection, or "nasus", on their head and greatly reduced, nonfunctional mandibles (see Fig. 7.7b,c). When threatened, they can eject from this snout a sticky and irritating fluid composed largely of terpenes—a chemical defense more effective than the powerful mandibles of lower termite soldiers. The Nasutitermitinae—particularly the genus *Nasutitermes*—dominate Neotropical forests in terms of both abundance and number of species. They are also very important in the Asian region and New Guinea, but are relatively less abundant in Africa. Most species in this subfamily are wood-feeders but some have other specialties. In the Neotropics, species of *Syntermes* forage at night, making circular cuts several millimeters in diameter in dead leaves and carrying the pieces back to their underground nests. In the Amazonian rain forest, 20–50% of the leaves in the litter layer show evidence of feeding by these termites (Martius 1994). In Southeast Asia, columns of *Hospitalitermes* individuals forage high in the forest canopy for lichens, bryophytes, and algae, which are carried back to the nest as food balls (Miura & Matsumoto 1998). The high nutritional value of this diet (relative to wood) apparently makes the added risk of foraging in the open worthwhile, while the need to forage over a wide area makes the covered foraging trails used by most other termites impractical. Aggressive soldier termites lead and protect the foraging columns of *Hospitalitermes*; the high ratio of soldiers to workers may partly compensate for the lack of cover.

Other Nasutitermitinae, such as the Neotropical *Subulitermes*, are soil-feeders. Soil-feeders dominate the rain forest termite community but are very difficult to study and this general designation probably hides a great deal of variation in what is actually eaten and where. In addition to fragments of dead plant material and humus, the guts of soil-feeders contain fine roots and fungal mycelium, as well as soil minerals. Soil-feeding termites must pass huge quantities of soil through their guts to obtain enough digestible organic matter and they probably have profound effects on soil chemistry and structure. The other pantropical subfamily of higher termites, the Termitinae, includes both soil- and wood-feeding species, and is important in all regions, although relatively less so in the

Neotropics. Different branches of the subfamily have penetrated the soil-feeding niche in Africa and Asia.

The other two subfamilies of the Termitidae have more restricted distributions. The largely soil-feeding Apicotermitinae are most diverse and abundant in West Africa, but this subfamily is also well represented in the Neotropics and a few species are found in most of the Oriental region. It has not reached Madagascar, New Guinea, or Australia. The remaining subfamily, the Macrotermitinae ("big termites"), is confined to Africa, Madagascar, and Asia, and has reached neither the Neotropics, New Guinea, nor Australia. Members of this subfamily are distinguished by the fact that they cultivate a fungus in order to degrade the cell walls in their food materials and thus increase the digestibility of the cellulose. The termites grow the fungus—usually a species of *Termitomyces* (Basidiomycotina)—on specialized structures known as "combs" built from their feces, largely the undigested cell walls of plant material, and then feed on the older parts of the comb after fungal degradation. The fungus itself also serves as a food source for some species (Hyodo et al. 2003). Molecular studies have shown that this fungus-farming mutualism arose only once, in Africa, from where there have been several independent migrations to Asia (Aanen et al. 2002). This association allows the Macrotermitinae to make extremely efficient use of a wide range of relatively undecomposed plant materials. The fungus, in turn, receives preconditioned plant material as food in the form of termite feces, an equable microenvironment for growth, and control of potential competitors. The Macrotermitinae are most abundant in drier forests and become less important with increasing rainfall, perhaps because continuous dampness favors free-living fungi and reduces the benefits derived from the symbiosis. It seems a curious coincidence that the fungus-growing leaf-cutter ants are confined to the New World and the fungus-growing termites to the Old World, but despite the obvious parallels, the diets of the two groups show little overlap, and there seems to be no reason why they could not coexist in the same forest.

The importance of termites in soil processes means that the biogeographical differences in the composition of rain forest termite communities described above may translate into real differences between regions in ecosystem processes, such as decomposition, energy flow, and nutrient cycling. With regard to ecosystem function, the most important differences are likely to be the low diversity of soil-feeding termites in Southeast Asia, Madagascar, New Guinea, and Australia, and the absence of fungus-growing Macrotermitinae from the Neotropics, New Guinea, and Australia (Davies et al. 2003). However, while there is some evidence for differences in the importance of termites between regions, with much lower energy flow through termites in Borneo than Africa (Eggleton et al. 1999), this will not necessarily lead to differences in ecosystem processes, since other invertebrates and free-living microbes may compensate for the missing termites.

## Social wasps

The term "wasp" is applied to all members of the vast order Hymenoptera that are not bees or ants, and includes species with a huge range of life histories. Relatively few wasps are truly social, but these species are a conspicuous (and often unpleasant) feature of the tropics, particularly in relatively open and disturbed habitats, where nests are nearer ground level. These are also the habitats where most research has been done, and the role of social wasps in undisturbed rain forests is virtually unstudied. Social wasps, similarly to the ants and social bees, have often made a huge investment in their nests and have evolved the behavior and weapons to defend it. Indeed, stings from social wasps probably kill more people than any other wild animal in the tropics. The most aggressive of all wasps may be *Polybioides raphigastra* (Vespidae, Polistinae). Found in the rain forests of Malaysia, it will attack humans without provocation up to 20 m (65 feet) from the nest, which is constructed inside a hollow tree. Like bees, but unlike most other wasps, the sting of this species is left in the victim, where it releases chemicals that attract more wasps (Sledge et al. 1999).

The truly social wasps belong in two subfamilies of the family Vespidae: the Polistinae, which includes the cosmopolitan paper wasps (*Polistes*) and a great diversity of other, mostly tropical species, and the Vespinae—the larger species of which are commonly known as hornets—which are confined to tropical Asia and New Guinea and to the northern temperate zone. The paper wasps have been very extensively studied because they build naked combs in exposed sites, which are easy to observe. However, their colonies are relatively small and short-lived, even in the tropics. In contrast, many other Polistinae and the Vespinae build large nests protected by an envelope of chewed plant fibers, which can, at least in some species, last for several years. The nests of *Polybia scutellaris* in Brazil can reach up to 1 m in diameter and persist for as long as 25 years (Wilson 1972). In many species of social wasps, including all Vespinae, new nests are founded by one or more queens without the help of workers, but in the New World *Polybia* and related genera, and some species in the Old World genus *Ropalidia*, the nests are founded by swarms, as in the honeybees and stingless bees.

Most wasps are basically carnivorous, with animal protein as the main larval food, but the adults also consume nectar, sap, and fruit juice for energy. Pollen is not a wasp food, except in the nonsocial "pollen wasps" (Masaridae), which provision their larval cells with pollen and nectar. A great variety of insects are taken as prey by different wasp species and, although the overall impact has not been quantified, it is likely they are important predators of herbivorous insects in the rain forest. The wasps around flowering plants are often hunting other flower visitors, but they also feed on exposed nectar and, in a few cases, are significant pollinators (Corlett 2004). Several species of social wasps are also known to harvest honeydew produced by aphids and other sucking bugs, which they "milk" and defend from competing ants (Dejean & Turillazzi 1992).

## Bees

Although butterflies are the best-known group of insects, the honeybee, *Apis mellifera*, is almost certainly the best-known single species. Honeybees and the other highly social bees, notably the 500 stingless bee species, make up only a small fraction of the total of 20,000 or so bee species, most of which are solitary—that is, they nest and forage as single individuals (Michener 2000). In contrast to most other insect groups, bee diversity is not highest in the rain forest, but appears to reach a maximum in dry, warm temperate regions, such as the southwestern United States. The social bees, however, are concentrated in the wet tropics, and their numerical dominance there may be a major reason for the lower diversity of solitary species.

Bees originated around 140 million years ago in the Cretaceous, evolving from wasps that became specialized on pollen instead of animal prey as a source of protein-rich food for their larvae. Their later diversification proceeded in parallel with the speciation of flowering plants over the same period. While bees came to depend on plants for both protein-rich pollen and energy-rich nectar, many plant species became dependent on bees for cross-pollination. The evolutionary histories of bees and flowers have thus been intertwined from near the origins of both groups. Today, bees are the most important group of pollinators in all tropical rain forests. In a lowland dipterocarp forest in Sarawak, Borneo, for instance, 32% of the 270 plant species studied were pollinated by social bees and an additional 18% by nonsocial bees (Momose et al. 1998). At the La Selva field station in Costa Rica, bees were the main pollinators of at least 42% of the tree species studied and were frequent visitors to many others (Fig. 7.8) (Kress & Beach 1994).

Rain forest bees range in size from the 1.5 mm-long workers of some stingless bees to the giant mason bee, *Megachile pluto*, from the Northern Moluccas, in which the females reach lengths of almost 40 mm (1.6 inches) (Michener 2000). Most are black or brown, but many are boldly striped, while the orchid bees of the Neotropics rival the flowers they visit in their shiny green and blue colors. Most bee species are solitary, some form communal nests, while others show a gradient of increasing cooperation and specialization, culminating in the permanent highly social colonies of the honeybees and stingless bees, which consist of a single queen and her worker daughters.

Despite their ancient origin and powers of flight, there are some distinct differences between the bee faunas of different rain forest regions. The biggest differences, as is so often the case, are between the Old and New Worlds. Neotropical rain forests have the most diverse bee faunas, possibly because of the absence (until recently) of the honeybee genus, *Apis*, with its ability to recruit huge numbers of foragers and thus dominate large patches of flowers.

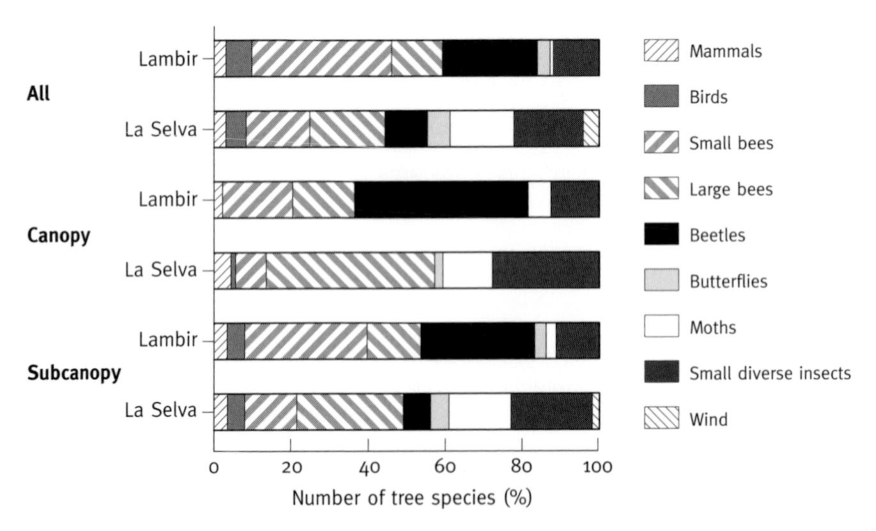


**Fig. 7.8** Bees are important pollinators at two rain forest sites: Lambir in Borneo and La Selva in Costa Rica. Shown are the percentages of trees pollinated by major classes of pollinators, on all trees, canopy trees, and subcanopy trees. (From Turner 2001b.)

Indeed, Southeast Asian rain forests, which support two or three species of honeybee at any one locality, have only half as many bee species as Neotropical sites (Roubik 1989). Another factor in this difference, however, may be the phenomenon of community-wide mass flowering at multiyear intervals in Southeast Asian rain forests, which favors large perennial colonies that are able to forage over very large areas, and then either migrate or store enough resources to get them through the lean periods (Corlett 2004).

## Large bees

Large bees are not only a conspicuous element of the rain forest fauna but are also very important as pollinators of a variety of large flowers, both in the canopy and the understorey. Bees 20 mm or more in length are found in several groups, including the orchid bees and honeybees, which are considered separately below. In the Neotropics, the most important of the large bee pollinators are in the family Apidae, including the large, hairy, fast-flying solitary bees in the genus *Centris* (Michener 2000). Most species of *Centris* use oil, rather than nectar, as a supplementary larval food. This oil is collected by the female bees from the flowers of various plants, particularly in the largely New World family Malpighiaceae, which produce it as a floral reward instead of nectar. Although both oil-collecting bees and oil-producing flowers occur in other parts of the world, it is only in the Neotropics that they are an important component of rain forests.

Another group of large-bodied bees, the carpenter bees, in the pantropical genus *Xylocopa*, have been described as important pollinators of rain forest trees in the Malay Peninsula. In more recent studies in Sarawak, however, carpenter bees foraged mostly in open habitats and along the forest edge, and were rare in the undisturbed forest (Momose et al. 1998). The common name of the *Xylocopa* bees refers to the female bee's excavation of its nest in solid wood or in the hollow internodes of bamboo, for which it is provided with exceptionally powerful jaws. Species of *Megachile* (Megachilidae), including the giant mason bee mentioned above, are another prominent group of relatively large bees in Asian forests. Overall, however, large solitary bees appear to be less important as rain forest pollinators in Southeast Asia than in the Neotropics, perhaps because such bees are particularly vulnerable to long-term fluctuations in the availability of floral resources (Corlett 2004).

## Orchid bees

The common name of the orchid bees (Apidae, Euglossini) refers to the role of the male bees in pollinating many of the larger flowers of orchids of the Neotropics, but could equally well reflect the brilliant colors of many of these bees themselves, which include metallic blues, greens, reds, or coppers. These bees and the plants that depend on them are confined to the Neotropics, with most species found in rain forests (Cameron 2004). They range in size from 8.5 mm to almost 30 mm in length, and their scientific name refers to their characteristic long tongues. Although in the same family as the highly social stingless bees and honeybees, most orchid bees are solitary or only temporarily social.

Fig. 7.9 (a) Long-tongued euglossine bee approaching a flower, in Brazil. Note the extended tongue. (Courtesy of Rob Bierregaard.) (b) Stingless bee (*Trigona hockingsi*) nest, from northeast Queensland, Australia. (Courtesy of Andrew Dennis.)

Both male and female euglossines visit flowers to collect nectar and pollen like other bees, but the males alone also visit certain species of orchids, gesneriads, and aroids to gather fragrances (Fig. 7.9a). More than 650 Neotropical orchid species in 55 genera provide these fragrances as their only reward for flower

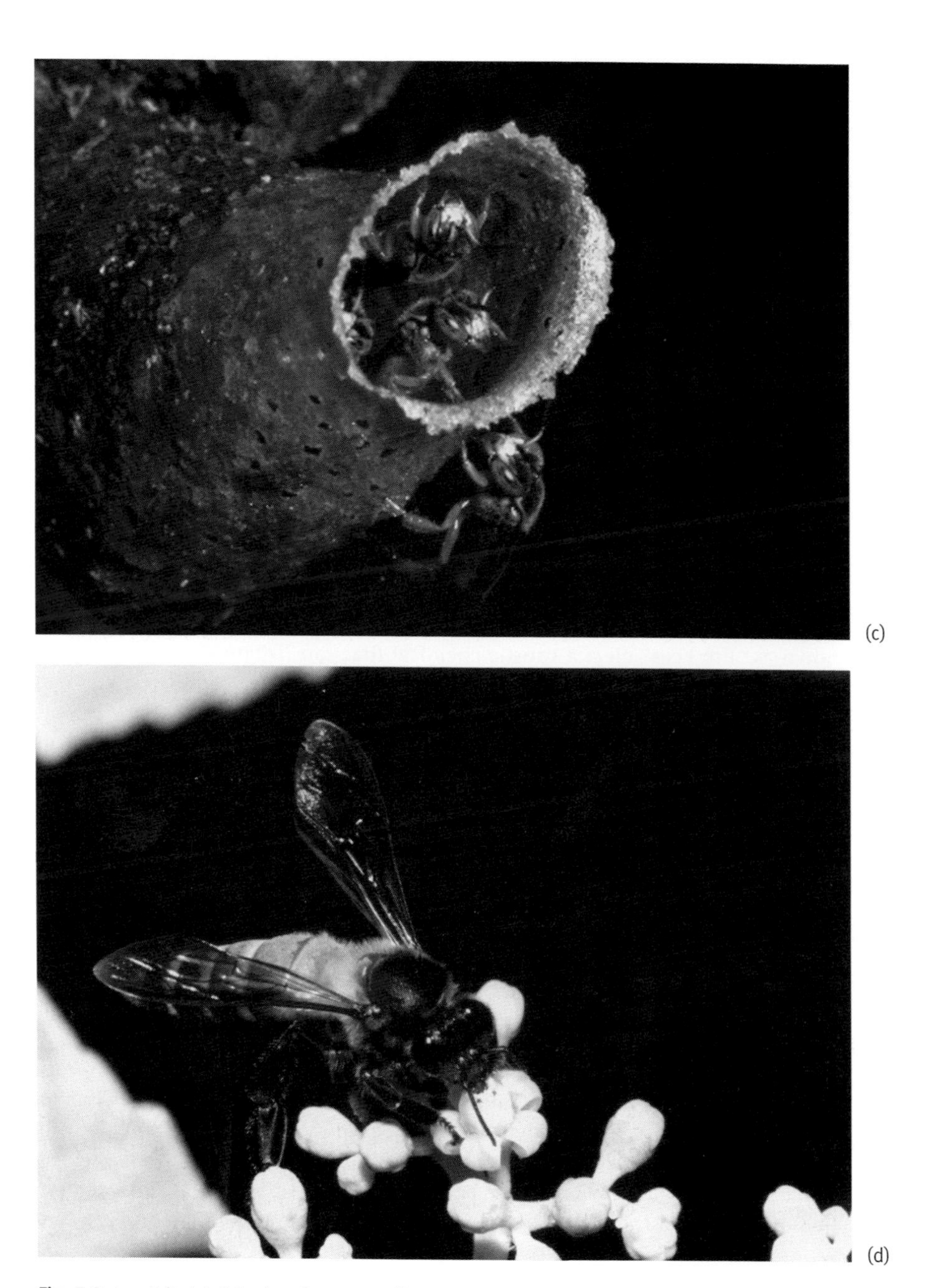


**Fig. 7.9** (*cont'd*) (c) Stingless bees guarding a nest entrance, in Costa Rica. (Courtesy of Dan Perlman.) (d) Giant honeybee (*Apis dorsata*) visiting flowers in the Western Ghats of India. (Courtesy of N.A. Aravind.)

visitors. The complex shapes of the orchid flowers manipulate the bees so that the pollinia (waxy masses of pollen) of different species are usually precisely placed on different parts of the bee's body (see Plate 7.5, between pp. 150 and 151). Although these fragrant chemicals are obviously extremely important to the male euglossines, which have special structures on their legs to collect and store them, their exact role is still not understood. In other bee species, fragrances are released by males as a means of attracting females during courtship behavior (D.W. Roubik, personal communication).

## Stingless bees

Most individual bees seen in tropical rain forests are stingless bees (tribe Meliponini) and they are probably the most important single group of flower visitors and pollinators (Fig. 7.9b,c). These bees form permanent colonies with a social system very similar to that of the better-known honeybees. The largest species, the Neotropical *Melipona fuliginosa*, is slightly bigger than the common honeybee, and will kill honeybees in nest raids. Most stingless bees are smaller than honeybees and many are less than 5 mm (0.2 inches) in length. Colonies range in size from fewer than 100 individuals in some Neotropical *Melipona* species to more than 100,000 individuals in some species of *Trigona*. Most species nest in preexisting cavities in tree trunks or the ground. Although they lack functional stings, the nests of some are defended aggressively by worker bees. They crawl over human intruders, biting and pulling at hairs. Some species known as "fire bees" eject a caustic chemical that can irritate skin.

Stingless bees are an ancient group and occur in all major rain forest regions, but they are most diverse in Central and South America, where there are an estimated 300 species, compared with around 70 in Southeast Asia, 60 in Africa, and 14 in Australia (Velthuis 1997). Many species can coexist at a single rain forest site, with up to 64 species co-occurring at sites in the Neotropics and up to 22 in Southeast Asia. Coexisting species appear to reduce competition by adopting different foraging strategies. Some species forage individually on scattered floral resources, while others can rapidly recruit hundreds or even thousands of coworkers to a large flowering tree. Recruitment ability varies widely among stingless bees, but in some species it appears to be as effective as in the honeybees discussed below. Some stingless bees are known to leave a trail of scent marks as a guide to followers, but in other species some form of three-dimensional communication is apparently used for recruitment (Nieh 2004). Some stingless bees defend clumps of flowers against other bees, while others depend on locating them rapidly. A few Neotropical species, including *Trigona necrophaga* in Central America, have abandoned the key bee innovation of feeding pollen to their larvae, and instead use animal protein collected from the carcasses of animals (Serrão et al. 1997). Some species even keep scale insects in their nest as a source of nectar and wax.

## Honeybees

The true honeybees, in the genus *Apis*, occur naturally only in the Old World tropics, with three exceptions. Two cavity-nesting species, *A. mellifera* and *A. cerana*, extend north into temperate Eurasia, and the giant Himalayan honeybee, *A. laboriosa*, is found in the temperate zone of the Himalayas. The common domesticated honeybee, *A. mellifera*, also occurs naturally throughout tropical Africa and has now been widely introduced into other regions of the world for honey production. There are at least nine *Apis* species in the Asian tropics, occurring as far east as Sulawesi and the Philippines. Prior to the introduction of the domestic honeybee, there were no honeybees in New Guinea, Australia, or the Neotropics.

In addition to a group of at least five medium-sized, cavity-nesting honeybee species, which includes the familiar *A. mellifera*, there are two much smaller species, *A. florea* and *A. andreniformis*, and two very large species (2 cm long), *A. dorsata* (see Fig. 7.9d) and *A. laboriosa* (Michener 2000). Southeast Asian rain forests typically support one species from each group. Instead of using the hollow trees favored by the medium-sized honeybees, the nests of both the small and the large species consist of a single exposed vertical comb of hexagonal wax cells. In the small species, these nests are hidden low down among dense foliage, but the huge nests of *A. dorsata* and *A. laboriosa*, sometimes 1 $m^2$ (10 sq. feet) or more in area, hang conspicuously from rock faces or the branches of emergent trees, depending for defense on their inaccessibility and the formidable stings of their occupants. These nests are often aggregated, with up to 100 in a single huge tree.

Tropical honeybees move their nest sites more readily than do their temperate counterparts. This habit is taken to the extreme in *A. dorsata*, which makes long-distance migrations of 100 km (60 miles) or more. In the seasonal areas of tropical Asia, such as Sri Lanka, these migrations are annual and take advantage of the predictable seasonal patterns of flowering. In the lowland dipterocarp forests of Sarawak, in contrast, *A. dorsata* colonies appear only during the mass flowering episodes that occur at irregular, multiyear intervals (Itioka et al. 2001). Only 11% of the 305 plant species studied at Lambir Hills, Sarawak, are pollinated by *Apis* bees, but these species, which are mostly large canopy trees, include a high proportion of the total available floral resources during mass-flowering episodes. *A. dorsata* bees become active before dawn and may continue flying after sunset. Several of the tree species they pollinate have open flowers at these times, when no other social bees are foraging, suggesting that the relationship between mass flowering and honeybee migration in Southeast Asian rain forests is an ancient one.

The success of *A. dorsata* and other honeybees in the rain forest depends to a large extent on the efficiency with which individual worker scouts recruit large numbers of their sisters to newly discovered sources of food. The "waggle dances" with which *A. mellifera* communicates the direction of a food source relative to the sun, as well as its distance and quality, are well known (Dyer 2002). Similar dances have been observed in several other *Apis* species and are assumed to occur in all. In *A. florea*, the dances are carried out on the expanded horizontal base of the comb, while *A. dorsata* dances on the vertical comb itself. The ability to exploit large flowering trees by rapid recruitment, coupled with large colony sizes and wide flight ranges, makes *Apis* bees very important components of Old World rain forest ecosystems.

Honeybees have been introduced into many other regions of the world, altering the organization of rain forest insect communities and pollination relationships. European strains of *A. mellifera* have been deliberately transported all over the world in the last few centuries, but are poorly adapted to the lowland tropics and have invaded rain forests only on islands, such as Hawaii and Mauritius, and in Australia, where they are now the most common visitors to some mass-flowering rain forest trees (Williams & Adam 1997). It was not until queens of an African strain of *A. mellifera* were imported to Brazil in 1956 to increase honey production and subsequently escaped that honeybees became established in Neotropical rain forests. These "Africanized" bees have spread throughout South and Central America, and are estimated to visit the flowers of at least a quarter of the plant species at Barro Colorado Island in Panama (Roubik et al. 2003). The bees are abundant in the mosaic of forest fragments and agricultural habitats created by human activities, but are much less common in undisturbed tracts of primary forest (Dick et al. 2003). They differ from European strains of honeybees in several important ways, including their ability to nest in a greater variety of smaller cavities, their willingness to emigrate when the local resources are exhausted, and their much more aggressive defense of nest sites (Winston 1993). This latter characteristic, which presumably evolved in response to nest predation by vertebrates in Africa (including, for several millennia, humans), has lead to a number of human deaths and the popular name "killer bees".

*Apis mellifera* is not the only honeybee species that has been deliberately introduced outside its natural range. The closely related and widely domesticated Asian hive bee, *A. cerana,* is now established on many islands east of Sulawesi and has recently colonized New Guinea. Occasional colonies of this and other honeybee species have been found aboard ships in Australian ports, and there is a clear risk of further introductions both there and in the Neotropics. Introduced honeybees have been blamed for a dramatic decline in the native solitary bees of Hawaii. Although there is no evidence yet for a similar decline in bee diversity in Neotropical or Australian rain forests (Paini 2004), there is no doubt that introduced honeybees do compete with native bees for nectar and pollen, particularly in plants with flowers that are cup-like or flat in shape, or which only have a short floral tube (Roubik & Wolda 2001). Only time will tell if these introductions are the ecological disaster that many ecologists fear them to be.

## Conclusions and future research directions

This chapter has touched only briefly on the most well-known groups of insects: the butterflies and social insects. Other important groups such as beetles, plant-sucking bugs, flies, and moths must await further investigation before cross-continental comparisons can be made. Although the butterfly communities of the various regions each include unique and visually distinctive groups of species, these differences between the regions do not appear to have any major ecological importance. In contrast, there are at least three major differences between the social insects of the different rain forest regions that have considerable ecological significance. First, leaf-cutter ants in the Neotropics selectively harvest large amounts of biomass and cultivate fungi in underground chambers. Second, the aggressive foraging of army ants in the Neotropics and Africa removes animal life from local areas of the forest floor and creates foraging opportunities for many other species. And third, honeybee species in the Asian and African tropics forage as a group and dominate many flowering plants, excluding other insects from those resources. The major differences between the termite communities of different regions are probably also of considerable ecological significance, but more studies are needed to identify which differences are important and why.

The recent introduction of Africanized honeybees into the American tropics is of grave conservation concern, but it is also a chance to examine the interaction between species of two divergent biogeographical regions (Wilms & Wiechers 1997; Roubik 2000; Roubik & Wolda 2001). Although no one could reasonably suggest carrying out such an experiment for scientific reasons, now that honeybees are in both the New World and the Australian region, we can study their impact on other insects and native plant species (Paini 2004). Massive changes in the insect fauna of the New World have been predicted as a result of the introduction of honeybees. At the least, numerous species of native bees, flies, beetles, and butterflies would be expected to decline in abundance as their nectar and pollen sources are taken over by honeybees. From existing reports, it appears that honeybee densities are greatest in areas disturbed by selective logging, farming, and other human disturbances, and that honeybees are present but less abundant in primary rain forest. If this is always the case, we may not be able to distinguish the specific impact of honeybees on active insect communities from the more general impact of direct human activities. It has also been shown that honeybees may have a positive impact on the survival of forest trees in highly disturbed landscapes by providing pollination services to isolated trees that the native pollinators do not visit (Dick et al. 2003).

Honeybee densities are currently being manipulated in many Asian and African forests through the collection of honey from bee colonies. When honey is recovered carefully from the colony, the colony can recover. However, if the harvesting is done carelessly and the queen is killed, the colony may not recover. In areas where honeybee populations have decreased due to the overharvesting of honey, it would be valuable to determine if fruit set has declined in certain plant species through lack of pollination or, alternatively, perhaps other insect species have increased in abundance and are carrying out pollination activities.

The ecological significance of leaf-cutter ants could potentially be investigated in a comparative manner by clipping the leaves of African or Asian trees in a manner comparable to the damage inflicted on a Neotropical forest by these ants. While such an experiment is almost certainly impractical, it is at least worth contemplating. An alternative and more practical approach would be to experimentally eliminate leaf-cutter ants from an area and monitor the impact on the surrounding forest. It would be expected that the trees most heavily harvested by the ants would respond positively, while the numerous subterranean animals that live in association with the leaf-cutter ants would suffer. This manipulation of leaf-cutter ant colonies is already being carried out by farmers who poison colonies that are harvesting their plantation trees. The farmers know from experience that their orange trees do not do well with ants clipping off their leaves and flowers.

One of the most challenging groups to investigate, and one of great ecological significance, is the termites. With their subterranean foraging behavior and nests, they are not readily suited to observation and manipulation. One group that can be observed and investigated is the processional termites of Southeast Asia. These termites can be manipulated by damaging trails and removing entire colonies. Using such manipulations, the role of at least these highly specialized termites in forest processes could potentially be investigated.

Insects remain one of the great, unknown forces in tropical rain forests. The uneven distribution of honeybees, army ants, termite families, and leaf-cutter ants highlight the differences among rain forests. But other groups of invertebrates, such as soil spiders, seed-eating beetles, and canopy thrips, might be found to have an enormous role in ecological processes once they are more completely investigated and compared across continents.

## Further reading

Abe T. & Higashi M. (2001) Isoptera. In *Encyclopedia of Biodiversity*, Vol. 3 (ed, Levin S.A.). Academic Press, San Diego, CA, pp. 581–611.

Corlett R.T. (2004) Flower visitors and pollination in the Oriental (Indomalayan) region. *Biological Reviews* **79**, 497–532.

DeVries P.J. (1997) *The Butterflies of Costa Rica and their Natural History, II Riodinidae*. Princeton University Press, Princeton, NJ.

DeVries P.J. (2001) Butterflies. In *Encyclopedia of Biodiversity*, Vol. 1 (ed, Levin S.A.). Academic Press, San Diego, CA, pp. 559–573.

Hölldobler B. & Wilson E.O. (1990) *The Ants*. Springer-Verlag, Berlin.

Hölldobler B. & Wilson E.O. (1994) *Journey to the Ants: a Story of Scientific Exploration*. Harvard University Press, Cambridge, MA.

Michener C.D. (2000) *The Bees of the World*. Johns Hopkins University Press, Baltimore, MD.

Roubik D.W. (1989) *Ecology and Natural History of Tropical Bees*. Cambridge University Press, Cambridge, UK.

# The Future of Rain Forests

In this book, we have examined the unique features of rain forests in many regions of the world. Although we have focused on the intact rain forests, in almost every case these rain forests have been damaged by human activities—in the worst cases, they are in the process of being utterly destroyed. In this final chapter, we will consider the threats to rain forests and strategies to protect them. Whenever possible, we will try to highlight threats and possible solutions that are unique to each rain forest area.

The major threats to tropical rain forests are the clearance and fragmentation of the forest, and the overexploitation of plant and animal species in the areas that remain (Myers 1986; Laurance et al. 2001; Primack 2002). Invasions by exotic species are also a growing threat, particularly on oceanic islands, and the impact of global climate change is likely to increase in significance over the next few decades. Many threatened rain forest species face two or more of these threats, speeding their way to destruction and hindering efforts to protect them. Often, these threats develop so rapidly and on such a large scale that efforts to save species and representative examples of rain forests are difficult. The threats also build upon one another in a spiral of destruction. For example, logging companies build roads to take logs out of the forest, and then timber workers and local people use this road network to intensively hunt animals for the camp mess hall. Later, landless people will move in on the logging roads and establish farms and ranches. These people often cut down the remaining trees and burn them to create a nutrient-rich ash to fertilize their crops. When such agriculture is practiced near selectively logged forests, the fire can easily spread from the fields into the forest—particularly during dry periods—creating thousands of fires over a large area, devastating forests, and producing a persistent smoky haze over an entire region.

## Different forests, different threats

Just as rain forests differ both within and between the major rain forest regions, so does the amount of rain forest remaining and the nature of the threats to its continued survival. In many parts of the world, tropical rain forests have already been largely destroyed by human activity, with lush forests converted into barren grasslands and scrub. Yet despite the extent of the destruction that has occurred already, there are still vast areas of rain forest in several parts of the tropics. To protect these remaining rain forests, it is important to understand not only the biological differences between each forest as detailed in the preceding chapters, but to also account for the different economic and social factors threatening forests in each area of the world (Newman 1990; Rudel & Roper 1996; Bawa & Dayanandan 1998; WRI 2000).

In the discussion that follows, we have made use of quantitative data from a variety of sources. These figures are the best available, but in many cases they give a very misleading appearance of accuracy. Few, if any, tropical countries have accurate data on, for example, present and past forest cover, current deforestation rates, or numbers of cattle. For some countries, corruption, political instability, or simply the inaccessibility of rural areas mean that all available statistics must be viewed with suspicion, including those on human populations, birth rates, and exports of forest products. In a rapidly changing world, even the best statistics are soon out of date. These inaccurate, unreliable, or outdated figures are often combined and manipulated further to produce new numbers that may or may not reflect the reality on the ground. We have used these figures because there is no alternative, but readers should be aware of their limitations.

## Population growth and poverty in Africa and Madagascar

The rain forest countries of Africa and the island of Madagascar share two features that distinguish them from the other major rain forest regions: they have among the highest population growth rates in the world, as a result of each woman bearing an average of 5–7 children (United Nations Population Division 2001), and they have among the lowest per capita incomes, with most people living on less than US$1 a day (Table 8.1) (UNDP 2002). Threats to rain forests in these countries stem largely from the poverty of this rising population dependent upon subsistence agriculture and herding. Among individual countries within Africa, higher deforestation rates are associated with a higher percentage of crop area, a high density of cattle, and a high rural population density (Bawa & Dayanandan 1997, 1998).

The rain forests of West Africa have already largely gone, transformed by fire, cattle grazing, agriculture, and the gathering of firewood. By some estimates, these are the most fragmented tropical forests in the world (Minnemeyer 2002). Vast areas of rain forest still remain in Central Africa, but these now face the same threats. Ironically, the very poverty that produces pressure on forest lands has also protected some areas. Large areas of rain forest remain relatively intact in the Democratic Republic of the Congo (the DRC, formerly Zaire), for example, because the road and river networks there do not provide adequate access to loggers, commercial hunters, and landless migrants (Minnemeyer 2002). In fact, in many areas, the roads are in worse condition and the human population is lower than in 1960 when the DRC became independent—which is good news for the forests, at least in the short term, because poor infrastructure still restricts the development of large-scale logging operations. As a result, the largest remaining areas of rain forest in Africa are in the DRC, the Republic of the Congo, and Gabon, along the Congo River basin. Throughout the accessible forests of the region, however, subsistence and commercial hunting of wildlife for "bushmeat" is intense, with meat from wild game an important source of

**Table 8.1** Some statistics relevant for the future of rain forests in six countries. Note the different dates to which they apply.

| | Brazil | DRC | Indonesia | PNG | Madagascar | Australia |
|---|---|---|---|---|---|---|
| Area of natural forest* (thousands km$^2$) (2000) | 5,389 | 1,351 | 951 | 305 | 114 | 1,535 |
| Percentage of original forest cover remaining (1996) | 66 | 60 | 65 | 85 | 13 | 64 |
| Percentage of frontier forest (1996) | 42 | 16 | 28 | 40 | 0 | 18 |
| Annual change in forest cover (1990–2000) | −0.4 | −0.4 | −1.5 | −0.4 | −1.0 | – |
| Annual timber production (1999–2001) (millions m$^3$) | 234 | 69 | 124 | 9 | 10 | 29 |
| Number of cattle (millions) (1996–98) | 163 | 1 | 12 | 0.1 | 10 | 27 |
| Human population (millions) (2002) | 175 | 54 | 218 | 5 | 17 | 20 |
| Fertility (children per woman) (2000) | 2.2 | 6.7 | 2.3 | 4.3 | 5.7 | 1.8 |
| Per capita GDP (US$) (2000) | 4,624 | – | 986 | 989 | 239 | 23,893 |

* Includes all forest types. In the cases of Madagascar and Australia, most of this is not tropical rain forest.
DRC, Democratic Republic of the Congo; GDP, gross domestic product; PNG, Papua New Guinea.
From WRI (2000, 2003).

protein for rural populations and sometimes preferred to that from domestic animals even in urban areas.

Until recently, the density of people was far greater in the rain forest regions of Africa than the other major rain forest areas, although this has changed with the increasing industrialization of parts of Southeast Asia and tropical America. In much of the interior Congo River area, human population densities are 2–20 people per square kilometer in the river valley areas, with widespread cultivation of irrigated crops and tree plantations (Fig. 8.1) (Times Books 1994). Densities of 0.4–2 people per square kilometer are seen even in the hill areas away from the main river channels. Thus, even though around 60% of the DRC's original forest remained in 1996, only 16% was considered to be "frontier forest"—that is, large blocks of forest in largely undisturbed condition, with relatively intact animal populations and ecosystem processes (Bryant et al. 1997). With a projected human population of 200 million by the year 2050 (United Nations Population Division 2001), the future of the DRC's rain forests does not look good. The greatest threat in the immediate future is that the expansion of logging activity—half the forest of the Congo basin has already been allocated for logging—will open up previously isolated areas to hunters and migrant farmers, catalyzing the destruction of the rain forest (Sizer & Plouvier 2000).

Madagascar faces these same problems of poverty and population growth, but its biota is in an even more desperate situation because more than 87% of its rain forests have already been destroyed (Bryant et al. 1997; WRI 2000). Human population densities of 20–100 people per square kilometer occur throughout most of the island, including areas adjacent to the rain forest areas themselves; in Madagascar, no "frontier forests" are left. Many people keep large herds of cattle and burn the grasslands annually to improve forage quality. Madagascar has 0.61 cattle per person, which is somewhat less than Brazil (0.93 cattle per person), but is much greater than the DRC (0.02), Indonesia (0.06), or Papua

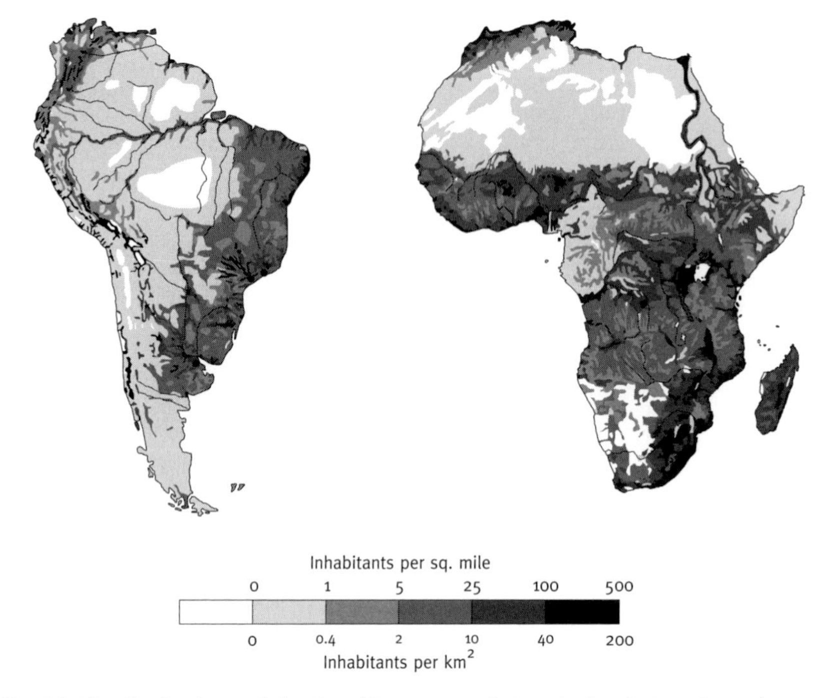


**Fig. 8.1** The distribution and density of human populations in South America and Africa, illustrating the much greater density of human populations in rain forest regions in Africa than in rain forest regions in South America. (From Times 1994.)

New Guinea (very low). One bright spot, however, is the absence—so far—of large-scale commercial logging of rain forests in Madagascar.

## Logging and cash crops in Asian rain forests

In contrast to Madagascar, Southeast Asia's rain forests are strongly affected by commercial logging. A century ago, Southeast Asia was still largely covered with forests, and as recently as 1950 Thailand and the Philippines—which today are the countries with the least forest cover in the region—still had more than half their forests (Fig. 8.2). Today, more than half of the region's forest is gone, and Southeast Asia has by far the highest rates of forest loss and degradation in the tropics (Achard et al. 2002). Moreover, the rate of clearance increased from the 1980s to the 1990s (DeFries et al. 2002) and is probably still increasing. Large areas of forest with a more or less intact biota are now confined to Borneo, Sumatra, and Sulawesi, all of which are increasingly threatened (Jepson et al. 2001). Although poverty and population growth have played an important role in the destruction of the region's rain forests, the high rates seen today are mainly due to a combination of the region's large logging industry, extracting wood for industrial use, and the conversion of forest to cash crops (such as oil palm, rubber, and cocoa) or industrial timber plantations (Bawa & Dayanandan 1997).

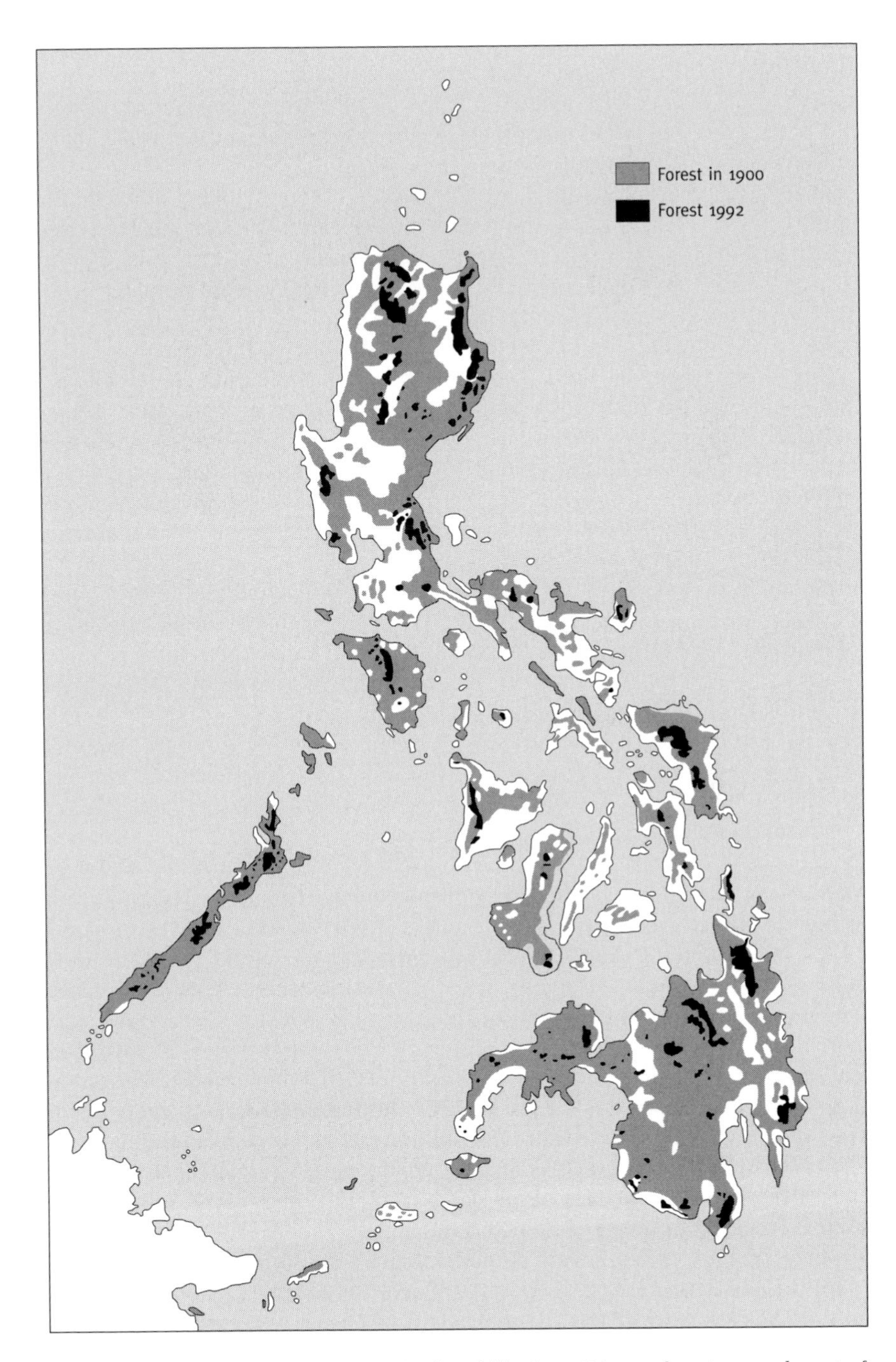


**Fig. 8.2** Loss of forest cover over time in the Philippines. Primary forest covered most of the Philippines in 1900 (gray), but only a small amount of primary forest was left by 1992 (black). (After Heaney & Regalado 1998.)

Indonesia has most of the region's surviving rain forests, but it has also replaced the Philippines as the region's new rain forest disaster area (e.g. Jepson et al. 2001; Curran et al. 2004). Indonesia is the world's largest supplier of plywood and other processed wood products, creating a huge demand for wood that is largely met by illegal logging (Barber et al. 2002). In the year 2001, Indonesia exported US$5.6 billion worth of wood products—more than any other tropical country (FAO 2002). Exports from the smaller forest area of Malaysia were worth $2.7 billion. In contrast, exports from Brazil, a country with a much greater forested area than Indonesia, were worth $3.2 billion, although Brazil also has a large domestic consumption. At the far end of the scale, the DRC only exported $11 million worth of forest products, despite its huge forest area.

Timber extraction, by itself, does not destroy a forest, but in Indonesia it commonly catalyzes destruction by providing road access to both small farmers and giant agribusinesses (Whitmore 1999). The same industrial conglomerates control much of the logging, wood processing, and plantation industries in Indonesia, so the link between logging and deforestation is often direct, with the former just the first stage in the conversion of rain forest into a plantation monoculture. Logging is followed by burning to clear the remainder of the forest, and if the fires burn out of control beyond the boundaries of the concession area, it is no loss to the plantation owner. Two-thirds of the plantations on former forest land in Indonesia consist of oil palm, which covered an estimated 3 million ha (7.5 million acres) in the year 2000 (Glastra et al. 2002). Global demand for palm oil is expected to double in the next 20 years, with half of the new plantation land required to expand production likely to be in Indonesia. Most of this will come from the conversion of lowland rain forests in Sumatra, Kalimantan (Indonesian Borneo), and Papua, the Indonesian half of New Guinea.

An additional factor contributing to Indonesia's rapid loss of forest has been the movement of people into remote rain forest regions. In contrast to Africa and Madagascar, where such population movements have been largely spontaneous, in Indonesia they were the deliberate result of development policies supported by the international finance community, including the World Bank and other international funding organizations. These organizations funded programs such as Indonesia's transmigration program, which, until it ended in 1999, transported large numbers of people from the densely settled central islands of Java, Bali, and Madura to remote rain forest locations in Borneo, Sumatra, and New Guinea (Primack 2002). Depending on the services and training available, these people often practice plantation agriculture or shifting cultivation. Linked to this movement of people, there was a 20% increase in Indonesia's cropland between 1984 and 1994, among the highest of any large tropical country (WRI 2000). Large-scale movements of people within the Indonesian Archipelago continue today, but they are now driven largely by poverty and political instability.

The archipelago nature of Southeast Asia, which facilitated isolation and speciation in the past, now allows the movement of people, the extraction of forest resources, and agricultural development. These movements of people and goods are leading inevitably to forest destruction. Indonesia in particular is a country with massive logging and plantation industries that simultaneously possesses huge political, economic, and social problems, with a repeating series of internal conflicts. The rate of forest loss is accelerating, with recent estimates putting it at 20,000 km$^2$ per year—about 2% of the remaining forest (Barber et al. 2002). Indonesia's curse is that it has just enough political stability to destroy its forests.

**Roads and development in the Brazilian Amazon**

In the Amazon region, the situation is somewhat different from that in either Africa or Southeast Asia. For one thing, the human population densities were, until recently, very low: less than 1 person per square kilometer over most of the basin, except along the main rivers where densities were in the range of 2–10 people per square kilometer (Times Books 1994). Population pressure, therefore, was not the principal source of threats to Amazonian habitats. Yet Brazil's rain forest now faces many of the same threats as in Indonesia: logging, clearance for agriculture, and an influx of settlers, but with the added impact of a huge cattle industry looking for cheap rangeland (Laurance et al. 2001). The increasing deforestation in the Amazon region during the 1980s and 1990s was greatly facilitated by the development of a new road system, financed by multibillion dollar loans from the World Bank and other international lenders (see Plate 8.1, between pp. 150 and 151). These new roads, coupled with government policies intended to encourage immigration and economic development in the Amazon region, have resulted in a rapid growth in the population of the region, from 2.5 million in 1960 to 20 million in 2000 (Laurance et al. 2002). Deforestation, logging, and forest fires are all concentrated along the new roads.

The area of rain forest lost each year in Brazil is enormous, but because the Amazon forest is so large and transportation is still difficult over most of the region, the rate of deforestation in percentage terms is still comparatively low, at about 0.5% per year. This will certainly increase, however, if the Brazilian government goes ahead with its plans for a major expansion of the present road network, greatly increasing the accessibility of areas that are now remote from transportation (Fig. 8.3) (Laurance et al. 2002). Around 7500 km (4600 miles) of highways will be paved and the network of unpaved roads will also be expanded, opening up new areas to deforestation, logging, and fire. Much of this infrastructure is justified by the production of soybeans for export, a crop grown

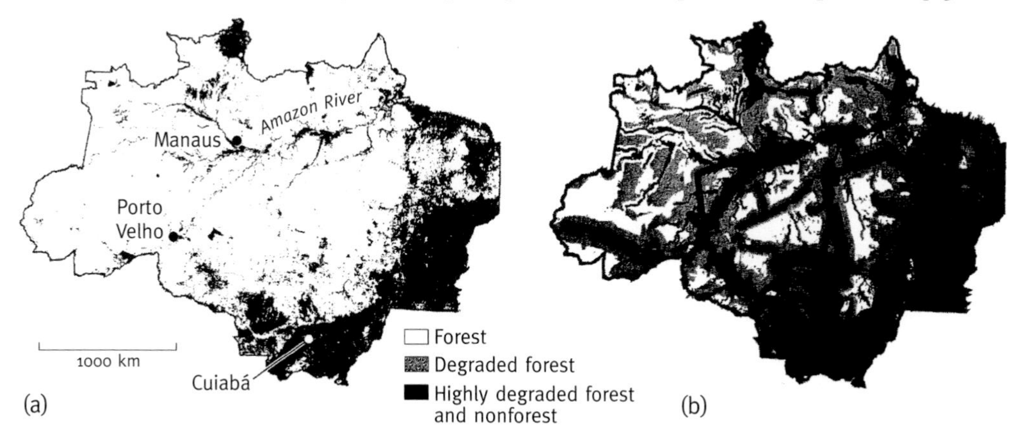


**Fig. 8.3** (a) A current map of the Brazilian Amazon showing forest (white) and deforested, degraded, and savanna areas (black). Note that deforestation tends to be found along rivers and roads and in eastern and southern populated areas. (b) When Brazil completes its proposed network of new roads by the year 2020, the amount of pristine forest cover (white) far from roads is predicted to be dramatically reduced, with much of the land lightly and moderately degraded (gray) and deforested, highly degraded, or converted to savanna (black). If strong conservation measures are implemented by the government, the levels of degradation and deforestation could be somewhat reduced. (From Laurance et al. 2001.)

by wealthy agribusinesses that employ very few people (Fearnside 2002). Although environmental protection is improving in Amazonia, the government does not currently have the capacity to control illegal deforestation, logging, and mining across this vast area. Regulations that exist on paper have often been impossible to enforce on the forest frontier.

### New Guinea: paradise lost?

The western half of the island of New Guinea is the Indonesian province of Papua (previously Irian Jaya), whose future prospects are linked to those of that troubled nation. As rain forest resources are depleted in western Indonesia, attention is being increasingly focused on the vast rain forests of Papua. Logging, exploitation of wildlife, fires, and clearance for cash crops, such as oil palm, are all increasing problems in Papua, although huge areas of intact rain forest remain. Recent newspaper reports suggest that Papuan rain forests are now the major source of Indonesia's illegal log exports, with the military and military-linked companies heavily involved. As in neighboring Papua New Guinea, the political naivety of the traditional landowners is exploited to obtain logging concessions at little or no cost. The ongoing construction of roads in previously inaccessible areas will inevitably accelerate this illegal exploitation.

Until recently, the rain forests of Papua New Guinea, which occupies the eastern half of the island, appeared to have a much brighter future, with the relatively low human population density in the lowlands and a unique system of clan control of forest lands creating barriers to large-scale logging or conversion to cash crops. However, over the last two decades, Papua New Guinea has experienced a massive boom in the worst type of commercial logging, causing severe, but still localized, environmental damage (Sizer & Plouvier 2000; World Bank 2002). Corruption has been a massive problem, with logging concessions awarded in return for bribes to senior officials. The traditional landowners were often paid very little for the logging rights and in many cases did not fully understand the agreements they were signing. A recent moratorium on new logging concessions may signal an improvement in the situation, but the future of Papua New Guinea's rain forests is still uncertain.

An additional problem, as in many areas of Africa, has been overharvesting of wildlife for food and other products. The traditional land tenure system makes the establishment of formal protected areas very difficult in Papua New Guinea, so alternative approaches to conservation will have to be developed. Several huge protected areas have been established in neighboring Papua, where the provincial government has stronger powers over land, but these appear to have been relatively ineffective in limiting exploitation. High birth rates (averaging 4.3 children per woman) pose a problem for the future, with the human population of Papua New Guinea expected to grow from its present 4.8 million to 11 million by 2050 (WRI 2000).

### Australia: paradise regained?

Australia is the only developed country in the world with a significant area of tropical rain forest, so it is not surprising that it now has the world's best-protected rain forests. After a long period of exploitation, followed by an epic struggle between competing interest groups in the 1980s, Australia has now protected most of its remaining tropical rain forests, including all of the larger blocks, in the 8940 $km^2$ Wet Tropics of Queensland World Heritage Area (McDonald & Lane 2000; Anon 2002). Despite the absence of such spectacular wildlife as primates, elephants, and hornbills, the protected rain forests have been a hugely successful tourist attraction. Packaged with the Great Barrier Reef in "rain forest and reef" tours, the rain forest attracts large numbers of Australian and international tourists. The region still has many real environmental problems, but most of these seem relatively minor in comparison with the massive threats to rain forests elsewhere. One exception is the threat from climate change, since the concentration of the Australian endemic rain forest vertebrates in upland areas makes them especially vulnerable to global warming (Hilbert et al. 2001; Meynecke 2004).

## The major threats

In the preceding section, we described in brief some of the major threats that face each rain forest region. In the following section, we will elaborate on each of these threats and show how they affect rain forest biodiversity.

### Clearance for agriculture

Few if any rain forest plant and animal species can survive a complete removal of the forest. Newly created open habitats are invaded by nonforest species, often from outside the region, and their richness of species is much lower than that of the rain forest they replace (Corlett 2000). On a global scale, the majority of rain forest destruction still results from small-scale cultivation of crops by poor farmers. Some of these farmers have lived in or near the rain forest for generations, but many have been forced to remote, undeveloped forest lands to practice shifting agriculture out of desperation and poverty (see below). Some of this land is converted to permanent farm plantations and pastures (Fig. 8.4), but much of the area returns to scrub or secondary forest following a brief period of cultivation, due to weed problems, low soil fertility, and a lack of secure land tenure. This secondary woody vegetation can support some of the more tolerant rain forest species, with the diversity gradually increasing if the area is permanently abandoned by the farmers (Dunn 2004). More often, the more fertile areas are cleared again for repeated cultivation, while the rest of the landscape is degraded by fire, hunting, and firewood collection (Fig. 8.5). More than 2 billion people cook their food with firewood, so this impact can be of major significance. In the DRC, for instance, it has been estimated that the volume of fuelwood harvested each year is 200 times greater than the volume of the commercial timber harvest (Leslie et al. 2002).

In an increasing proportion of the tropics, clearance by peasant farmers to meet subsistence needs is now dwarfed by clearance by large landowners and commercial interests, to create pasture for cattle ranching or to plant cash crops, such as oil palm and soybeans. Cattle ranching has been particularly important in tropical America, while plantations of tree crops are the major cause of

(a)

(b)

**Fig. 8.4** Cattle are an important factor in rain forest destruction. (a) Cattle in a cut-over forest in Bolivia. (Courtesy of Margie Mayfield.) (b) Cattle being herded across a stream in Uganda. (Courtesy of Tim Laman.)

deforestation in much of Southeast Asia and are increasing elsewhere (Bawa & Dayanandan 1997, 1998). In Brazil, soybean cultivation by large agribusinesses is expanding in the Amazon region, adding a new threat to the forest (Fearnside 2001, 2002). Soybeans are particularly damaging as a crop because their high value is used to justify massive transportation infrastructure and because

(a)

(b)

**Fig. 8.5** Agriculture is a major contributor to rain forest destruction. (a) In this case, indigenous people in the Brazilian Amazon have cut down trees and burned them in preparation for planting their crops. Here, a local chief stands in front of land that has been cleared. (Courtesy of Milla Jung). (b) In other cases, logging occurs first, then the brush is burned by local people to ready the ground for planting. (Courtesy of Mark Brenner.)

they displace small farmers. Commercial agriculture is generally worse for biodiversity than clearance by peasant farmers, because large areas are maintained under a uniform crop cover. Tree crops are somewhat better than pasture or soybeans, because their structure and microclimate is more similar to that of forest, but even then, only a few tolerant rain forest species can survive in this artificial habitat.

(c)

(d)

**Fig. 8.5** *(cont'd)* (c) Rain forests are also cleared for plantation agriculture, as shown by the this very young oil palm plantation in Borneo. (Courtesy of Richard Primack). (d) A farmer harvesting fruit in a cocoa plantation in Ghana. (Courtesy of Sarah Frazee, Conservation International.)

## Fragmentation

In addition to outright clearance, tropical forests are being threatened by fragmentation. Habitat fragmentation is the process whereby a large, continuous

**Fig. 8.6** After logging and the creation of open land for cash crops, such as tea, the landscape is a mixture of forest fragments and fields, as shown in this photo from the East Usambara Mountains of Tanzania. (Courtesy of Norbert Cordeiro.)

area of habitat is both reduced in area and divided into two or more fragments. When rain forest is cleared for cultivation, a patchwork of forest fragments is often left behind (Fig. 8.6). In many cases, these are sites that escape immediate clearance because they are too steep, swampy, or infertile for cultivation (Corlett 2000). Patterns of land ownership or legal restrictions may also prevent complete clearance, and forest patches may also be deliberately retained as sources of timber and firewood, as shelters for cattle, and as protection for the local water supply. These rain forest fragments are often isolated from one another by a highly modified landscape. Fragmentation almost always occurs during a severe reduction in habitat area, but it can also occur when the total area is reduced to only a minor degree if the original habitat is divided by roads, railroads, canals, power lines, fences, oil pipelines, fire lanes, or other barriers to the free movement of species.

Extensive experimental studies have been carried out, particularly in the Amazon rain forest, to determine the effects of rain forest fragmentation (Laurance & Bierregaard 1997; Laurance & Williamson 2001). These studies have shown that when a habitat is fragmented, the potential for dispersal and colonization of plants and animals is reduced. Many bird, mammal, and insect species of the forest interior will not cross even very short distances of open area. Even a narrow dirt road can be a barrier to some bird species (Laurance et al. 2004). If they do venture into the open, they may find predators, such as hawks, owls, flycatchers, and cats, waiting on the forest edge to catch and eat them. When animal movements are reduced by habitat fragmentation, plants with fleshy fruits that depend on these animals for dispersal are also affected. As species go extinct within individual fragments through natural successional and population processes, new species will be unable to arrive due to barriers to colonization, and the number of species present in the habitat fragment will decline over time.

Forest fragmentation also changes the microenvironment at the fragment edge, and this can have a significant impact on species composition. Some of the more important edge effects include changes in light, temperature, wind, humidity, and incidence of fire (Schelhas & Greenberg 1996; Laurance & Bierregaard 1997). When the forest canopy is removed, the ground is exposed to direct sunlight. It becomes much hotter during the day and, without the canopy to reduce heat and moisture loss, much colder at night and generally less humid. These effects will be strongest at the edge of the habitat fragment and decrease toward the interior of the fragment (Fig. 8.7). In studies of Amazonian forest fragments, microclimate effects were evident up to 100 m (300 feet) into the forest interior (Laurance & Bierregaard 1997). Since species of plants and animals are often precisely adapted to temperature, humidity, and light levels, changes in these factors will eliminate many species from forest fragments. Habitat fragments are also susceptible to fires spreading from nearby cultivated fields into the forest interior, in the process eliminating sensitive species.

Studies of fragments have also revealed the crucial importance of the disturbed matrix around the fragments in determining how many and which species survive (Laurance & Bierregaard 1997). Species that are able to live in and move across the matrix will increase in abundance in small, isolated fragments, while other species decline. The more similar the matrix habitats are to the original rain forest, the weaker are the edge effects in forest fragments, and the more species can survive. Pasture or low-growing crops are the worst matrix types, while tree crops or secondary forest regrowth can act as a buffer and protect the fragment from the external microclimate. These observations suggest ways in which the conservation value of fragments could be increased by planting buffer zones or encouraging regrowth in the surrounding area.

The overall effect of all these fragment processes is that all forest fragments tend to lose species, with both the rate of loss and the precise species that are lost depending on fragment size, fragment isolation from other forest areas, and the nature of the matrix. Species loss from rain forest fragments of the size most commonly left in agricultural landscapes—typically 1–500 ha (2–1200 acres) in area—is rapid, with the most sensitive species disappearing within a few years. In real—rather than experimental—fragments, the situation is made worse by hunting and firewood collection, as well as the continued erosion of fragments by fires and new clearance activity (Corlett 2000). A surprising diversity of the more tolerant rain forest species can persist, however, even in tiny fragments, particularly if they are protected from further exploitation and damage.

## Logging

Huge areas of rain forest are damaged every year through commercial logging (Fig. 8.8). Most of this is highly selective, with only one or a few high-value tree species harvested per hectare. Logging intensities are higher, however, in Southeast Asian dipterocarp forests, because hundreds of similar dipterocarp species can be grouped into a small number of categories for marketing (Whitmore 1998) (see Chapter 2). In these forests, dipterocarp trees form the majority of large trees, or even all of the large trees. Because the logs are straight and of

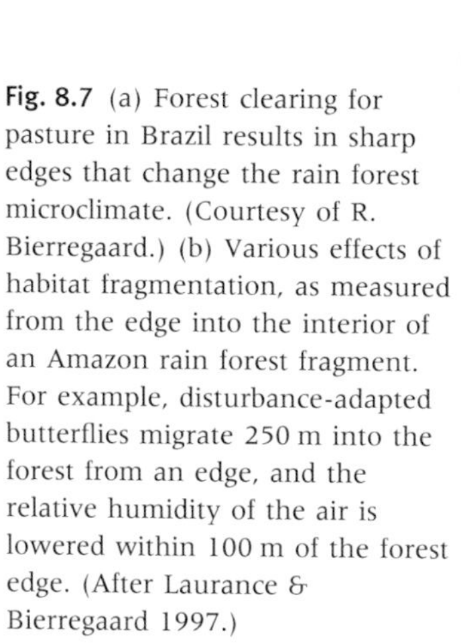
**Fig. 8.7** (a) Forest clearing for pasture in Brazil results in sharp edges that change the rain forest microclimate. (Courtesy of R. Bierregaard.) (b) Various effects of habitat fragmentation, as measured from the edge into the interior of an Amazon rain forest fragment. For example, disturbance-adapted butterflies migrate 250 m into the forest from an edge, and the relative humidity of the air is lowered within 100 m of the forest edge. (After Laurance & Bierregaard 1997.)

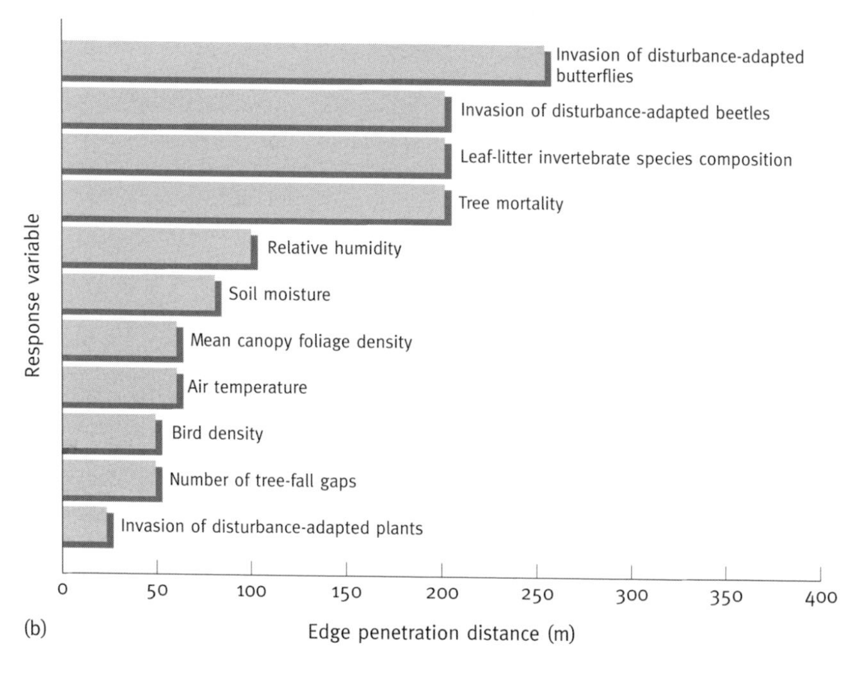


considerable length, and the wood is often light in weight, dipterocarp logs are in great demand by the timber industry to be used for inexpensive construction and plywood manufacture. A few comparisons can be used to contrast the major forest regions of the world for logging intensity. In Southeast Asia, there are 195 major timber species, many of which are dipterocarps, in contrast with only 13 species for all of tropical South America and only seven species for Africa

(a)

(b)

(c)

**Fig. 8.8** Logging contributes to rain forest destruction in every area of the world. (a) Logging truck overturned on a highway in Panama, perhaps as a metaphor for the dangerous and problematic timber industry. (Courtesy of Mark Brenner.) (b) Logging in a peat swamp forest, in Borneo. (Courtesy of Richard Primack.) (c) Logging yard showing the massive scale of the logging industry. (Courtesy of Olivier Langrand, Conservation International.)

(Whitmore 1998). Due to this greater percentage of commercial trees in the forest, harvesting intensity can be much greater in Asian forests; logging in Asian dipterocarp forests removes an average of 33 $m^3$ per hectare, in contrast to an average of 13 $m^3$ per hectare in Africa and only an average of 8 $m^3$ per hectare in American forests (Whitmore 1998). Another reason for the low logging intensity in American forests is the preponderance of trees with heavy timber (more than 700 kg per cubic meter), which are much less desirable as commercial timber. Thus, although Brazil has a huge annual timber production, most is consumed locally and its forest product exports are relatively modest (Table 8.2). Of the top 10 timber-exporting countries in the world, only two are tropical countries, and both of these are from Asia: Indonesia and Malaysia. In contrast annual timber production in tropical Africa is much lower, due to low extraction rates and a smaller area being logged. India has an enormous total timber production but most of this comes from dry forests and plantations, and only very little of it is exported.

This pattern is set to change with the rapid exhaustion of accessible forests in Asia and increasing investment by transnational logging companies in Africa, the Neotropics, and Papua New Guinea (Sizer & Plouvier 2000). In addition, the growing domestic and Asian markets for rain forest timber are far less fussy about the species, size, and quality of logs than the traditional markets in Europe, North America, and Japan. This means that more species are saleable and at a smaller size, so the initial logging intensity is greater and accessible areas are often re-logged later for smaller timber and less valuable species. Most logging, however, is still "mining" of huge, ancient trees in virgin forest. Only the most valuable species can justify the large investment in machinery and infrastructure that logging in remote areas requires. This logging is "selective" only in terms of

**Table 8.2** Total area of natural forest, annual wood production, export value of forest products, and net export value (exports minus imports) for selected tropical forest countries. Roundwood production represents all trees cut in the country, whereas the export value represents only those products that are exported. Note that only a fraction of India's large timber production comes from rain forest.

| | Natural forest area (thousands km$^2$) (2000) | Annual roundwood production (millions m$^3$) (1999–2001) | Forest products export value (millions US$) (1999–2001) | Forest product exports minus imports (millions US$) (1999–2001) |
|---|---|---|---|---|
| **Asia** | | | | |
| India | 315 | 296 | 73 | –850 |
| Indonesia | 951 | 124 | 5142 | 4005 |
| Malaysia | 175 | 20 | 2944 | 1952 |
| Myanmar | 336 | 38 | 255 | 232 |
| Thailand | 98 | 26 | 752 | –347 |
| **Africa** | | | | |
| Cameroon | 238 | 11 | 456 | 435 |
| Congo | 220 | 2 | 117 | 115 |
| DRC | 1351 | 69 | 14 | 10 |
| Gabon | 218 | 3 | 341 | 337 |
| **South America** | | | | |
| Bolivia | 530 | 3 | 25 | –12 |
| Brazil | 5382 | 234 | 2739 | 1876 |
| Colombia | 495 | 12 | 93 | –250 |
| Peru | 646 | 9 | 85 | –113 |
| Venezuela | 486 | 5 | 52 | –242 |
| Madagascar | 114 | 10 | 19 | 10 |
| Papua New Guinea | 305 | 9 | 203 | 192 |

DRC, Democratic Republic of the Congo.
From WRI (2003).

the logs actually removed from the site; the process of finding, cutting, preparing, and extracting these logs is devastating and leaves the whole forest looking like a war zone. Often forests are degraded over large areas just to extract a few trees per hectare, with most of the trees knocked down and the soil exposed and compacted by heavy machinery. Despite this devastation, many studies have shown that most wildlife can survive a single cycle of selective logging (Johns 1997; Fimbel et al. 2001). Recovery is slow, but it is possible. The major impact is not the extraction of timber from the site, but the opening up of the forest by the construction of roads and other infrastructure. Improved access brings in hunters and encourages recurrent cycles of logging. Landless farmers often move into the area and cut down the remaining trees for agriculture. Recently logged forests are also far more likely to burn than those that have not been logged or were logged long ago. The combination of large, well-equipped logging camps with chainsaws, bulldozers, and trucks, landless farmers, politically connected logging companies, and a huge international market creates a powerful force for deforestation that is difficult, if not impossible, to stop.

### Fire

Under normal conditions, tropical rain forests do not burn. The presence of charcoal buried in the soil beneath many rain forest areas shows that fires are possible under extreme conditions of drought, but in the absence of other human impacts such conditions were clearly very rare. Closed-canopy rain forests are remarkably resistant to drought, remaining immune to fires even after months without rain (Uhl et al. 1988). Because fires were so rare, rain forest trees have not evolved resistance to fire and most species have very thin bark. Natural rain forest fires must therefore have been rare, but catastrophic, events.

Fires are no longer either natural or rare in the rain forest, but they are still catastrophic (Cochrane 2003b). Fire is the most convenient tool for clearing rain forest and preventing regrowth, but it is not an easy tool to control. Out-of-control fires burned an estimated 20 million ha (50 million acres) of forest in Southeast Asia and the Neotropics during the unusually dry conditions caused by the 1997–98 El Niño event. These fires primarily affect logged forests, because the canopy has been opened and logging waste increases the fuel supply (Siegert et al. 2001). Forest fragments are also vulnerable, because structural changes at the edges increase the fuel load while exposure to wind and sunlight reduces humidity. The fires in turn kill trees and leave huge amounts of flammable dead wood, greatly increasing the risk of recurrence. Rain forest recovery after a single fire is a slow process, aided by the seed sources provided by surviving trees and unburned "islands" of forest. Recurrent fires rapidly remove these seed sources and encourage the invasion of grasses and vines, further increasing flammability. The result is a landscape that becomes vulnerable to fire after weeks, rather than months, without rain. In much of the rain forest region of Southeast Asia, for example, the fires that were an exceptional occurrence a few decades ago, and then an El Niño-associated disaster in the 1980s and 1990s, are now becoming an annual event.

### Hunting

Even where rain forests still appear intact, many rain forest animals are being threatened with extinction. The current round of extinctions is the latest episode in the long-term elimination of vulnerable species from the landscape. As people arrived for the first time in each rain forest region, they drove the large native species to extinction, largely by hunting (Martin 2001; Brook & Bowman 2004). In the Americas, Madagascar, New Guinea, and the Hawaiian Islands, the arrival of people signaled the decline and eventual extinction of numerous animal species. The extinction of large mammals appears to have been much less common in Africa and Asia where animals evolved with people. But even in these areas, animals are now facing extinction due to increasing hunting by people.

In all rain forests areas, people have hunted and harvested the food and other resources they need in order to survive (Fig. 8.9). As long as human populations were small and the methods of collection unsophisticated, this hunting and harvesting could be sustainable. Animal populations could be lowered in size, particularly near villages, but some animals would survive, often at distance from human settlements. In recent decades, human populations in rain forest areas have increased, and their methods of harvesting wildlife have become dramatically more efficient. In addition, previously isolated human populations,

**Fig. 8.9** A hunter has just killed a bearded pig in Borneo. (Courtesy of Richard Primack.)

which hunted for their own use, are now connected to regional and international markets whose demand is insatiable. Hunting has been transformed from a subsistence activity into a commercial enterprise (Walsh et al. 2003). This has lead to an almost complete depletion of larger animals from many rain forests, leaving strangely "empty" forests, with animal densities reduced by 90% or more (Redford 1992; Robinson et al. 1999; Wilkie & Carpenter 1999). People only appreciate how "empty" a typical modern rain forest is when they visit a rain forest that is extremely remote or that has been vigorously protected, such as Tikal National Park in Guatemala or Barro Colorado Island in Panama. At these locations animals are at a high density, and they are readily seen because they are not afraid of people.

In traditional societies, restrictions were often imposed to prevent overexploitation of natural resources (e.g. Bottoms 2000). For example, the rights to specific harvesting territories were rigidly controlled; hunting in certain areas was banned; there were often prohibitions against taking females, juveniles, and undersized individuals; certain seasons of the year and times of the day were closed for harvesting; and certain efficient methods of harvesting were not allowed. These restrictions, which allowed certain traditional societies to exploit communal resources on a long-term, sustainable basis, are almost identical to the hunting restrictions imposed on game animals in industrialized nations.

In much of the world today, however, traditional social constraints are no longer effective and resources are exploited opportunistically. If a market exists for a product, local people will search the nearby forest to find and sell it. Whether people are poor and hungry or rich and greedy, they will use whatever methods are available to secure that product. Whole villages in Vietnam are organized to search as a group through sections of national parks, picking up every animal and plant that can be eaten or sold. In rural areas, the traditional controls that regulate the extraction of natural products have generally weakened. For example, former taboos in Africa about killing and eating chimpanzees have been abandoned as community structure has fallen apart. Where there has been substantial human migration, civil unrest, or war, controls may no longer exist. In countries beset with civil conflict, such as Indonesia, Colombia, the Democratic Republic of the Congo, and Rwanda, firearms have come into the hands of rural people. The breakdown of social controls and food distribution networks in these situations leaves the resources of the natural environment vulnerable to whoever can exploit them. The most efficient hunter can kill the most animals, sell the most meat, and make the most money for himself and his family.

Throughout the world, guns are now used instead of blowpipes, spears, or arrows for hunting in the tropical rain forests; motor vehicles are used to drive along roads into the forest; spotlights are used to hunt at night. Hunting with snares is particularly common in Africa, due to the abundance of large mammals foraging on the ground. The small game that dominated the catch of traditional subsistence hunters has been replaced by large-bodied primates, ungulates, and rodents, which find a ready market in towns and logging camps (Chapman & Peres 2001). Overall, such game meat has emerged as the second most valuable forest product after wood. In some places, game meat is actually the forest product with the highest value. The extreme overhunting of animals throughout tropical forests regions, both for local consumption and for sale in towns, has been labeled the "bushmeat crisis", and has become a major focus of a number of prominent conservation organizations (Bennett & Robinson 2000; Chapman & Peres 2001). The situation is particularly critical in tropical Africa and the Neotropics. Commercial hunting is less serious a problem in Asia only because past overharvesting has made it less profitable today. Even in Asia, however, almost all protected areas have a hunting problem.

Besides the exploitation of animals for meat, there is also a large trade in living rain forest animals for display in zoos, use by medical researchers, or for private citizens to keep as exotic pets. This trade includes primates, birds, amphibians, and a wide range of other animals. Besides a surprisingly large legal trade, billions of dollars are involved in the illegal trade of wildlife or wildlife products from tropical forest habitats. A black market links poor local people, corrupt customs officials, rogue dealers, and wealthy buyers who do not question the sources that they buy from (Webster 1997). This trade has many of the same characteristics, the same practices, and sometimes the same players, as the illegal trade in drugs and weapons. Confronting these illegal activities has become a job for international law enforcement agencies.

Throughout this book, we have emphasized the interrelationships between plants and animals in the rain forest. These interactions mean that the hunting of forest vertebrates has impacts far beyond the survival of the species that are harvested (Wright 2003). Hunters favor larger species and it is these that are responsible for dispersing the seeds of many rain forest plants. Plants with small fruits and small seeds will find a wide range of dispersal agents in even the most degraded landscapes, but larger, bigger seeded fruits are consumed by fewer dispersers, with the largest depending on the few species of large birds and mammals that are most vulnerable to hunting (Corlett 1998, 2002). And it is not only seed dispersal that is affected. Selective hunting of carnivores can lead to proliferation of their prey, while the elimination of the prey threatens any carnivores that are not hunted themselves. Vertebrates that consume leaves or seeds influence the competitive balance between plant species, so their elimination will change the plant community. These various interactions between species are complex and little understood, so the long-term impacts of hunting and the animal trade are impossible to predict.

## Invasive exotics

A further threat to degraded and fragmented rain forests is invasion by exotic species, to the detriment of native species. So far, exotic species have been a problem mainly on oceanic islands. As humans spread beyond the edges of the continental shelves, they came into contact with island faunas that had no experience of terrestrial predators and no resistance to continental diseases. The result was an expanding front of extinctions, eventually encompassing more than 1000 species of birds and hundreds of species of reptiles, as well as many small mammals and countless insects, snails, and other invertebrates (Martin & Steadman 1999). Although many of these species were eliminated by overexploitation or habitat destruction, exotic species also played a major role. The great voyaging canoes in which the Polynesian voyagers first reached the remote Hawaiian Islands brought not only people, but also their crop plants, domestic animals, and assorted stowaways. As the Polynesian settlers cleared the lowland rain forest and hunted valuable species to extinction, their pigs and chickens, along with stowaway rats, geckos, and skinks, spread into the forest that was left. European settlers, from the eighteenth century onwards, introduced a huge array of additional exotic plants and animals. No species was deliberately introduced to the rain forest, but enough invaded it to cause massive damage to the native flora and fauna (Loope et al. 2001).

There are still rain forests in the Hawaiian Islands, but they are now very different from those seen by the first Polynesian voyagers 1500 years ago. Exotic pigs root through the understorey, destroying native plants, feeding on exotic earthworms, and dispersing the seeds of exotic plants, such as the strawberry guava (*Psidium cattleianum*) and the banana poka (*Passiflora mollissima*). The black rat (*Rattus rattus*) climbs into the trees, eating seeds, snails, and insects, as well as the eggs and nestlings of the surviving native birds. Flocks of Japanese white-eyes move through the canopy, feeding on insects, fruits, and nectar. Introduced honeybees visit the flowers of native trees, while introduced predatory ants and yellowjacket wasps hunt both native and exotic invertebrates. The exotic rosy wolfsnail (*Euglandina rosea*) stalks the few native snails that have survived the onslaught of the rats. Native moths are attacked by exotic parasites introduced to control pests of lowland crops, and bird-biting mosquitoes (*Culex quinquefasciatus*) transmit avian pox and malaria to susceptible native birds.

The same story, with varying intensity, was repeated on other rain forest islands, such as Mauritius in the Indian Ocean and many islands in the Caribbean. Continental rain forests, in contrast, have proved much more resistant, for reasons that are still being debated. It has been argued that island rain forests have simply received more intense and more prolonged human impacts (Whittaker 1998). This seems unlikely to be the complete explanation, however, since even the most degraded and fragmented continental rain forests—such as those of Singapore—have generally proved resistant to exotic invasion (Teo et al. 2003). It appears that the low diversity of island biotas and the complete absence of some major groups of continental organisms make island rain forests particularly vulnerable to invasion, perhaps because some resources are incompletely utilized. However, there are now an increasing number of cases where species from one continent have invaded disturbed and fragmented rain forests on other continents. The invasion of "Africanized" Old World honeybees into Amazonian forest fragments is a striking example (see Chapter 7), as are the invasions of several Neotropical pioneer trees and shrubs (e.g. *Cecropia* spp., *Clidemia hirta,* and *Piper aduncum*) into Old World forest fragments (Peters 2001; Rogers & Hartemink 2001). After a brief initial period when control may be possible, species introductions, like extinctions, are forever, so this is a trend that can only get worse.

## The forces behind the threats

The threats above can be seen as symptoms of a smaller number of underlying problems, including population growth, poverty, inequity, and poor governance in the rain forest countries, and excessive consumption in the developed world.

### Population growth

There is one important way in which all rain forests are alike: the threats to rain forests are all caused by an ever-increasing use of the world's natural resources by an expanding human population, particularly in species-rich tropical countries (see Table 8.1). Until the last few hundred years, the rate of human population growth was relatively slow, with the birth rate only slightly exceeding the mortality rate. The greatest destruction of rain forests has occurred over the last 150 years, during which the human population has gone from 1 billion in 1850, to 2 billion in 1930, and reached 6.5 billion in 2004. World population will reach an estimated 9 billion by the year 2050, with most of this increase occurring in tropical countries (Fig. 8.10a) (Cohen 2003). Human numbers have increased because birth rates have remained high while mortality rates have declined as a result of both modern medical discoveries (specifically the control of disease) and the presence of more reliable food supplies. Population growth has slowed in the industrialized countries of the world, but it is still very high in the rain forest countries of tropical Africa and Madagascar, in Papua New Guinea, and in some parts of Latin America and Asia (United Nations Population Division 2001). Birth rates are falling in other rain forest countries, with Brazil and Thailand near the long-term replacement rate of 2.1 children per woman, and Indonesia approaching this level. Even in these countries, however, the youthful age profile of the population will ensure that it keeps growing rapidly for several more decades before it begins to stabilize. Moreover, internal migration, as in Indonesia and Brazil, can concentrate people at the forest frontier where they do most damage.

People use natural resources, such as fuelwood, wild meat, and wild plants, and they convert vast amounts of natural habitat into farmland and grazing land. In urban areas, land is used for homes and industry. Because some degree of resource use is inevitable, population growth is partially responsible for the loss of rain forests (Fig. 8.10b) (Cincotta & Engelman, 2000). All else being equal, more

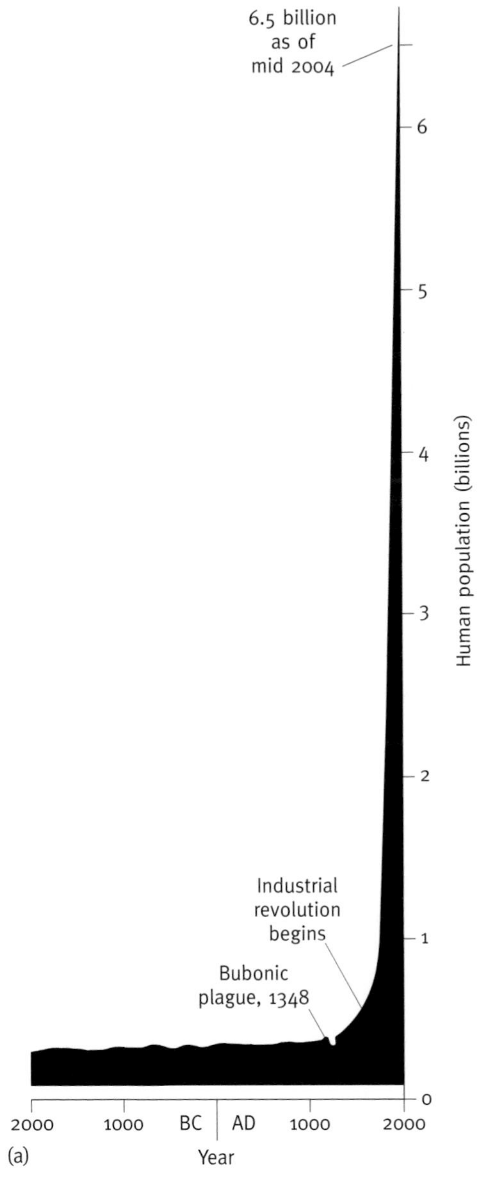


**Fig. 8.10** (a) The human population has increased spectacularly since the seventeenth century. At current growth rates, the population will double in less than 40 years. (From Primack 2002.)

people equals less forest. Some scientists have argued strongly that controlling the size of the human population is the key to protecting the natural areas of the world (Hardin 1993; Myers 1998). However, population growth is not the only cause of the loss of rain forests: an increasing use of resources by individual citizens is also responsible. People want a more varied diet, larger homes, and more possessions—and satisfying these wants has impacts on the environment.

(b)

**Fig. 8.10** *(cont'd)* (b) A driving force in rain forest destruction is the need for tropical countries to provide for the needs of a growing population, as illustrated by this village photo from Borneo. (Courtesy of Richard Primack.)

## The phenomenon of "shifted cultivators"

In many tropical countries (and temperate countries also) there is extreme inequality in the distribution of wealth, with the majority of the wealth (money, good farmland, livestock, timber resources, etc.) owned by a small percentage of the population. Brazil, for instance, has one of the least equal distributions of income in the world, with the richest 20% of the population receiving two-thirds of the total national income, while the poorest 20% receive less than 3%. As a result, in these countries poor rural people with no land or resources of their own are forced to cut down forests and hunt wildlife (Kummer & Turner 1994; Skole et al. 1994). In several countries, large landowners and business interests use armed gangs to force local farmers off their land, sometimes backed up by the government, the police, or the army. Farmers are also forced to move when an increasing population can no longer be supported by the existing farmland. This is particularly true when crop yields are declining due to decreases in soil fertility.

These farmers often have no choice but to move to remote, undeveloped areas and attempt to eke out a living through a type of temporary agriculture known as shifting cultivation. In this kind of subsistence farming, sometimes also referred to as "slash and burn" agriculture, plots of rain forest are cut down, burned away, and the cleared patches are farmed for two or three seasons (see Fig. 8.5a). The patch is abandoned as soil fertility declines and weeds become more difficult to control, and more forest must be cleared. This type of agriculture is often practiced in such areas because the farmers are unwilling or unable to spend the time and money necessary to develop more permanent forms of agriculture on land that they do not own and may not occupy for very long. Political instability, lawlessness, and war also force farmers off their land and into remote, undeveloped areas where they feel safer (Homer-Dixon et al. 1993). Rather than being called "shifting cultivators", however, these newly arrived people would be more accurately described as "shifted cultivators", in order to distinguish them from the traditional farmers who have long inhabited rain forest areas. In the past, when rural populations were low, shifting cultivation did not destroy the forest, because the forest had time to recover from farming and only the best lands were used by people who had learned, over many generations, which crops and practices were most successful locally. However, now that the land is used more intensively by a higher population, with farming even on steep slopes, it has led to the loss of forest cover and soil over large areas. These "shifted cultivators" are often ignorant of the crops and agricultural practices most appropriate to the unfamiliar landscapes they colonize, leading to crop failures and land degradation, and forcing these people to move on again.

### Poor governance in rain forest countries

Damage to biodiversity can be limited by the enforcement of laws that protect threatened habitats and species, that regulate the exploitation of nonendangered species, and that limit the adverse environmental impacts of new developments. This approach has been used most effectively to protect the remaining rain forests in Australia, but has also worked to a varying degree in many less wealthy parts of the tropics. The reasons why it is not working everywhere are many and varied, but one widespread problem, which is repeated in many rain forest countries, is the poor quality of governance at the national, provincial, and local level. Until recently, this was a problem that many involved in conservation were too embarrassed to mention, preferring to see the villains as wealthy landowners, uncaring bankers, and greedy multinational corporations, or to blame everything on pervasive poverty. Now, however, bad governance is increasingly being seen as one of the root causes of both conservation and development problems (Balmford & Whitten 2003; Smith et al. 2003; Laurance 2004). Good governance will not save the rain forest, but it may be a necessary condition for the rain forest to be saved.

The term governance covers all aspects of the way a country is governed, including government policies, laws, and regulations, and the ways in which these are implemented in practice. Good governance includes such characteristics as the rule of law, transparency, efficiency, effectiveness, and accountability. Bad governance is the lack of these characteristics. Bribery and corruption are often the most visible symptoms of bad governance, but inefficiency and ineffectiveness can be just as damaging. In relation to rain forest habitats and species, the major problem is sometimes outdated laws—usually a relict of colonial administration—but even good laws are often ineffectively enforced by weak forestry and conservation departments. Poorly paid staff can be bribed or simply ignored. In many rain forest countries, decisions on forest exploitation are made by a small group of powerful people in, or closely linked to, the government, who see rain forests as a short-term source of personal revenue, rather than a resource to be managed in the long-term national interests (Sizer & Plouvier 2000). At the local level, major landowners and powerful local families are effectively above the law and can have a major influence on conservation-related decisions, while at the forest frontier there may no law at all.

Poor governance encourages an exploitive "mining" mentality and discourages long-term investment. Why should a peasant farmer invest labor in developing a permanent farm that may be taken away at any moment? Why should a timber company invest scarce capital in sustainable forest management when its patron in the capital city may be removed in the next election or coup? Good governance encourages a longer term view of costs and benefits, but only when all participants have confidence that the bad days will not return.

Deforestation can be considered to result, in part, from a market failure in which consumers in developed countries do not pay for the environmental costs of deforestation. The costs of deforestation are borne by the producing country, especially the people living in the local area. For example, people in developed countries drink orange juice (perhaps from Brazil) and eat chocolate produced in tropical countries (perhaps West Africa), but do not pay for the loss of forest resources or the damage to the nearby streams caused by soil erosion. If the environmental costs of deforestation were included in the cost of producing tropical products, the prices would be much higher (Pearce & Brown 1994; Bawa & Dayanandan 1998). The only practical way to insure that consumers pay for the environmental costs is by taxing the producers and using the money to mitigate the impacts of production. A tax on logs extracted from the rain forest, for instance, could be used to fund an effective forestry service that would enforce laws and minimize environmental damage. However, a fair and equitable taxation system is one of first victims of poor governance, so rain forests continue to be exploited and cleared without producers and consumers paying for the damage.

## The responsibility of developed countries

The inequities that are responsible for the phenomenon of shifted cultivators in rain forest countries are played out on an international scale as well. Just as a few individuals consume the lion's share of the resources in some tropical countries, some nations in the world—predominantly industrialized nations, with the United States at the top of the list—consume a disproportionate share of the world's energy, minerals, wood products, and food (Fig. 8.11) (Myers 1997), and therefore have a disproportionate impact on the environment. Each year, the average US citizen uses 241 times more gasoline, 24 times more meat, and almost 3000 times more paper products than the average citizen of the Democratic Republic of the Congo (WRI 2000). Even compared to Brazil, a country with a fairly well-developed economy, an average US citizen uses 17 times more gasoline and seven times more paper products than an average Brazilian citizen. This excessive consumption of resources by citizens of the developed world is not sustainable in the long run, and in the short term it directly and indirectly contributes to the loss of rain forests. If this pattern is adopted by the expanding middle class in the developing world, it will require an increased level of utilization of natural resources. The affluent citizens of developed countries must confront their excessive consumption of forest resources and reevaluate their life styles at the same time as they offer aid to curb population growth and protect rain forests in the developing world.

**Fig. 8.11** In 1 year, an average US family of four with a typical American life style consumes approximately 3785 liters (1000 gallons) of oil, shown here in barrels, for fueling its two cars and heating its home. The same family burns about 2270 kg (5000 lb) of coal, shown here in a pile in the right foreground, to generate the electricity to power the lights, refrigerators, air conditioners, and other home appliances. The air and water pollution that results from this consumption of resources directly harms biological diversity, and creates an "ecological footprint" that extends far beyond their home. Regulating such consumption must be addressed in a comprehensive conservation strategy at individual, local, regional, national, and global scales. (Courtesy of Robert Schoen, Northeast Sustainable Energy Association.)

This prospect is not quite as radical an idea as one might think. During the 1980s, Costa Rica had one of the world's highest rates of deforestation as a result of the conversion of its forests into cattle ranches (Downing et al. 1992). Much of the beef produced on these ranches was sold to the United States and other developed countries to produce inexpensive hamburgers. Adverse publicity resulting from this "hamburger connection", followed by consumer boycotts, led major restaurant chains in the United States to stop buying beef from tropical ranches. Even though the story was oversimplified and deforestation continued in Costa Rica, the boycott was important in making people aware of the international connections promoting deforestation. Many other such connections exist, involving products such as wood, paper, palm oil, soybeans, chocolate, coffee, bananas, orange juice, and parrots. Consumers need to be aware of the environmental impacts of purchasing tropical products. Such purchases can accelerate rain forest destruction, or contribute to preservation in the case of a sustainable industry.

Developed countries fuel rain forest loss not only through their consumption of products, but also by financing the companies directly involved in rain forest exploitation. Financial institutions in the developed world could—but rarely do—exert a great deal of influence on the practices of logging and plantation companies. The same applies to the potential influence of international institutions, such as the World Bank and the International Monetary Fund, at the government level. Good governance and "the environment" now feature prominently on the websites and publications of these institutions, but it is too soon to judge how much these issues influence important decisions. There is an additional problem with the economic reforms that these institutions promote. The effect is often to open up the economy to foreign investment, without any strengthening of the government's capacity to reduce the resulting environmental damage (Sizer & Plouvier 2000).

We must be careful, however, not to exaggerate the influence of the developed world on rain forest exploitation. Although North America, Europe, and Japan continue to be very important markets for rain forest-related products, many rain forest countries also have huge domestic markets. The combined population of Indonesia and Brazil, for example, is larger than that of the USA. Moreover, developing countries, such as China, India, Thailand, and the Philippines, are also major importers. The same story applies to sources of investment. A large fraction of recent capital investment in the rain forest logging industry, for instance, has come from companies based in Malaysia (Sizer & Plouvier 2000). These alternative markets and alternative sources of investment have inevitably reduced the leverage that consumers, bankers, and governments in the developed world have with the producers. A Malaysian company logging forests in Gabon for export to China will not be very concerned about the environmental sensitivities of the American, Japanese, or European public. Unfortunately, the countries that are likely to see the greatest increase in demand for rain forest-associated products over the coming decades are also the ones least likely, on current evidence, to make environmental considerations a major factor in their decisions.

## Global climate change

There is one human impact that affects all forests, tropical or temperate, no matter where they are located: global climate change that stems from human activities such as use of fossil fuels. Today, atmospheric concentrations of carbon dioxide ($CO_2$) and other greenhouse gases are increasing so much as a result of human activity that scientists believe they are already affecting the Earth's climate. The term global warming is used to describe the rising temperatures resulting from human activities; and global climate change refers to the complete set of climate characters that will change, including patterns of temperature, wind, and rainfall. During the past 100 years, global levels of $CO_2$, methane, and other trace gases have been steadily increasing, primarily as a result of burning fossil fuels—coal, oil, and natural gas (Gates 1993; IPCC 2001). Carbon dioxide concentration in the atmosphere has increased from 290 parts per million (ppm) to 370 ppm over the last 100 years, and it is projected to double somewhere in the latter half of the twenty-first century. Clearing forests to create farmland, logging,

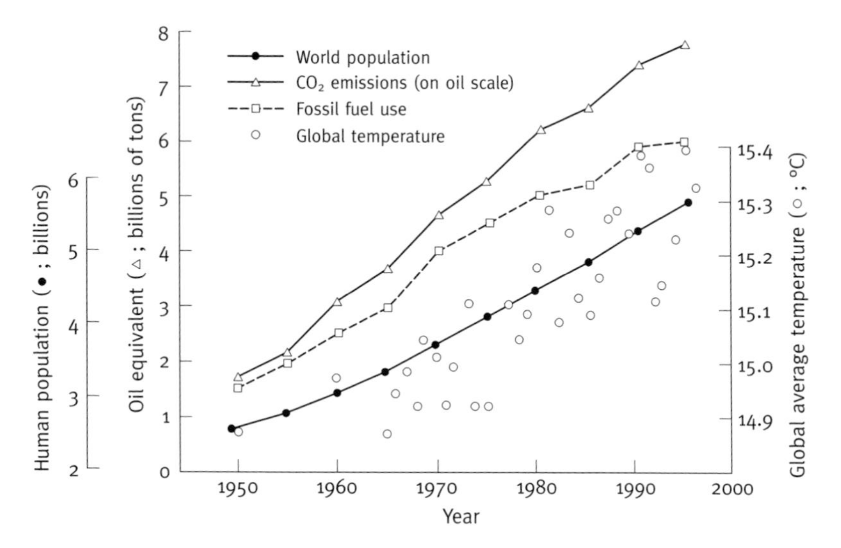


**Fig. 8.12** Over the last 50 years, human populations have increased dramatically. Increased fossil fuel use and forest destruction have led to greater release of carbon dioxide ($CO_2$) into the atmosphere. Most scientists believe the observed increase in global temperature is caused by increased atmospheric concentrations of $CO_2$ and other greenhouse gases. (From Houghton et al. 1996.)

and burning firewood for heating and cooking also contribute to rising concentrations of $CO_2$. The contribution of tropical deforestation and forest degradation to this rise has been estimated as between 15 and 40%, with the most recent estimates in the lower part of this range (DeFries et al. 2002; Houghton 2003). The remainder largely comes from the burning of fossil fuels. The contribution from tropical forest clearance varies from year to year, with the El Niño year of 1997 marked by a large jump in atmospheric $CO_2$ concentrations as a result of the burning of forest and peat in Indonesia (Page et al. 2002). Deforestation also releases other, more potent, greenhouse gases, particularly methane and nitrous oxide, and it has been estimated that these could add 6–25% to the impact from $CO_2$ emissions alone (Fearnside & Laurance 2003).

Most scientists believe that these increased levels of greenhouse gases have affected the world's climate already, and that these effects will increase in the future (Fig. 8.12). An extensive review of the evidence supports the conclusion that world climate has warmed by around 0.6°C during the last century (IPCC 2001). There is a growing consensus among meteorologists that the world climate will increase in temperature by an additional 1.4–5.8°C over the next 100 years as a result of increased levels of carbon dioxide and other gases. The increase could be even greater if $CO_2$ levels rise faster than predicted; it could be slightly less if all countries agreed to reduce their emissions of greenhouse gases. The increase in temperature will be greatest at high latitudes and over large continents (Fig. 8.13) (IPCC 2001). Temperatures are predicted to increase by 2–4°C over most continental areas currently occupied by rain forests, with a somewhat greater increase over the northern Amazon basin. Many scientists also predict an increase in extreme weather events associated with this warming, such as hurricanes, flooding, and regional drought (Karl et al. 1997).

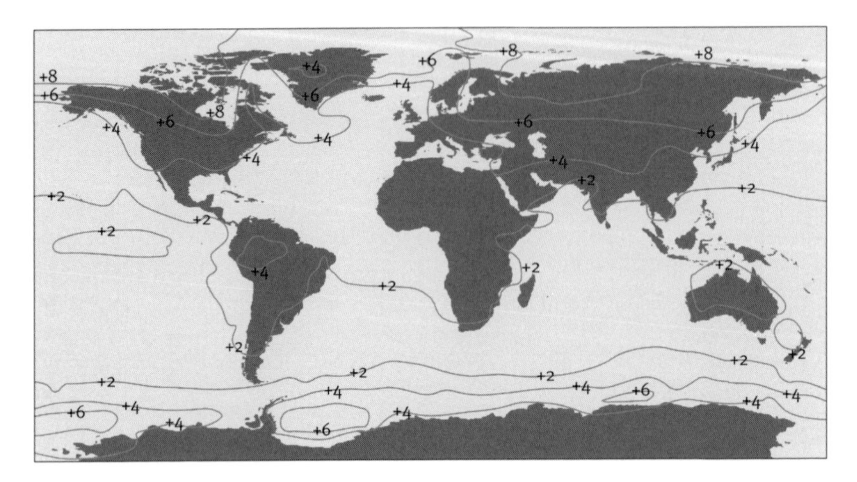


**Fig. 8.13** Complex computer models of global climate predict that temperatures will increase significantly from the late twentieth century to the late twenty-first century, when $CO_2$ levels are projected to double. Predicted temperature increases, shown in °C, are greatest over continents and at high latitudes (i.e. closer to the poles). (From IPCC 2001.)

There is abundant evidence that global climate change is already affecting biological communities in the northern temperate zone (Jensen 2004). For example, birds are laying their eggs earlier in the year and plants are flowering earlier than they did in the past. In the tropics, there is evidence that rain forest tree species have been increasing their growth rates, mortality rates, and recruitment rates over the last few decades, presumably as a result of increased temperatures and $CO_2$ concentrations (Phillips & Gentry 1994; Baker et al. 2004; Phillips et al. 2004). There is also increasing evidence that lowland rain forest plants are growing at or near their maximum temperature tolerances at present, so further increases in temperature may eliminate many temperature-sensitive species and reduce the growth rates of others (Clark et al. 2003; D.B. Clark, personal communication, 2003). Even slight increases in temperature reduce carbon uptake, and thus growth, by increasing nighttime respiration. Changes in rainfall are harder to predict, but many tropical areas will likely be drier than at present. The combination of higher temperatures and lower rainfall will make the forests more susceptible to fire, hastening their replacement by open woodlands and grasslands. The survival of species may depend on their ability to migrate to higher elevations or their ability to persist near to rivers or other wetlands. Global climate change may have especially harmful effects on species that depend on seasonal weather patterns to coordinate reproduction (Bawa & Markham 1995). In Southeast Asian forests, the mass flowering of dipterocarp species could be disrupted, with widespread and potentially devastating consequences for the animals that are dependant on this cycle (Curran & Leighton 2000).

Global climate change may be particularly damaging to montane forests. There is evidence that the vegetation zonation on tropical mountains is strongly controlled by temperature. A 3°C warming would result in temperature zones moving 500 m (1600 feet) vertically up the mountain, permitting lowland plants to migrate upwards and eliminating the species in the highest zones (Foster 2001). Some global climate models also predict a reduction in low-level cloudiness, with a potentially severe impact on the "cloud forests" that depend for their existence on frequent mists.

Although global climate change and rising $CO_2$ levels have, by themselves, the potential to radically alter tropical rain forest communities, they are particularly ominous when combined with increasing levels of direct human disturbance and fragmentation. In large areas of continuous forest, the more mobile species will be able to migrate in response to climate change, but few species can migrate across the barriers formed by agricultural or urban areas. If rain forests persist only as isolated national parks, and these parks are then exposed to a changing climate, many species may be lost as the local environment becomes unsuitable and migration routes are blocked.

## How bad is it?

Quantifying the destruction and degradation of tropical rain forest is a much more difficult task than one might at first expect. Definitions of forest types differ greatly between countries, as does the reliability of national statistics on their extent. Definitions of destruction also differ: conversion of forest to pasture or crops is obviously destruction, but vast areas of rain forest have suffered varying degrees of degradation that fall short of total clearance. In particular, large areas have been selectively logged, with varying numbers of trees remaining. It some cases, recovery of the forest is possible if the area is protected from further abuse, but in others it may not be, due to soil damage, proliferation of vines, and the invasion of exotic species. There are also huge areas of "empty forest", from which all but the smallest vertebrates have been harvested by hunters, but which appear as intact in forestry statistics (Redford 1992). Remote sensing data from satellites provides a way of standardizing measurements across the tropics, but it is not normally possible to distinguish among the various closed-canopy forest types or to detect any but the most extreme forms of forest degradation. Also, all surveys based on satellite imagery so far have relied on relatively small samples, covering 10% or less of the total tropical forest area, so their accuracy is questionable (Matthews 2001). Moreover, the situation is changing very rapidly, with rates of forest loss sensitive to a multitude of factors, including the El Niño cycle, the state of local and global economies, and political events in individual countries. For all these reasons, the numbers given below should be viewed with caution. Most refer to either all closed-canopy forest—including deciduous forests that would not meet our definition of "rain forest"—or all tropical forest, including open woodlands.

The original extent of closed-canopy tropical forests has been estimated at 16 million $km^2$ (6 million sq. miles), based on current patterns of rainfall and temperature (Myers 1984, 1986; Sayer & Whitmore 1991). A survey based entirely on satellite imagery estimated that approximately 11.5 million $km^2$ remained in 1990, and that this was lost at a rate of 58,000 $km^2$ per year—0.52%—between 1990 and 1997 (Achard et al. 2002). A further 23,000 $km^2$ per year (0.2%) was visibly degraded. This survey could not detect selective logging, unless it had catalyzed further degradation, and it was completed before the exceptional El

Niño fires of 1997–98. The rates of both forest loss and forest degradation were twice as high in Asia (including New Guinea in this case) as in Africa, with tropical America showing the lowest overall rates of both (see Table 8.1). These average rates, however, mask huge differences within regions, with deforestation "hotspots", such as Acre in Brazilian Amazonia, parts of Madagascar, and central Sumatra, experiencing deforestation rates of more than 4% per year.

Estimates of tropical forest losses from other sources are generally somewhat higher than those of Achard et al. (2002), but comparisons are difficult because of differences in methodology and definitions. The total area of natural tropical forest—including open woodland—lost per year may be as high as 160,000 $km^2$ (Matthews 2001). Despite the difficulty in obtaining accurate numbers, a general consensus exists that tropical deforestation rates are alarmingly high and are growing. There are no reliable estimates based on ground surveys for how much additional area is logged, since most logging in tropical forests is illegal (Smith 2002). An indication of the extent of illegal logging comes from comparisons of trade statistics between exporting and importing countries, with China, for instance, importing 103 times as much timber from Indonesia in the year 2000 as Indonesia reported exporting (Johnson 2002)!

In many tropical countries of the world, particularly on islands and in locations where human population density is high, most of the original forest habitat has already been destroyed. More than 50% of the forest has gone in many Old World tropical countries (WRI 2000). In tropical Asia, fully 65% of the primary

**Table 8.3** Loss of forest habitat in some countries of the Old World tropics. All data are from 1996. Note that the definition of forest habitat is slightly different from that used in Tables 8.1 and 8.2.

| Country | Current forest remaining (×1000 ha) | Percentage of habitat lost | Percentage of current forest as frontier forest |
|---|---|---|---|
| **Africa** | | | |
| Democratic Republic of the Congo | 135,071 | 40 | 16 |
| Gambia | 188 | 38 | 0 |
| Ghana | 1,694 | 91 | 0 |
| Kenya | 3,423 | 82 | 0 |
| Madagascar | 6,940 | 87 | 0 |
| Rwanda | 291 | 84 | 0 |
| Zimbabwe | 15,397 | 33 | 0 |
| **Asia** | | | |
| Bangladesh | 862 | 92 | 4 |
| India | 44,450 | 80 | 1 |
| Indonesia | 88,744 | 35 | 28 |
| Malaysia | 13,007 | 36 | 14 |
| Myanmar (Burma) | 20,661 | 59 | 0 |
| Philippines | 2,402 | 94 | 0 |
| Sri Lanka | 1,581 | 82 | 12 |
| Thailand | 16,237 | 78 | 5 |
| Vietnam | 4,218 | 83 | 2 |

From WRI (2000).

forest habitat has been lost (Table 8.3). In Asia, the two biologically rich countries of Malaysia and Indonesia still have about half of their original forest area, but others have lost most of theirs, including the Philippines (only 6% left in 1996), Thailand (22%), Sri Lanka (18%), and Vietnam (17%). Even in Malaysia and Indonesia, much of the remaining forest has been selectively logged, with only about 20% still in its original condition. Sub-Saharan Africa has similarly lost about 65% of its original forests, with losses being most severe in the West African countries of Nigeria (only 11% left), Ghana (9%), Ivory Coast (10%), and Sierra Leone (10%). The biologically rich Central African nation of the Democratic Republic of Congo, is relatively better off, still having about half of its forests, though much of these forested lands have been strongly altered by human activity. In South America, the percentage of the original forest cover remaining is fairly high for Brazil (66% in 1996), Peru (87%), and Colombia (54%), but lower percentages remain in many Central American and Caribbean countries, such as Costa Rica (35%), Cuba (29%), and Jamaica (36%). Recent rates of deforestation (1990–2000) vary considerably among countries, with particularly high annual rates of over 2% reported in such tropical countries as the Philippines (2.1%), Thailand (2.9%), El Salvador (6.1%), Guatemala (2.2%), and Nicaragua (3.2%) (WRI 2003). There is also a huge amount of regional variation within countries in the rate of forest loss.

The percentage of the original forest area remaining and the rate of deforestation give only part of the picture. As a result of habitat fragmentation, farming, logging, and other human activities, very little frontier forest—intact blocks of undisturbed forest large enough to support all aspects of biodiversity—remains in most of Asia and Africa (WRI 2000). In the New World, the situation is somewhat better; an estimated 42% of Brazilian forest and 59% of Venezuelan forest was frontier forest in 1996. In Papua New Guinea, 85% of the original forest cover is still present, with 40% remaining as frontier forest in 1996. However, the percentage of frontier forest declines alarmingly each year, and these estimates are undoubtedly already out of date.

At the current rate of loss, there will be very little tropical forest left by the year 2050, except in relatively small protected areas. The situation is actually grimmer than these projections indicate because the world's population is still increasing and poverty is on the rise in many developing tropical countries, putting ever greater demands on the dwindling supply of rain forest. Further, many countries have proved unable to protect and manage their present system of protected areas. These predictions are not based on mere speculation: the almost complete loss of rain forests has already occurred in many areas of the world over the last few decades, including the Philippines, much of southern Borneo, much of Sumatra, West Africa, Central America, and western Ecuador.

The loss of rain forest can be illustrated by the Atlantic Coastal Forest of Brazil, an area with a high endemism. Fully half of its tree species are endemic to the area—that is, they are found there and nowhere else in the world—and the region supports a number of rare and endangered animals, including the golden lion tamarin (*Leontopithecus rosalia*). Clearance of the Atlantic Forest started with the arrival of Portuguese colonists, 500 years ago, and has continued since, with much of the damage occurring during the last few decades, as the remaining forest was cleared for sugarcane, coffee, orange, and cocoa production. Today, less than 9% of the original forest remains. The surviving forest is not in one large block but is divided into isolated fragments that may be unable to support long-term viable populations of many wide-ranging species. The single largest patch is only 7000 $km^2$ and is highly disturbed in places. Only 3000 $km^2$ of forest are included in protected areas, and many species in this region are at severe risk of extinction (Brooks & Balmford 1996).

The coastal region of western Ecuador provides another example. This region originally was covered by a rich forest filled with endemic species. It was minimally disturbed by human activity until 1960. At that time, roads were developed and almost all the forests were cleared to establish human settlements and oil palm plantations. One of the last surviving fragments is the 1.7 $km^2$ Rio Palenque Science Reserve. This tiny conservation area has 1025 recorded plant species (Gentry 1986), of which around 30 species are not known to occur anywhere else.

## Rain forest extinctions

In previous chapters, we have drawn attention to the extinctions of large vertebrates that accompanied the first arrival of modern humans in New Guinea, Australia, tropical America, and Madagascar (Brook & Bowman 2004). In this chapter, we have covered the massive recent losses in rain forest area and the continuing threats from hunting, logging, and agricultural development. Although tropical rain forests only covered 6–7% of the Earth's land surface, it is generally accepted that they supported at least 50% of the species of plants and animals (Primack 2002). There is therefore a consensus among rain forest ecologists and conservationists that if the current destruction of the rain forest continues it will inevitably lead to a new wave of mass extinctions, dwarfing anything caused by previous generations of humans.

Local extinctions of rain forest species are very common; there are few areas of rain forest left in Asia or Central America, for instance, that still support all the animals that were present a century ago. Large mammals, in particular, have been eliminated from many forest areas by hunting (Bennett & Robinson 2000). The Southeast Asian island of Singapore, which was almost entirely covered in rain forest 200 years ago, has lost around half of its species as a result of deforestation and many of the survivors are threatened (Brook et al. 2003). However, documented global extinctions—the complete loss of a rain forest species from the Earth—are still rare. It is true that extinctions among most invertebrate groups, where many species are undescribed, would simply not be noticed. With plants, although most species are already described the distributions of tropical species are poorly known, which makes it very difficult to assess their current status. Birds and the more conspicuous mammal groups, in contrast, are relatively well studied and catastrophic extinctions among these groups would not remain undetected. Such extinctions among rain forest birds and mammals have so far been very few, except on islands. Indeed, the great majority of documented recent extinctions of rain forest organisms have taken place in the Caribbean or on islands in the Indian and Pacific Oceans. Why have there apparently been so few extinctions from continental rain forests?

No area illustrates the apparent paradox of massive forest loss but few global extinctions better than the Atlantic Forest of Brazil: most of the forest has gone, but no forest bird species has yet become extinct. One study used the well-known relationship between the number of species and the total area of habitat available to estimate how many species would be expected to go extinct as a result of forest loss (estimated as 88% for this study) (Brooks et al. 1999a). The predicted number of extinctions was 51 species, which is 41% of the total fauna of forest-dependent birds. This is very similar to the number of species (45) that are currently considered endangered, suggesting a simple explanation for the apparent discrepancy: extinction is not instantaneous. Bird species can persist as tiny populations in forest fragments for decades after most of their habitat has been lost, but extinction is still inevitable unless immediate action is taken to save them. Similar studies on birds in Southeast Asia have confirmed this conclusion: calculations based on loss of habitat predict correctly or even underestimate the numbers of species currently threatened and likely to become extinct in the future (Brooks et al. 1999b).

Extrapolating these results from birds to other groups of organisms would be unwise. Birds tend to have larger ranges than mammals, which will make them less vulnerable to global extinction when large areas of forest are cleared. Birds also tend to be more mobile. The species most vulnerable to immediate extinction as a result of deforestation will be those that are endemic to a small area of forest, as seems to be the case with many species of plants (Gentry 1986). Many plant species are restricted to specialized habitats, such as rock outcrops, unusual soil conditions, and river edges. On the other hand, plants and many invertebrates are more likely to survive than vertebrates in the tiny forest fragments that are commonly left after agricultural clearance. The good news is that we still have time to save most of the amazing biodiversity of tropical rain forests. The bad news is that this is only a temporary respite. Numerous species are already committed to extinction unless comprehensive action to protect their habitats starts immediately.

## Solutions

There is currently a vigorous—at times fiery—debate among conservationists about the best ways to conserve tropical biodiversity. Major areas of disagreement include the relative effectiveness of traditional national parks and community-based conservation initiatives (Terborgh 1999; Schwartzman et al. 2000; Bruner et al. 2001; Terborgh et al. 2002); the role of indigenous people in protected areas (Galetti 2001; Terborgh & Peres 2002); and the effectiveness of forest certification and similar initiatives in promoting the sustainable use of tropical forest (Leslie et al. 2002; Gullison 2003; Rametsteiner & Simula 2003). Much of the debate could be characterized—or caricaturized—as "use it or lose it" against "use it and lose it", although most participants fall somewhere between these extremes. There are multiple factors causing the decline of rain forest, and these threats vary from country to country and continent to continent. The threats also vary from place to place within continents and countries. Therefore the methods required to protect rain forest need to be tailored to each situation. The relative effectiveness of each method will also depend, to a varying extent, on social and political factors that are usually beyond the control of conservationists. The ongoing destruction of the world's tropical rain forests is not one problem, so we should not expect there to be only one—or even a few—solutions.

### National parks and other protected areas

This book has been about the ecology of more or less intact rain forest communities: rain forests with giant trees, primates, carnivores, herbivores, birds, bats, and insects. With a few exceptions, such rain forests survive today only in regions with very few human inhabitants or in national parks and similar areas set aside for their protection. An extra billion people in the tropics over the next 15 years, coupled with the expanding exploitation of rain forest resources, will insure that isolation alone will not provide protection much longer. The single most important strategy for protecting intact rain forest communities is therefore to establish protected areas, and then to effectively manage them. In most cases, these protected areas will be national parks in which rain forest protection is the principal objective. Throughout much of the modern world, national governments play the leading role in rain forest preservation, although state and local governments, local communities, conservation organizations, and even private individuals are important in some areas. National parks are the single largest source of protected lands in many countries. For example, Costa Rica's national parks and other protected areas cover 700,000 ha (1.75 million acres) or almost 14% of the nation's total land area (WRI 1998). Outside the parks, deforestation is proceeding rapidly, and soon the parks may represent the only undisturbed habitat in the country.

Costa Rica's parks are a success story, but the picture is less clear in some other parts of the tropics. Many of Indonesia's protected areas are subject to virtually uncontrolled logging, hunting, and, in some areas, forest clearance (Curran et al. 2004). Most rain forest protected areas lie between these extremes, however. They often have huge problems, including poaching (see Plate 8.2, between pp. 150 and 151), encroachment of park boundaries, and invasive species, but both the vegetation and fauna are usually in much better condition inside parks than outside (Bruner et al. 2001). There is clearly a need not only to create additional protected areas and extend existing ones, but also to enhance the protection they receive.

Rain forest parks come in all different sizes and shapes. The massive—but almost completely unmanaged—national parks of the Brazilian Amazon represent one extreme (Peres & Terborgh 1995). Such huge parks are probably the only way to preserve complete rain forest ecosystems, including the full range of habitats and species. The window of opportunity for establishing large parks is rapidly closing, as loggers and settlers move into new areas, so the completion of a pantropical network of representative protected rain forest parks is the most urgent priority in rain forest conservation. Large parks alone will not be enough, however, and smaller rain forests reserves—down to a few hectares in size—can also play a valuable role in an overall conservation strategy, protecting species and habitats that are not represented in the larger parks (Turner & Corlett 1996). Indeed, in much of the tropics, there are no large areas of intact rain forest left to protect and small reserves are the only way to save what is left of the rain forest biota. Such small parks may need to be carefully managed to prevent the extinction of species present in low numbers and species affected by habitat fragmentation.

Bukit Timah Nature Reserve in Singapore is an excellent example of such a small reserve that provides long-term protection of numerous species due to its careful management. This 164 ha reserve represents 0.2% of the original forested area on Singapore and has been isolated from other forests since 1860. Yet Bukit Timah and the adjacent reservoir area still support more than half of Singapore's original flora and fauna (Corlett & Turner 1997; Corlett 2000). Another example is the Parque Natural Metropolitano within Panama City, which has 186 recorded bird species in 192 ha of semideciduous rain forest. These small reserves have lost many of the most spectacular organisms mentioned in this book—such as tigers, jaguars, harpy eagles, and hornbills—and will certainly lose more species in coming decades, but both support diverse communities of the smaller and more tolerant animal species. Small reserves located near populated areas also make excellent conservation and nature study centers that further the long-range goals of rain forest conservation by developing public awareness of important issues. Both Bukit Timah Nature Reserve and the Parque Natural Metropolitano are a few minutes' drive from densely populated urban areas and are used in both the formal education of school and university students and the informal education of adult visitors.

Once rain forests are legally protected, funds must be available to support park management, through salaries and vehicles in particular. Many tropical parks are large and look impressive on paper, but on the ground they lack the personnel and infrastructure needed to protect and manage them. In accessible areas in politically stable countries, tourism can provide a source of income for park management. In many cases, however, national parks in developing countries will need external financial support from international donor agencies, such as the World Bank, or partnerships with zoos and conservation organizations in developed countries, such as the World Wildlife Fund. It is almost always cheaper to protect endangered species in their natural habitat rather than in zoos or other captive facilities, so investments in park establishment and management represent a cost-effective strategy.

Experience throughout the tropics has shown the importance of local support when parks are established. The costs of establishing new protected areas in inhabited regions are typically borne largely by the local people, who lose access to resources within the park boundaries, and who may be displaced from their homes and farms (Ferraro 2002). In theory these costs may be offset by benefits, both financial, from jobs and tourism, and ecological, such as clean water and reduced soil erosion. In practice, however, the financial benefits often go largely to outsiders and the ecological benefits are rarely obvious enough to compensate for the losses. Not surprisingly, the establishment of parks and other protected areas has often bred resentment among local communities, making the task of park management much more difficult. The impact on local communities needs to be considered at the planning stage, with active steps taken to insure that the cost of the park is not paid by those least able to afford it.

The exact type of conservation strategy will depend on the country involved. For example, conservation projects in India require a major social component to address the related problems of rural poverty and a high density of people. In countries such as Costa Rica, rain forest preservation is linked to the rapidly rising tourist industry. In Brazil, the rights of indigenous people need to be considered. In Papua New Guinea, the complex pattern of traditional clan rights over land may make the whole idea of national parks impractical, without some form of co-management with local communities.

### Regulating exploitation

Parks will not be enough. However successful we are in expanding the present coverage of protected areas and insuring their proper management, they will inevitably be too few, too small, and too unrepresentative to preserve all of the rain forest biodiversity. Most rain forest regions will continue to have a much larger area of forest that is not inside parks. Whether this is in legally designated "production forests" or just blank areas on the map, regulating the exploitation of rain forest outside the parks can make a major contribution to the protection of rain forest diversity. It is also important to remember that even the unregulated exploitation of timber and wildlife, as long as it maintains forest cover, protects much more biodiversity than clearance for pasture or crops. The aim of regulating exploitation outside parks should not therefore be to prevent it, but rather to insure that it is done without the irreversible loss of species and habitats.

Reducing the adverse impacts of the logging industry is—or should be—a priority in most rain forest countries, not only because of the massive direct damage logging does to the rain forest, but also because of its role in catalyzing further degradation and deforestation. When very little forest is left, as in Thailand, a total ban on logging is simpler and more effective than attempts at regulation. Most rain forest countries, in contrast, need the income from logging and the employment that it and the wood-processing industry can provide. As a result, a great deal of research effort has gone into devising improved methods of logging that cause less environmental damage. These methods, known collectively as reduced impact logging or RIL, involve the application of a series of guidelines designed to minimize damage to soils and the next generation of commercial trees, as well as nontarget species of plants and animals (Putz et al. 2000). RIL includes such practices as: training of workers; protection of forest on steep slopes and along streams; careful planning of the road network to reduce soil damage; a pre-logging inventory to set cutting limits; directional felling of trees to avoid injuring neighboring trees; and pulling trees out of the forest using cables to minimize vegetation damage and soil disturbance. A number of studies have now shown that the application of RIL can potentially benefit everybody, reducing not just environmental damage, but also the financial costs of logging (Putz et al. 2000; Boltz et al. 2003; Pearce et al. 2003). Why then do destructive logging practices persist almost everywhere in the tropics?

One major reason why RIL is not more widely adopted is that the long-term benefits do not accrue to the logger, who is very rarely the owner of the forest. Currently, most rain forest logging is either illegal or involves only a short-term concession, so the logger derives no financial benefit from protecting soils and future generations of trees. The logger's interest is to take out as many logs as quickly and cheaply as possible. Some aspects of RIL, such as the training of workers, careful planning of roads, and directional felling of trees, make sense in any logging operation, but others, such as the exclusion of steep slopes and streamside forests, merely cut profits. The forest owner—the state in most rain forest countries—would undoubtedly benefit from the strict application of RIL guidelines, but enforcing them on logging companies requires a well-trained, adequately paid, and highly motivated team of forest officers, and few rain forest countries have this. This is not an excuse for abandoning all attempts at controlling logging practices, however, since any improvement in the present situation would have immediate benefits. The potential for the consumer nations to influence logging practices is discussed below.

Regulating the exploitation of wildlife is another important step. Until recently, most attention was focused on the exploitation of rare and endangered species, such as tigers, rhinoceroses, primates, and parrots. In countries with effective border controls, the control of international trade in these species and the products made from them may be the most effective approach. However, for many species, the internal trade is more important—witness, for example, the huge numbers of parrots and primates kept as pets in rain forest countries—while many rain forest countries, such as Indonesia, have borders that are far too long and porous for an effective enforcement of a trade ban.

Bushmeat is sold by weight, rather than rarity, so the bushmeat trade is a much broader threat to biodiversity than the trade in specific endangered species. Regulating the bushmeat trade is being increasingly recognized as a critical conservation step (Bennett & Robinson 2000). The Malaysian state of Sarawak has taken a very positive step in this direction by banning the sale of wild meat. Malaysia and Papua New Guinea have successfully regulated the sale of ammunition and guns. At the international level, the United Nations organized a conference on the control of the small arms trade in 2001, as a means of reducing access to cheap guns. However, effective action on this issue has been blocked by the United States, which has argued that access to small arms by private individuals is a legitimate right. Where bushmeat is the major source of protein for the local population, as it is in much of the African rain forest area, the provision of alternative sources of protein will also be essential. The current importance of bushmeat protein in rural Africa is not an argument for continued exploitation, since current harvest levels are not sustainable (Fa et al. 2003).

## Traditional societies and sustainable development

The solutions proposed so far—protected areas and regulated exploitation ("fences and fines")—are those used successfully in much of the developed world. There have been successes in the tropics too, but also enough failures to make alternative approaches to conservation worth considering. Many of these failures have resulted from a lack of attention to local people living in or near to rain forest areas (Western et al. 1994; Primack et al. 1998; Redford & Sanderson 2000). To much of the general public in the developed world, rain forests existed until recently without human influence. Although remote tropical rain forests may be designated as "wilderness" by governments and conservation groups, in fact they usually have at least a small, sparse human population. Many tropical areas of the world have had a particularly long association with human societies because the tropics have always been free of glaciation and are particularly amenable to human settlement (Mercader 2003). We are, after all, a tropical species. The present mixture and relative densities of plants and animals in many biological communities may even reflect the historic activities of people in those areas, such as selective hunting of certain game animals, fishing, and planting useful species (Redford 1992). The great biological diversity of the tropics has coexisted with human societies for thousands of years, although in many areas, such as New Guinea, Australia, Madagascar, and South America, this peaceful coexistence followed a period of rapid extinction among large vertebrates after the initial arrival of humans.

Local people practicing a traditional way of life in rural areas, with relatively little outside influence in terms of modern technology, are variously referred to as tribal people, indigenous people, native people, or traditional people (Dasmann 1991; Richards 1996b). These established indigenous people need to be distinguished from more recent settlers, who may not be as concerned with sustaining the health of the biological community or as knowledgeable about the species present. Worldwide, there are approximately 250 million indigenous people in more than 70 countries, occupying 12–19% of the Earth's total land surface (Redford & Sanderson 2000). An important goal of many conservation efforts is to allow traditional people living in an area to maintain their culture and livelihood, and to protect rain forests at the same time. This type of integration of a local economy with nature preservation has been the objective of the UNESCO's Man and the Biosphere (MAB) Program and other similar projects.

Many traditional societies have strong conservation ethics; they may be subtler and less clearly stated than Western conservation beliefs, but they do tend to affect people's actions in their day-to-day lives, perhaps more than Western beliefs (Gómez-Pompa & Kaus 1992; Western 1997). In some cases, these people benefit from interacting with outside conservationists to help them develop and articulate their own ideas about conservation. Local people with a concern for the natural environment sometimes even take the lead in protecting biological diversity from destruction by outside influences (Fig. 8.14). The destruction of communal forests by government-sanctioned logging operations has been a frequent target of protests by traditional people throughout the world (Poffenberger

**Fig. 8.14** Buddhist priests in Thailand offer prayers and blessings to protect communal forests and sacred groves from commercial logging operations. (Courtesy of the Project for Ecological Recovery, Bangkok.)

1990). In India, followers of the Chipko movement hug trees to prevent logging (Gadgil & Guha 1992). In Borneo, the Penan, a small tribe of hunter-gatherers, have attracted worldwide attention with their blockades of logging roads entering their traditional forests. Certain isolated forests are maintained by local people because of their perceived importance to the culture and religious beliefs of the people living nearby. In these "sacred forests", hunting and collecting of plants may be prohibited by local custom (Folke & Colding 2001). Empowering such local people and helping them to obtain legal title to their traditionally owned lands is often an important component of efforts to establish locally managed protected areas in developing countries.

One approach that has been tried in many areas is to include the economic needs of local people in local conservation management plans, with the aim of helping both the people and the reserves. Such compromises, known as integrated conservation–development projects (ICDPs), are seen by some experts as conservation strategies worthy of serious consideration (Alpert 1996; Maser 1997; Primack et al. 1998; Salafsky et al. 2001). In theory, these projects encourage rural communities to conserve biodiversity, either by helping them to use it sustainably or by providing alternative sources of income. UNESCO's MAB Program, probably the most well-known program, includes among its goals the maintenance of "samples of varied and harmonious landscapes resulting from long-established land use patterns" (Batisse 1997). The MAB Program recognizes the role of people in shaping the natural landscape, as well as the need to find ways in which people can sustainably use natural resources without degrading the environment. The MAB research framework, applied in its worldwide network of 440 designated biosphere reserves in 97 countries, integrates natural science and social science research. It includes investigations of how biological communities respond to different human activities, how humans respond to changes in their natural environment, and how degraded ecosystems can be restored to their former conditions. Increasingly, such projects are being evaluated and monitored to determine if they are meeting their conservation goals and if the project funds have been spent responsibly.

One specific example of such a rain forest conservation project is the Community Baboon Sanctuary in eastern Belize, created by a collective agreement among a group of villages to maintain the tropical forest habitat required by the local population of howler monkeys (*Alouatta pigra*) (Fig. 8.15) (Horwich & Lyon 1998). Ecotourists visiting the sanctuary must pay a fee to the village organization and additional fees if they stay overnight and eat meals with a local family. Conservation biologists working at the site have provided training for local nature guides, a body of scientific information on the wildlife, funds for a local natural history museum, and business training for the village leaders. In such cases, improving the economic conditions of people's lives will need to be part of the overall strategy of preserving tropical rain forests. In addition, the ecotourists from developed countries who visit the projects may become advocates for rain forest conservation when they return to their own countries. And increasingly, people are visiting rain forests parks and projects within their own country and are emerging as a voice for conservation.

Although working with indigenous people and other long-established local communities has resulted in a number of successful conservation initiatives, many conservationists have doubts about the long-term viability of this approach (Terborgh et al. 2002; Balmford & Whitten 2003). Such projects may be difficult

(a)

Fig. 8.15 (a) Aerial bridges allow howler monkeys to cross over roads and gaps in the forest. These bridges become popular viewing points for tourists. (b) The Community Baboon Sanctuary in Bermudian Landing, Belize, is attempting to preserve a network of forest corridors (stippled areas) along the Belize River and between fields. (Photograph courtesy of R.P. Horwich and J. Lyon.)

(b)

to implement and maintain due to social and economic issues, and may even have a negative net impact on biodiversity if more people are attracted to the area by economic opportunities (Oates 1999; van Schaik & Rijksen 2002). Many observers have concluded that ICDPs have, in general, been more successful at promoting development than at enhancing conservation, and that in many cases they have achieved neither objective (Terborgh et al. 2002; Ferraro 2002). The ability of human populations to coexist with wildlife for millennia was a result of low population densities and limited technology: conditions that cannot be maintained in an increasingly globalized planet. It is not clear if traditional societies will be able to pass on either their cultures or their sustainable methods of rain forest exploitation to a new generation that has been exposed to Western consumer culture and is equipped with the technologies of the developed world. Moreover, many of the threats to rain forest come not from traditional societies with long-established ties to the land, but rather from "shifted cultivators" on the forest frontier. The strict protection of national parks by park personnel may be the best hope in these situations.

Another alternative increasingly being discussed involves international conservation groups making direct payments to individual landowners and local communities that protect critical ecosystems, in effect paying the community to be good land stewards (Ferraro 2001; Ferraro & Kiss 2002, 2003). Such conservation payments have the advantage of greater simplicity than programs that attempt to link conservation and economic development. Direct financial incentives are common in the developed world, but uncertain land tenure and weak legal systems make this approach more difficult to implement in many rain forest countries. However, the same problems apply to alternative approaches, such as ICDPs, and common sense suggests that paying for what you want (i.e. protection of rain forest and its fauna) will be cheaper and more effective than paying for something indirectly related to it (such as rural development). Moreover, since both land values and incomes are typically very low at the forest frontier, direct conservation payments can be surprisingly affordable (Ferraro & Kiss 2002, 2003). It is too early to say whether direct payments will have a wide application in rain forest conservation, but the widespread failure or, at best, partial success of alternative approaches to resolving conflicts between biodiversity conservation and the needs of local people suggests that it should at least be tried.

## Rain forest conservation efforts in the developed world

A recent review of the costs and benefits of tropical conservation concluded that the costs are borne largely by local people in the tropical countries—the rural poor rather than the urban elite—while the benefits are shared globally (Balmford & Whitten 2003). A logical conclusion from this is that the developed world, as the wealthiest of the beneficiaries, should pay far more for tropical conservation. How can this best be done? As discussed above, the influence of the developed world in rain forest countries has been reduced by the growth of alternative markets and sources of capital. Moreover, poor governance in many tropical countries limits the effectiveness of aid and debt relief as a way of influencing conservation practices. The leaders and citizens of these countries are also resentful of lectures from the developed world on the importance of saving the rain forest, seeing them as, at best, hypocritical, and at worst, neocolonial interference.

However, even in an increasingly globalized, multipolar world, coordinated and focused pressure from the developed countries can have a beneficial influence, particularly when it is used in support of the growing conservation movements in the rain forest countries themselves.

Stopping trade is the answer only for species that are directly threatened by harvesting for export. The multibillion dollar trade in endangered species and their products is controlled by the CITES Convention—the Convention on International Trade in Endangered Species of Wild Fauna and Flora. More than 160 countries are signatories to this convention. Signatories are committed to preventing or restricting trade in around 5000 animal species and 28,000 plant species, including elephants, rhinoceroses, tigers, and birds of paradise, as well as most rain forest primates. Periodic news reports of large seizures at ports and airports in the developed world show that there is still a big market there for wildlife products. At the other end of the scale are naïve tourists to tropical countries who attempt to bring back souvenir curios made from endangered species. Governments in the developed world can help reduce the impact of this evil trade by greater efforts in enforcement at home, by pressuring other importer nations to abide by the Convention, and by providing financial support and technical assistance to the source countries.

Boycotts of the major rain forest-related products, such as rain forest timber and palm oil or soybeans grown on deforested land, are unlikely to be effective, because of the many alternative markets for these products, and may even be counterproductive, by further reducing the influence of the developed world. Instead, the aim should be to work with the producers to reduce environmental damage. One way of doing this is by paying a premium for products that have been produced without damaging the rain forest. An example of an initiative that deserves support from consumers in the developed world is forest certification. Certification of a forest, and the wood products harvested from it, is intended to show the consumer that the forest is being managed sustainably, so that his or her purchase is not contributing to environmental degradation. Certification gives the producers the right to use a trademarked label. There are several international certification schemes, with the one run by the Forest Stewardship Council (FSC) most widely known, as well as a range of national schemes in consuming and producing countries (Rametsteiner & Simula 2003). Requirements for certification vary between schemes, but all include reduced impact logging and compliance with local laws. An increasing number of both individual consumers and major industrial buyers of timber and wood products now insist on certification. Over 1 million $km^2$ of forests are now certified—more than 3% of the world's forest area—but less than one-tenth of this is in the tropics. The major problem for tropical producers is that the costs of meeting certification standards are rarely justified by the premium paid for certified products. An additional problem is that domestic markets for timber in many rain forest countries are far larger than export markets, so that international pressures have little effect (Leslie et al. 2002). However, while forest certification is clearly not the complete answer to the problem of unsustainable logging of tropical rain forests, it can help by providing models for best forestry practice, which other producers can then be encouraged to copy.

Forest certification works because timber, furniture, and paper are easily recognizable rain forest products, so the consumer can be given a clear choice. However, many of the products most directly linked to rain forest loss are invisible in the end-products purchased by the consumer, so mobilizing consumer support is a lot more difficult. Developed countries import huge amounts of palm oil and soybean, for instance, but few people knowingly buy either product at the supermarket. Palm oil is hidden in a wide variety of processed foods, as well as in animal feed, soaps, detergents, cosmetics, and candles. Rain forest soybeans reach consumers as chicken, pork, and beef, or in a similar variety of food and industrial products. In this case, the pressure needs to be on the private companies that import the raw products to insure that they have been produced in a way that does not contribute to deforestation. As with certified timber, however, the direct impact is likely to be small.

Governments in the developed world can have an influence on rain forest conservation through both their foreign aid budgets and their purchasing power. Governments are major consumers of wood and other products and should insure that their purchases do not damage the environment in the producer country. Aid and debt relief to tropical countries should be linked to explicit measures of conservation performance, as they are for human rights. It should be possible for citizens of both donor and recipient countries to find out exactly what standards are being set and whether or not they are being met. Making progress in the sustainable management of timber extraction, or the control of rain forest fires, a condition for development aid is no more "neocolonial interference" than requiring an end to torture and arbitrary arrests. Governments have an obligation to reflect the values of the citizens who elect and finance them.

More aid from the developed world needs to be targeted specifically at rain forest conservation, with investment in existing protected areas and the enforcement of existing laws the first priority in most countries (Walsh et al. 2003). Training and technical support will also be valuable, particularly in areas, such as the management of visitors to protected areas, in which the developed world has a great deal of experience. Could the success of the Wet Tropics of Queensland World Heritage Area as a tourist destination be repeated in other countries?

National governments can also influence rain forest conservation through their participation in international agreements and conventions. The success of the CITES Convention, mentioned above, in limiting—if not yet halting—the trade in endangered species shows what is possible when many countries work together. No environmental problem is more urgently in need of international cooperation than the control of climate change. The Kyoto Protocol—from which the USA has, unfortunately, withdrawn—has two major implications for rain forest conservation. First, as discussed above, tropical rain forests are as vulnerable to climate change as other ecosystems, and the Kyoto agreement—flawed though it is—is currently the best available hope for slowing the rise in greenhouse gas concentrations. Second, the Kyoto Protocol, under the so-called Clean Development Mechanism, offers the possibility of financial incentives for reducing the carbon emissions that result from deforestation and forest degradation. If the many technical issues can be resolved, it is possible that the monetary value of the carbon stored in trees will tip the balance in favor of the sustainable use of rain forests (Fearnside 2002; Swingland 2003).

Rain forest researchers in the tropics tend to have a love–hate relationship with the giant environmental pressure groups that dominate the public debate on rain forest protection in the developed world, such as the World Wide Fund for Nature (WWF), The Nature Conservancy, and Conservation International. On the one hand, these groups are seen as using simplistic slogans to raise funds for their own support, without either educating the general public on the complex realities of rain forest conservation or being very effective in saving rain forest biodiversity on the ground. They are also seen as excessively competitive among themselves, rather like the nineteenth-century colonial powers squabbling over spheres of influence. On the other hand, everyone who has worked in the tropics has seen specific situations where one or other of these groups is making a real difference in support of local conservation efforts. The big environmental nongovernment organizations (NGOs) are also often very effective at lobbying governments and private companies in the developed world. Unless you are a very wealthy individual or have a great deal of spare time on your hands, you will be more effective acting through one of these groups than on your own, but you should first look behind the sound bites to see what they are actually doing in terms of protecting rain forest biodiversity.

It has been estimated that the total annual expenditure on biodiversity conservation worldwide—by governments, international agencies, and NGOs—amounts to around US$10 billion (Myers 2003). This total is considerably larger than the estimated annual cost of protecting all the global biodiversity hotspots, including several of the major centers of rain forest diversity (Mittermeier et al. 1999). A reordering of funding priorities could go a long way towards meeting the financial needs of rain forest conservation.

## Conclusions and future research directions

Each rain forest of the world has a unique flora and fauna that is a product of the biogeographical history of that region, and which represents a unique community-level response to the local climate, soils, and topography. Each rain forest community also faces different types and levels of current threats from increasing human activity. The enormous forest regions of the Amazon, the Congo basin, and New Guinea were protected until recently by their size, lack of development, and remoteness. The Amazon and New Guinea also had relatively low human populations. Large tracts of these forests will survive until at least the middle of this century, but their long-term future is in doubt due to planned road networks and the pressure to open new areas for logging and cash crops. In Africa, the long association of wildlife with humans, and the ability of many species to live in fragmented landscapes, may provide some hope, but already many large animals such as primates, elephants, and antelopes are being locally eliminated by intensive hunting. Primary forests in most of everwet Southeast Asia are already disappearing because of increasing logging and agricultural development, and even some national parks are threatened.

The situation is not yet hopeless. Even in the worst hit regions, the majority of the rain forest biota still survives, in small protected areas, in unprotected fragments of primary forest on sites that are too steep, too wet, or too infertile to be worth clearing, in logged forests, in secondary forest on abandoned land, and in woody regrowth along streams and fences (Corlett 2000). Many of the more tolerant rain forest species can also make use of agricultural systems that include trees, such as shade coffee plantations, although few can survive in industrial monoculture plantations. More species will survive if bigger areas are fully protected and if unprotected areas are managed sustainably. International support is needed to insure that financing is available for the establishment and subsequent management of protected area systems and to encourage the use of best practices, such as RIL, in exploiting unprotected areas. It makes no sense to expect the world's poorest countries to pay for the protection of the world's richest ecosystems, when the benefits are global. Support for these conservation activities can come from countries of the developed world, conservation organizations, and multinational development banks, such as the World Bank. Individual citizens can contribute to this effort by visiting rain forests with local guides, by buying timber and other products obtained from sustainably managed forests, and by joining and donating funds to conservation organizations. And individual citizens and the conservation organizations to which they belong should insist that governments in the developing world protect and manage their rain forests in a responsible manner.

The major goal of conservation efforts should be to protect representative examples of intact rain forest communities so that when the current, crazy round of forest clearing finishes, there will be core areas of forest out of which species can migrate and establish new forest communities (Mittermeier et al. 1999). The ability of the forest to eventually heal following damage by logging, farming, and cattle is illustrated by the Maya forest region of Mexico and Central America (Primack et al. 1998). This area was highly fragmented and cleared over large areas by the Maya civilization during the first millennium to support large rural populations and growing cities. When the Maya civilization collapsed around 1000 AD, possibly due to some combination of exhaustion of its soil and forest resources and warfare, the human population declined drastically, and the forest was able to expand and recover from its remaining fragments. If one flies over the region today, one is struck by the extensive area of forest (Fig. 8.16). There

**Fig. 8.16** A thousand years ago, Maya farms and cities occupied a wide area of the Central American lowlands, with no apparent loss of species. Today the ruined cities are overgrown by tropical forests. (Courtesy of James Nations and Richard Primack.)

is little evidence that any species went extinct in the region as a result of this transformation of the landscape by Maya agriculture. Some species, however, such as the quetzal (*Pharomachrus mocino*), a bird with beautiful, long feathers, may have still not completely rebounded from past overuse and continued human impact in the region. Although this example suggests that rain forests may be more resilient than is often thought, it is important to realize that recovery is a very slow process, taking centuries rather decades. Moreover, a complete recovery will not be possible if native species have become extinct and exotic species have become established. The Maya had no guns, chain saws, or bulldozers and did not move species between continents.

In this book we have argued that comparative studies of rain forests in different parts of the world represent an important area for further research. We believe that in addition to their scientific importance such studies will help with conservation and management, by revealing key similarities and differences and thus the specific needs and vulnerabilities of each rain forest. Investigators such as Paul Richards (1996a) visited many parts of the world in the 1950s and 1960s to examine the variation in rain forest vegetation. In the 1970s, David Pearson undertook a comparative study of bird communities on different continents. This cross-continental approach has been extended in recent years by various researchers and organizations, most notably the Institute for Tropical Forest Science, which has established a worldwide network of 50 ha (125 acre) tropical forest plots to look at patterns of tree species distribution and forest dynamics (Condit 1998). Cross-continental comparisons of both bird and primate communities have also been undertaken, highlighting examples of both convergence and nonconvergence in community structure (Karr 1976b; Pearson 1977; Fleagle & Reed 1996).

What is needed now are community-oriented studies that include multiple taxonomic levels: for example, to what extent do parrots, small primates, squirrels, and marsupials in each rain forest overlap in their ecology and compete with or substitute for each other? Similar questions could be asked of the large, vertebrate-consuming carnivores: in what ways are cats, snakes, and eagles competitors and ecological equivalents? Another key topic is determining how rain forests respond to increasing levels of human impact, such as logging, fragmentation, hunting, and global warming. How do logging impacts differ between forests with different dominant families of timber-producing trees? Are forest fragments less isolated in Africa and Asia, where many mammals are willing to cross open areas, than in the Neotropics, where most are not? What is the impact on the plant community when hunters remove all of the large animals from the forest, and does this impact differ between rain forest regions? Is there a resulting failure of seed dispersal and regeneration, and are Madagascan forests more vulnerable because they have fewer dispersal agents? Will the reproductive cycle of Asian dipterocarp species be disrupted by global climate change?

In many cases, experimental manipulations may be the best way to investigate the significance of observed differences between rain forests. Manipulating rain forest communities on the scale needed may seem a daunting task, but large-scale experiments have been successfully used in the study of rain forest fragmentation (Laurance & Bierregaard 1997). For example, one of the most conspicuous differences among rain forests is the great diversity and abundance of epiphytes in New World forests in comparison with rain forests elsewhere. Manipulations in which the bromeliads and other epiphytes are experimentally removed from a section of a Neotropical forest, could potentially reveal the ecological importance of this group of plants in maintaining animal communities and mineral nutrient cycling. The experimental removal of dipterocarps from a Southeast Asian rain forest could be equally revealing, and is not as impractical as it sounds, since selective logging already removes the largest individuals. Such comparisons and experimental studies will be extremely valuable for understanding the unique features of each rain forest and how they will respond to increasing levels of human impact.

The world has many different rain forests, and the hope for each depends on a combination of the amount of forest remaining, the current threats to those forests, and the vulnerability of the species they still support. Protection of large areas of each rain forest type in well-managed national parks is the highest priority, followed by the sustainable management of forests outside these parks. If most species can survive until the present cycle of forest destruction ends, then rain forests will eventually expand from their remaining fragments and restore the damaged landscape. Biologists can play a role in these efforts by initiating comparative studies among rain forest regions that highlight the unique features of each area. Citizens can play a role by visiting rain forests, joining conservation organizations, and buying rain forest products that have been produced sustainably. If biologists and informed leaders can convey the message that each forest has special qualities, the public will support efforts to protect representative examples of the "many tropical rain forests".

## Further reading

Bruner A.G., Gullison R.E., Rice R.E. & da Fonseca G.A.B. (2001) Effectiveness of parks in protecting tropical biodiversity. *Science* **291**, 125–128.

Corlett R.T. (2000) Environmental heterogeneity and species survival in degraded tropical landscapes. In *The Ecological Consequences of Environmental Heterogeneity* (eds, Hutchings M.J., John E.A. & Stewart A.J.A.). Blackwell Science, Oxford, pp. 333–355.

Mittermeier R.A., Myers N., Gil P.R. & Mittermeier C.G. (eds) (1999) *Hotspots: Earth's Biologically Richest and Most Endangered Terrestrial Ecoregions*. CEMEX, Mexico City, Mexico.

Oates J.F. (1999) *Myth and Reality in the Rain Forest: How Conservation Strategies are Failing in West Africa*. University of California Press, Berkeley, CA.

Primack R.B. (2002) *Essentials of Conservation Biology*. Sinauer Associates, Sunderland, MA.

Primack R.B. (2004) *A Primer of Conservation Biology*. Sinauer Associates, Sunderland, MA.

Primack R.B., Bray D., Galetti H.A. & Ponciano J. (eds) (1998) *Timber, Tourists and Temples: Conservation and Development in the Maya Forest of Belize, Guatemala and Mexico*. Island Press, Washington, DC.

Primack R.B. & Lovejoy T.E. (eds) (1995) *Ecology, Conservation and Management of Southeast Asian Rainforests*. Yale University Press, New Haven, CT.

Redford K.H. & Sanderson S.E. (2000) Extracting humans from nature. *Conservation Biology* **14**, 1362–1364.

Terborgh J., van Schaik C., Davenport L. & Rao M. (2002) *Making Parks Work: Strategies for Preserving Tropical Nature*. Island Press, Washington, DC.

# References


Aanen D.K., Eggleton P., Rouland-Lefèvre C., Guldberg-Frøslev T., Rosendahl S. & Boomsma J.J. (2002) The evolution of fungus-growing termites and their mutualistic fungal symbionts. *Proceedings of the National Academy of Sciences of the United States of America* **99**, 14887–14892.

Abe T. & Higashi M. (2001) Isoptera. In *Encyclopedia of Biodiversity*, Vol. 3 (ed, Levin S.A.). Academic Press, San Diego, CA, pp. 581–611.

Abernethy K.A., White L.J.T. & Wickings E.J. (2002) Hordes of mandrills (*Mandrillus sphinx*): extreme group size and seasonal male presence. *Journal of Zoology* **258**, 131–137.

Abril A.B. & Bucher E.H. (2002) Evidence that the fungus cultured by leaf-cutting ants does not metabolize cellulose. *Ecology Letters* **5**, 325–328.

Achard F., Eva H.D., Stibig H.-J., Mayaux P., Gallego J., Richards T. & Malingreau J.-P. (2002) Determination of deforestation rates of the world's humid tropical forests. *Science* **297**, 999–1002.

Adam P. (1992) *Australian Rainforests*. Oxford University Press, Oxford.

Adrianjakarivelo V. (2003) Artiodactlya: *Potamochoerus larvatus*, bush pig, lambo, lambodia, lamboala, antsanga. In *The Natural History of Madagascar* (eds, Goodman S.M. & Benstead J.P.). University of Chicago Press, Chicago, IL, pp. 1365–1367.

Allmon W.D. (1991) A plot study of forest floor litter frogs. *Journal of Tropical Ecology* **7**, 503–522.

Alpert P. (1996) Integrated conservation and development projects. *BioScience* **46**, 845–855.

Altringham J.D. (1996) *Bats: Biology and Behaviour*. Oxford University Press, Oxford.

Altringham J.D. & Fenton M.B. (2003) Sensory ecology and communication in the Chiroptera.In *Bat Ecology* (eds, Kunz T.H. & Fenton M.B.). University of Chicago Press, Chicago, IL, pp. 90–127.

Alvarez Y., Juste J., Tabares E., Garrido-Pertierra A., Ibáñez C. & Bautista J.M. (1999) Molecular phylogeny and morphological homoplasy in fruitbats. *Molecular Biology and Evolution* **16**, 1061–1067.

Andresen E. (2002) Dung beetles in a Central Amazonian rainforest and their ecological role as secondary seed dispersers. *Ecological Entomology* **27**, 257–270.

Anon (2002) *Wet Tropics Management Authority: Annual Report 2001–2002*. Wet Tropics Management Authority, Cairns, Australia.

APG (Angiosperm Phylogeny Group) (2003) An update of the Angiosperm Phylogeny Group classification for the orders and families of flowering plants: APG II. *Botanical Journal of the Linnean Society* **141**, 399–436.

Ashton P.S. (1982) Dipterocarpaceae. *Flora Malesiana, Series I* **9**, 237–552.

Ashton P.S. (2003) Floristic zonation of tree communities on wet tropical mountains revisited. *Perspectives in Plant Ecology, Evolution and Systematics* **6**, 87–104.

Ashton P.S. and the CTFS Working Group (2004) Floristics and vegetation of the forest dynamics plots. In *Tropical Forest Diversity and Dynamism: Findings from a Large-scale Plot Network* (eds, Losos E. & Leigh E.G. Jr). University of Chicago Press, Chicago, IL, pp. 90–102.

Baillie I.C. (1996) Soils of the humid tropics. In *The Tropical Rain Forest: an Ecological Study* (eds, Richards P.W., Walsh R.P.D., Baillie I.C. & Greig-Smith P.). Cambridge University Press, Cambridge, UK, pp. 256–286.

Bakarr M.I., Bailey B., Omland M., Myers N., Hannah L., Mittermeier C.G. & Mittermeier R.A. (1999) Guinean forests. In *Hotspots: Earth's Biologically Richest and Most Endangered Terrestrial*

*Ecoregions* (eds, Mittermeier R.A., Myers N., Gil P.R. & Mittermeier C.G.). CEMEX, Mexico City, Mexico, pp. 239–253.

Baker R.R. (1978) *The Evolutionary Ecology of Animal Migration*. Hodder & Stoughton, London.

Baker T.R. and 17 other authors (2004) Increasing biomass in Amazonian forest plots. *Philosophical Transactions of the Royal Society of London, Series B* **359**, 353–365.

Balmford A. & Whitten T. (2003) Who should pay for tropical conservation, and how should the costs be met? *Oryx* **37**, 238–250.

Balslev H., Renato V., Paz y Miño G., Christensen H. & Nielsen I. (1998) Species count of vascular plants in one hectare of humid lowland forest in Amazonian Ecuador. In *Forest Biodiversity in North, Central and South America, and the Caribbean*, Vol. 21 (eds, Dallmeier F. & Comiskey J.A.). UNESCO, Paris, pp. 585–594.

Barber C.V., Matthews E., Brown D., Brown T.H., Curran L.M., Plume C. & Selig L. (2002) *State of the Forest: Indonesia*. World Resources Institute, Washington, DC.

Barker F.K., Cibois A., Schikler P., Feinstein J. & Cracraft J. (2004) Phylogeny and diversification of the largest avian radiation. *Proceedings of the National Academy of Sciences of the United States of America* **101**, 11040–11045.

Bates M. (1964) *Land and Wildlife in South America*. Time Incorporated, New York.

Batisse M. (1997) Biosphere reserves: a challenge for biodiversity conservation and regional development. *Environment* **39**, 6–15, 31–33.

Bawa K.S., Bullock S.H., Perry D.R., Coville R.E. & Grayum M.H. (1985) Reproductive biology of tropical lowland rain forest trees: 2. Pollination systems. *American Journal of Botany* **72**, 346–356.

Bawa K.S. & Dayanandan S. (1997) Socioeconomic factors and tropical deforestation. *Nature* **386**, 562–563.

Bawa K.S. & Dayanandan S. (1998) Causes of tropical deforestation and institutional constraints to conservation. In *Tropical Rain Forest: a Wider Perspective* (ed, Goldsmith F.B.). Chapman & Hall, London, pp. 175–198.

Bawa K.S. & Markham A. (1995) Climate change and tropical forests. *Trends in Ecology and Evolution* **10**, 348–349.

Beattie A.J. & Hughes L. (2002) Ant–plant interactions. In *Plant–Animal Interactions: an Evolutionary Approach* (eds, Herrera C.M. & Pellmyr O.). Blackwell Science, Oxford, pp. 211–235.

Beck J., Mühlenberg E. & Fiedler K. (1999) Mud-puddling behavior in tropical butterflies: in search of proteins or minerals? *Oecologia* **119**, 140–148.

Beehler B.M. & Dumbacher J.P. (1996) More examples of fruiting trees visited predominantly by birds of paradise. *Emu* **96**, 81–88.

Beisiegel B.M. (2001) Notes on the coati, *Nasua nasua* (Carnivora: Procyonidae) in an Atlantic Forest area. *Brazilian Journal of Biology* **61**, 689–692.

Bennett B. (2000) Ethnobotany of the Bromeliaceae. In *Bromeliaceae: Profile of an Adaptive Radiation* (ed, Benzing D.H.). Cambridge University Press, Cambridge, UK, pp. 587–608.

Bennett E.L. & Robinson J.G. (2000) Hunting for sustainability: the start of a synthesis. In *Hunting for Sustainability in Tropical Forests* (eds, Robinson J.G. & Bennett E.L.). Columbia University Press, New York, pp. 499–520.

Benzing D.H. (1990) *Vascular Epiphytes*. Cambridge University Press, Cambridge, UK.

Benzing D.H. (2000) *Bromeliaceae: Profile of an Adaptive Radiation*. Cambridge University Press, Cambridge, UK.

Berg C.C. (2003) Flora Malesiana precursor for the treatment of Moraceae 1: the main subdivision of *Ficus*: the subgenera. *Blumea* **48**, 167–178.

Berg C.C. & Wiebes J.T. (1992) *African Fig Trees and Fig Wasps*. North-Holland, Amsterdam, Netherlands.

Berghoff S.M., Maschwitz U. & Linsenmair K.E. (2003) Influence of the hypogaeic army ant *Dorylus (Dichthadia) laevigatus* on tropical arthropod communities. *Oecologia* **135**, 149–157.

Blake S. & Fay J.M. (1997) Seed production by *Gilbertiodendron dewevrei* in the Nouabale-Ndoki National Park, Congo, and its implications for large mammals. *Journal of Tropical Ecology* **13**, 885–891.

Bleiweiss R. (1998) Origin of hummingbird faunas. *Biological Journal of the Linnean Society* **65**, 77–97.

Bleiweiss R., Kirsch J.A.W. & Matheus J.C. (1997) DNA hybridization evidence for the principal lineages of hummingbirds (Aves: Trochilidae). *Molecular Biology and Evolution* **14**, 325–343.

Blundell A. (1999) Flowering of the forest. *Natural History* **108**, 30–38.

Blüthgen N. & Fiedler K. (2002) Interactions between weaver ants *Oecophylla smaragdina*, homopterans, trees and lianas in an Australian rain forest canopy. *Journal of Animal Ecology* **71**, 793–801.

Blüthgen N. & Reifenrath K. (2003) Extrafloral nectaries in an Australian rainforest: structure and distribution. *Australian Journal of Botany* **51**, 515–527.

Boltz F., Holmes T.P. & Carter D.R. (2003) Economic and environmental impacts of conventional and reduced-impact logging in tropical South America: a comparative review. *Forest Policy and Economics* **5**, 69–81.

Bonaccorso F.J. (1979) Foraging and reproductive ecology in a Panamanian bat community. *Bulletin of the Florida State Museum Biological Sciences* **24**, 359–408.

Bottoms T. (2000) Bama Country—Aboriginal homelands. In *Securing the Wet Tropics?* (eds, McDonald G. & Lane M.). Federation Press, Leichhardt, Australia, pp. 32–47.

Bourliere F. (1973) The comparative ecology of rainforest mammals in Africa and tropical America: some introductory remarks. In *Tropical Forest Ecosystems in Africa and South America* (eds, Meggers B.J., Ayensu E.S. & Duckworth W.D.). Smithsonian Press, Washington, DC, pp. 279–292.

Bowman D.M.J.S. (2000) *Australian Rainforests: Islands of Green in a Land of Fire*. Cambridge University Press, Cambridge, UK.

Braby M.F. (2000) *Butterflies of Australia: their Identification, Biology and Distribution*. CSIRO Publishing, Collingwood, Australia.

Brady S.G. (2003) Evolution of the army ant syndrome: the origin and long-term evolutionary stasis of a complex of behavioral and reproductive adaptations. *Proceedings of the National Academy of Sciences of the United States of America* **100**, 6575–6579.

Brokaw N.V.L. & Walker L.R. (1991) Summary of the effects of Caribbean hurricanes on vegetation. *Biotropica* **23**, 442–447.

Bronstein J.L. (1992) Seed predators as mutualists: ecology and evolution of the fig–pollinator interaction. In *Insect–Plant Interactions* (ed, Bernays E.A.). CRC Press, Boca Raton, FL, pp. 1–44.

Brook B.W. & Bowman D.M.J. (2004) The uncertain blitzkrieg of Pleistocene megafauna. *Journal of Biogeography* **31**, 517–523.

Brook B.W., Sodhi N.S. & Ng P.K.L. (2003) Catastrophic extinctions follow deforestation in Singapore. *Nature* **424**, 420–423.

Brooks T.M. & Balmford A. (1996) Atlantic forest extinctions. *Nature* **380**, 115.

Brooks T.M., Pimm S.L., Kapos V. & Ravilious C. (1999b) Threat from deforestation to montane and lowland birds and mammals in insular South-east Asia. *Journal of Animal Ecology* **68**, 1061–1078.

Brooks T.M., Tobias J. & Balmford A. (1999a) Deforestation and bird extinctions in the Atlantic forest. *Animal Conservation* **2**, 211–222.

Brosset A. (1990) A long-term study of the rain forest birds in M'Passa (Gabon). In *Biogeography and Ecology of Forest Bird Communities* (ed, Keast A.). SPB Academic, The Hague, Netherlands, pp. 259–274.

Brown E.D. & Hopkins M.J.G. (1996) How New Guinea rainforest flower resources vary in time and space: implications for nectarivorous birds. *Australian Journal of Ecology* **21**, 363–378.

Brown J.H. & Lomolino M.V. (1998) *Biogeography*. Sinauer Associates, Sunderland, MA.

Brown L. & Amadon D. (1968) *Eagles, Hawks and Falcons of the World*. Country Life Books, Feltham, UK.

Brown N., Press M. & Bebber D. (1999) Growth and survivorship of dipterocarp seedlings: differences in shade persistence create a special case of dispersal limitation. In *Change and Disturbance in Tropical Rainforest in South-East Asia* (eds, Newbery D.M., Clutton-Brock T.H. & Prance G.T.). Imperial College Press, London, pp. 123–131.

Bruce M.D. (1999) Family Tytonidae (barn-owls). In *Handbook of the Birds of the World*, Vol. 5. *Barn-owls to Hummingbirds* (eds, del Hoyo J., Elliott A. & Sargatal J.). Lynx Edicions, Barcelona, Spain, pp. 34–75.

Brühl C.A., Gunsalam G. & Linsenmair K.E. (1998) Stratification of ants (Hymenoptera, Formicidae) in a primary rain forest in Sabah, Borneo. *Journal of Tropical Ecology* **14**, 285–297.

Bruner A.G., Gullison R.E., Rice R.E. & da Fonseca G.A.B. (2001) Effectiveness of parks in protecting tropical biodiversity. *Science* **291**, 125–128.

Bryant D., Nielsen D. & Tangley L. (1997) *The Last Frontier Forests: Ecosystems and Economies on the Edge*. World Resources Institute, Washington, DC.

Bühler P. (1997) The visual peculiarities of the toucan's bill and their principal biological role (Ramphastidae, Aves). In *Tropical Biodiversity and Systematics* (ed, Ulrich H.). Zoologisches Forschungsinstitut und Museum Alexander Koenig, Bonn, pp. 305–310.

Bullock S.H., Mooney H.A. & Medina E. (eds) (1995) *Seasonally Dry Tropical Forests*. Cambridge University Press, Cambridge, UK.

Burney D.A., Robinson G.S. & Burney L.P. (2003) *Sporormiella* and the late Holocene extinctions in Madagascar. *Proceedings of the National Academy of Sciences of the United States of America* **100**, 10800–10805.

Byrne M.M. (1994) Ecology of twig-dwelling ants in wet lowland tropical forest. *Biotropica* **26**, 61–72.

Caine N.G. (2002) Seeing red: consequences of individual differences in color vision in callitrichid primates. In *Eat or be Eaten: Predator Sensitive Foraging among Primates* (ed, Miller L.E.). Cambridge University Press, Cambridge, UK, pp. 58–73.

Cameron S.A. (2004) Phylogeny and biology of Neotropical orchid bees (Euglossini). *Annual Reviews of Entomology* **49**, 377–404.

Cao M., Zhang J., Feng Z., Deng J. & Deng X. (1996) Tree species composition of a seasonal rain forest in Xishuangbanna, southwest China. *Tropical Ecology* **37**, 183–192.

Chadwick O.A., Derry L.A., Vitousek P.M., Huebert B.J. & Hedin L.O. (1999) Changing sources of nutrients during four million years of ecosystem development. *Nature* **397**, 491–497.

Chanderbali A.S., van der Werff H. & Renner S.S. (2001) Phylogeny and historical biogeography of Lauraceae: evidence from the chloroplast and nuclear genomes. *Annals of the Missouri Botanical Garden* **88**, 104–134.

Chapman C.A., Balcomb S.R., Gillespie T.R., Skorupa J.P. & Struhsaker T.T. (2000) Long-term effects of logging on African primate communities: a 28-year comparison from Kibale National Park, Uganda. *Conservation Biology* **14**, 207–217.

Chapman C.A. & Onderdonk D.A. (1998) Forests without primates: primate/plant codependency. *American Journal of Primatology* **45**, 127–141.

Chapman C.A. & Peres C.A. (2001) Primate conservation in the new millennium: the role of scientists. *Evolutionary Anthropology* **10**, 16–33.

Charles-Dominique P. (1986) Interrelations between frugivorous vertebrates and pioneer plants: *Cecropia*, birds and bats in French Guyana. *Tasks for Vegetation Science* **15**, 119–135.

Cheke R.A., Mann C.F. & Allen R. (2001) *Sunbirds: a Guide to the Sunbirds, Flowerpeckers, Spiderhunters and Sugarbirds of the World*. Yale University Press, New Haven, CT.

Cherrett J.M. (1989) Leaf-cutting ants. In *Tropical Rain Forest Ecosystems: Biogeographical and Ecological Studies* (eds, Lieth H. & Werger M.J.A.). Elsevier, Amsterdam, Netherlands, pp. 473–488.

Chesser R.T. (1995) Comparative diets of obligate ant-following birds at a site in northern Bolivia. *Biotropica* **27**, 382–390.

Cibois A. (2003) Mitochondrial DNA phylogeny of babblers (Timaliidae). *Auk* **120**, 35–54.

Cibois A., Slikas B., Schulenberg T.S. & Pasquet E. (2001) An endemic radiation of Malagasy songbirds is revealed by mitochondrial DNA sequence data. *Evolution* **55**, 1198–1206.

Cifelli R.L. (1985) South American ungulate evolution and extinction. In *The Great American Biotic Interchange* (eds, Stehli F.G. & Webb S.D.). Plenum Press, New York, pp. 249–268.

Cincotta R.P. & Engelman R. (2000) Biodiversity and population growth. *Issues in Science and Technology* **16**, 80–81.

Clark D.A., Piper S.C., Keeling C.D. & Clark D.B. (2003) Tropical rain forest tree growth and atmospheric carbon dynamics linked to interannual temperature variation during 1984–2000. *Proceedings of the National Academy of Sciences of the United States of America* **100**, 5852–5857.

Coates A.G. (1997) *Central America: a Natural and Cultural History*. Yale University Press, New Haven, CT.

Coates R.J. & Peckover W.S. (2001) *Birds of New Guinea and the Bismarck Archipelago. A Photographic Guide*. Dove Publications, Alderley, Queensland, Australia.

Cochrane E.P. (2003a) The need to be eaten: *Balanites wilsoniana* with and without elephant-dispersal. *Journal of Tropical Ecology* **19**, 579–589.

Cochrane M.A. (2003b) Fire science for rainforests. *Nature* **421**, 913–919.

Cogan M.M. (2001) *Skull Morphology in the Pteropodidae: Insights into Biomechanics, Diet, and Evolution*. PhD Thesis, University of Chicago, Chicago, IL.

Cohen J.E. (2003) Human population: the next half century. *Science* **302**, 1172–1175.

Colinvaux P.A., Irion G., Rasanen M.E., Bush M.B. & de Mello J. (2001) A paradigm to be discarded: geological and paleoecological data falsify the Haffer & Prance refuge hypothesis of Amazonian speciation. *Amazoniana* **16**, 609–646.

Collar N.J. (1997) Family Psittacidae (parrots). In *Handbook of the Birds of the World*, Vol. 4. *Sandgrouse to Cuckoos* (eds, del Hoyo J., Elliott A. & Sargatal J.). Lynx Edicions, Barcelona, Spain, pp. 280–477.

Condit R. (1998) *Tropical Forest Census Plots: Methods and Results from Barro Colorado Island, Panama, and a Comparison with Other Plots*. Springer-Verlag, Berlin.

Connell J.H. & Lowman M.D. (1989) Low diversity tropical and subtropical forests. *American Naturalist* **134**, 88–119.

Cook J.M. & Rasplus J.Y. (2003) Mutualists with attitude: coevolving fig wasps and figs. *Trends in Ecology and Evolution* **18**, 241–248.

Corlett R.T. (1998) Frugivory and seed dispersal by vertebrates in the Oriental (Indomalayan) region. *Biological Reviews* **73**, 413–448.

Corlett R.T. (2000) Environmental heterogeneity and species survival in degraded tropical landscapes. In *The Ecological Consequences of Environmental Heterogeneity* (eds, Hutchings M.J., John E.A. & Stewart A.J.A.). Blackwell Science, Oxford, pp. 333–355.

Corlett R.T. (2001) Seed dispersal in tropical rainforests: inter-regional differences and their implications for conservation. In *Tropical Ecosystems:*

*Structure, Diversity and Human Welfare* (eds, Ganeshaiah K.N., Uma Shaanker R. & Bawa K.S.). Oxford-IBH, New Delhi, India, pp. 366–369.

Corlett R.T. (2002) Frugivory and seed dispersal in degraded tropical East Asian landscapes. In *Seed Dispersal and Frugivory: Ecology, Evolution and Conservation* (eds, Levey D.J., Silva W.R. & Galetti M.). CABI Publishing, Wallingford, UK, pp. 451–465.

Corlett R.T. (2004) Flower visitors and pollination in the Oriental (Indomalayan) region. *Biological Reviews* **79**, 497–532.

Corlett R.T. & Lucas P.W. (1990) Alternative seed-handling strategies in primates: seed-spitting by long-tailed macaques (*Macaca fascicularis*). *Oecologia* **82**, 166–171.

Corlett R.T. & Turner I.M. (1997) Long term survival in tropical forest remnants in Singapore and Hong Kong. In *Tropical Forest Remnants: Ecology, Management and Conservation of Fragmented Communities* (eds, Laurance W.F. & Bierregaard R.O.). University of Chicago Press, Chicago, IL, pp. 333–345.

Cowlishaw G. & Dunbar R.I.M. (2000) *Primate Conservation Biology*. University of Chicago Press, Chicago, IL.

Cox P.A., Elmquist T., Pierson E.D. & Rainey W.E. (1991) Flying foxes as strong interactors in South Pacific island ecosystems: a conservation hypothesis. *Conservation Biology* **5**, 448–454.

Cracraft J. (2001) Gondwana genesis. *Natural History* **110**, 64–73.

Cranbrook, Earl of (1991) *Mammals of South-east Asia*, 2nd edn. Oxford University Press, Oxford.

Crisci J.V., Katinas L. & Posadas P. (2003) *Historical Biogeography: an Introduction*. Harvard University Press, Cambridge, MA.

Cristoffer C. (1987) Body size differences between New World and Old World, arboreal, tropical vertebrates: cause and consequences. *Journal of Biogeography* **14**, 165–172.

Cuevas E. (2001) Soil versus biological controls on nutrient cycling in terra firme forests. In *The Biogeochemistry of the Amazon Basin* (eds, McClain M.E., Victoria R.L. & Richey J.E.). Oxford University Press, Oxford, pp. 53–67.

Curran L.M. & Leighton M. (2000) Vertebrate responses to spatiotemporal variation in seed production of mast-fruiting Dipterocarpaceae. *Ecological Monographs* **70**, 101–128.

Curran L.M., Trigg S.N., McDonald A.K., Astiani D., Hardiono Y.M., Siregar P., Caniago I. & Kasischke E. (2004) Lowland forest loss in protected areas of Indonesian Borneo. *Science* **303**, 1000–1003.

Currie C.R., Wong B., Stuart A.E., Schultz T.R., Rehner S.A., Mueller U.G., Sung G.H., Spatafora J.W. & Straus N.A. (2003) Ancient tripartite coevolution in the attine ant–microbe symbiosis. *Science* **299**, 386–388.

Darwin C. (1862) *On the Various Contrivances by which British and Foreign Orchids are Fertilised by Insects, and on the Good Effects of Intercrossing*. J. Murray, London.

Dasmann R.F. (1991) The importance of cultural and biological diversity. In *Biodiversity: Culture, Conservation, and Ecodevelopment* (eds, Oldfield M.L. & Alcorn J.B.). Westview Press, Boulder, CO, pp. 7–15.

Davidson D.W., Cook S.C., Snelling R.R. & Chua T.H. (2003) Explaining the abundance of ants in lowland tropical rainforest canopies. *Science* **300**, 969–972.

Davies G. & Oates J. (eds) (1994) *Colobine Monkeys: their Ecology, Behaviour and Evolution*. Cambridge University Press, Cambridge, UK.

Davies R.G., Eggleton P., Jones D.T., Gathorne-Hardy F.J. & Hernandez L.M. (2003) Evolution of termite functional diversity: analysis and synthesis of local ecological and regional influences on local species richness. *Journal of Biogeography* **30**, 847–877.

Davies S.J.J.F. (2002) *Ratites and Tinamous*. Oxford University Press, Oxford.

Davis C.C., Bell C.D., Fritsch P.W. & Mathews S. (2002) Phylogeny of *Acridocarpus-Brachylophon* (Malpighiaceae): implications for tertiary tropical floras and Afroasian biogeography. *Evolution* **56**, 2395–2405.

Dayanandan S., Ashton P.S., Williams S.M. & Primack R.B. (1999) Phylogeny of the tropical tree family Dipterocarpaceae based on nucleotide sequences of the chloroplast rbcL gene. *American Journal of Botany* **86**, 1182–1190.

de Gouvenain R.C. & Silander J.A. (2003) Do tropical storm regimes influence the structure of tropical lowland rain forests? *Biotropica* **35**, 166–180.

de la Rosa C.L. & Nocke C.C. (2000) *A Guide to the Carnivores of Central America: Natural History, Ecology, and Conservation*. University of Texas Press, Austin, TX.

Dean W. (1995) *With Broadaxe and Firebrand: the Destruction of the Brazilian Atlantic Coastal Forest*. University of California Press, Berkeley, CA.

DeFries R.S., Houghton R.A., Hansen M.C., Field C.B., Skole D. & Townshend J. (2002) Carbon emissions from tropical deforestation and regrowth based on satellite observations for the 1980s and 1990s. *Proceedings of the National Academy of Sciences of the United States of America* **99**, 14256–14261.

DeGraaf R.M. & Rappole J.H. (1995) *Neotropical Migratory Birds: Natural History, Distribution and Population Change*. Comstock Publishing Associates, Ithaca, NY.

Dejean A. & Corbara B. (2003) A review of mosaics of dominant ants in rainforests. In *Arthropods of Tropical Forests* (eds, Basset Y., Novotny V., Miller S.E. & Kitching R.L.). Cambridge University Press, Cambridge, UK, pp. 341–347.

Dejean A. & Turillazzi S. (1992) Territoriality during trophobiosis between wasps and homopterans. *Tropical Zoology* **5**, 237–247.

del Hoyo J. (1994) Family Cracidae (chachalacas, guans and curassows). In *Handbook of the Birds of the World*, Vol. 2. *New World Vultures to Guineafowl* (eds, del Hoyo J., Elliott A. & Sargatal J.). Lynx Edicions, Barcelona, Spain, pp. 310–363.

Delissio L.J., Primack R.B., Hall P. & Lee H.S. (2002) A decade of canopy-tree seedling survival and growth in two Bornean rain forests: persistence and recovery from suppression. *Journal of Tropical Ecology* **18**, 645–658.

Dennis A.J. (2003) Scatter-hoarding by musky rat-kangaroos, *Hypsiprymnodon moschatus*, a tropical rain-forest marsupial from Australia: implications for seed dispersal. *Journal of Tropical Ecology* **19**, 619–627.

DeVries P.J. (1997) *The Butterflies of Costa Rica and their Natural History, II Riodinidae*. Princeton University Press, Princeton, NJ.

DeVries P.J. (2001) Butterflies. In *Encyclopedia of Biodiversity*, Vol. 1 (ed, Levin S.A.). Academic Press, San Diego, CA, pp. 559–573.

Dew J.L. & Wright P. (1998) Frugivory and seed dispersal by four species of primates in Madagascar's eastern rain forest. *Biotropica* **30**, 425–437.

Dick, C.W., Abdul-Salim K. & Bermingham E. (2003) Molecular systematic analysis reveals cryptic Tertiary diversification of a widespread tropical rain forest tree. *American Naturalist* **162**, 691–703.

Dick C.W., Etchelecu G. & Austerlitz F. (2003) Pollen dispersal of tropical trees (*Dinizia excelsa*: Fabaceae) by native insects and African honeybees in pristine and fragmented Amazonian rainforest. *Molecular Ecology* **12**, 753–764.

Dominy N.J. & Lucas P.W. (2001) Ecological importance of trichromatic vision to primates. *Nature* **410**, 363–366.

Downing T.E., Hecht S.B., Pearson H.A. & Garcia-Downing C. (eds) (1992) *Development or Destruction. The Conversion of Tropical Forest to Pasture in Latin America*. Westview Press, Boulder, CO.

Doyle J.J. & Luckow M.A. (2003) The rest of the iceberg. Legume diversity and evolution in a phylogenetic context. *Plant Physiology* **131**, 900–910.

Dransfield J. & Beentje H. (1995) *The Palms of Madagascar*. Royal Botanic Gardens, Kew, UK.

Dubost G. (1984) Comparison of the diets of frugivorous forest ruminants of Gabon. *Journal of Mammalogy* **65**, 298–316.

Ducousso M., Béna G., Bourgeois C., Buyck B., Eyssartier G., Vincelette M., Rabevohitra R., Randrihasipara L., Dreyfus B. & Prin Y. (2004) The last common ancestor of Sarcolaenaceae and Asian dipterocarp trees was ectomycorrhizal before the India–Madagascar separation, about 88 million years ago. *Molecular Ecology* **13**, 231–236.

Dudley R. & DeVries P.J. (1990) Tropical rain forest structure and the geographical distribution of gliding vertebrates. *Biotropica* **22**, 432–434.

Duellman W.E. & Pianka E.R. (1990) Biogeography of nocturnal insectivores: historical events and ecological filters. *Annual Review of Ecology and Systematics* **21**, 57–68.

Dumbacher J.P. & Fleischer R.C. (2001) Phylogenetic evidence for colour pattern convergence in toxic pitohuis: Müllerian mimicry in birds? *Proceedings of the Royal Society of London, UK, Series B* **268**, 1971–1976.

Dumont E.R. (2003) Bats and fruit: an ecomorphological approach. In *Bat Ecology* (eds, Kunz T.H. & Fenton M.B.). University of Chicago Press, Chicago, IL, pp. 398–429.

Dumont E.R. (2004) Patterns of diversity in cranial shape among plant-visiting bats. *Acta Chiropterologica* **6**, 59–74.

Dunn R.R. (2004) Recovery of faunal communities during tropical forest regeneration. *Conservation Biology* **18**, 302–309.

Dwiyahreni A.A., Kinnaird M.F., O'Brien T.G., Supriatna J. & Andayani N. (1999) Diet and activity of the bear cuscus, *Ailurops ursinus*, in North Sulawesi, Indonesia. *Journal of Mammalogy* **80**, 905–912.

Dyer F.C. (2002) The biology of the dance language. *Annual Review of Entomology* **47**, 917–949.

Edwards S.V. & Boles W.E. (2002) Out of Gondwana: the origin of passerine birds. *Trends in Ecology and Evolution* **17**, 347–349.

Eggleton P., Homathevi R., Jones D.T., MacDonald J.A., Jeeva D., Bignell D.E., Davies R.G. & Maryati M. (1999) Termite assemblages, forest disturbance and greenhouse gas fluxes in Sabah, East Malaysia. *Philosophical Transactions of the Royal Society of London, Series B, Biological Sciences* **354**, 1791–1802.

Eimerl S. & DeVore I. (1965) *The Primates.* Time Incorporated, New York.

Ellwood M.D.F. & Foster W.A. (2004) Doubling the estimate of invertebrate biomass in a rainforest canopy. *Nature* **429**, 549–551.

Ellwood M.D.F., Jones D.T. & Foster W.A. (2002) Canopy ferns in lowland dipterocarp forest support a prolific abundance of ants, termites, and other invertebrates. *Biotropica* **34**, 575–583.

Emmons L.H. (1995) Mammals of rainforest canopies. In *Forest Canopies* (eds, Lowman M.D. & Nadkarni N.M.). Academic Press, San Diego, CA, pp. 199–223.

Emmons L.H. (1997) *Neotropical Rainforest Mammals: a Field Guide.* University of Chicago Press, Chicago, IL.

Emmons L.H. (2000) *Tupai: a Field Study of Bornean Treeshrews.* University of California Press, Berkeley, CA.

Emmons L.H., Gautier-Hion A. & Dubost G. (1983) Community structure of the frugivorous–folivorous forest mammals of Gabon. *Journal of Zoology* **199**, 209–222.

Emmons L.H. & Gentry A.H. (1983) Tropical forest structure and the distribution of gliding and prehensile-tailed vertebrates. *American Naturalist* **121**, 513–524.

Ericson P.G.P., Irestedt M. & Johansson U.S. (2003) Evolution, biogeography, and patterns of diversification in passerine birds. *Journal of Avian Biology* **34**, 3–15.

Fa J.E., Currie D. & Meeuwig J. (2003) Bushmeat and food security in the Congo Basin: linkages between wildlife and people's future. *Environmental Conservation* **30**, 71–78.

FAO (2002) *Forest Products, 1996–2000.* Food and Agriculture Organization (FAO), Rome.

Fearnside P.M. (2001) Soybean cultivation as a threat to the environment in Brazil. *Environmental Conservation* **28**, 23–38.

Fearnside P.M. (2002) Avanca Brasil: environmental and social consequences of Brazil's planned infrastructure in Amazonia. *Environmental Management* **30**, 735–747.

Fearnside P.M. & Laurance W.F. (2003) Comment on "Determination of deforestation rates of the world's humid tropical forests". *Science* **299**, 1015.

Federle W., Maschwitz U., Fiala B., Riederer M. & Hölldobler B. (1997) Slippery ant-plants and skillful climbers: selection and protection of specific ant partners by epicuticular wax blooms in *Macaranga* (Euphorbiaceae). *Oecologia* **112**, 217–224.

Feer F. & Forget P.M. (2002) Spatio-temporal variations in post-dispersal seed fate. *Biotropica* **34**, 555–566.

Feinsinger P. (1976) Organization of a tropical guild of nectarivorous birds. *Ecological Monographs* **46**, 257–291.

Feinsinger P. & Colwell R.K. (1978) Community organization among Neotropical nectar-feeding birds. *American Zoologist* **18**, 779–795.

Ferguson-Lees J. & Christie D.A. (2001) *Raptors of the World.* Christopher Helm, London.

Ferraro P.J. (2001) Global habitat protection: limitations of development interventions and a role for conservation performance payments. *Conservation Biology* **15**, 990–1000.

Ferraro P.J. (2002) The local costs of establishing protected areas in low-income nations: Ranomafana National Park, Madagascar. *Ecological Economics* **43**, 261–275.

Ferraro P.J. & Kiss A. (2002) Direct payments to conserve biodiversity. *Science* **298**, 1718–1719.

Ferraro P.J. & Kiss A. (2003) Will direct payments help biodiversity? Response. *Science* **299**, 1981–1982.

Fiala B., Jakob A. & Maschwitz U. (1999) Diversity, evolutionary specialization and geographic distribution of a mutualistic ant–plant complex: *Macaranga* and *Crematogaster* in South East Asia. *Biological Journal of the Linnean Society* **66**, 305–331.

Fietz J. & Ganzhorn J.U. (1999) Feeding ecology of the hibernating primate *Cheirogaleus medius*: how does it get so fat? *Oecologia* **121**, 157–164.

Fimbel R.A., Grajal A. & Robinson J.G. (eds) (2001) *The Cutting Edge: Conserving Wildlife in Logged Tropical Forest.* Columbia University Press, New York.

Findley J.S. (1993) *Bats: a Community Perspective.* Cambridge University Press, New York.

Fish D. (1983) Phytotelmata: flora and fauna. In *Phytotelmata: Terrestrial Plants as Hosts of Aquatic Insect Communities* (eds, Frank J.H. & Lounibus L.P.). Plexus Publishing, Medford, NJ, pp. 1–27.

Fittkau E.J. & Klinge H. (1973) On biomass and trophic structure of the central Amazonian rain forest ecosystem. *Biotropica* **5**, 2–14.

Flannery T.F. (1995a) *Mammals of New Guinea.* Cornell University Press, Ithaca, NY.

Flannery T.F. (1995b) *Mammals of the South-West Pacific and Moluccan Islands.* Cornell University Press, Ithaca, NY.

Fleagle J.G. (1999) *Primate Adaptation and Evolution*, 2nd edn. Academic Press, San Diego, CA.

Fleagle J.G., Janson C. & Reed K.E (eds) (1999) *Primate Communities*. Cambridge University Press, Cambridge, UK.

Fleagle J.G. & Kay R.F. (1997) Platyrrhines, catarrhines, and the fossil record. In *New World Primates: Ecology, Evolution, and Behavior* (ed, Kinzey W.G.). Aldine de Gruyter, Hawthorne, NY, pp. 3–23.

Fleagle J.G. & Reed K.E. (1996) Comparing primate communities: a multivariate approach. *Journal of Human Evolution* **30**, 489–510.

Fleischer R.C. & McIntosh C.E. (2001) Molecular systematics and biogeography of the Hawaiian avifauna. *Studies in Avian Biology* **22**, 51–60.

Fleming T.H. (1986) Opportunism versus specialization: the evolution of feeding strategies in frugivorous bats. In *Frugivores and Seed Dispersal* (eds, Fleming T.H. & Estrada A.). Junk, Dordrecht, Netherlands, pp. 105–118.

Fleming T.H. (1988) *The Short-tailed Fruit Bat: a Study in Plant–Animal Interactions*. University of Chicago Press, Chicago, IL.

Fleming T.H. (1993) Plant-visiting bats. *American Scientist* **81**, 160–167.

Fleming T.H., Breitwisch R. & Whitesides G.H. (1987) Patterns of tropical vertebrate frugivore diversity. *Annual Review of Ecology and Systematics* **18**, 91–109.

Flenley J.R. (1998) Tropical forests under the climates of the last 30,000 years. *Climatic Change* **39**, 177–197.

Floren A. & Linsenmair K.E. (2000) Do ant mosaics exist in pristine lowland rain forests? *Oecologia* **123**, 129–137.

Folke C. & Colding J. (2001) Traditional conservation practices. In *Encyclopedia of Biodiversity*, Vol. 5 (ed, Levin S.A.). Academic Press, San Diego, CA, pp. 681–612.

Forget P.M., Hammond D.S., Milleron T. & Thomas R. (2002) Seasonality of fruiting and food hoarding by rodents in neotropical forests: consequences for seed dispersal and seedling recruitment. In *Seed Dispersal and Frugivory: Ecology, Evolution and Conservation* (eds, Levey D.J., Silva W.R. & Galetti M.). CABI Publishing, Wallingford, UK, pp. 241–256.

Forget P.M. & Vander Wall S.B. (2001) Scatter-hoarding rodents and marsupials: convergent evolution on diverging continents. *Trends in Ecology and Evolution* **16**, 65–67.

Foster P. (2001) The potential negative impacts of global climatic change on tropical montane cloud forests. *Earth Science Reviews* **55**, 73–106.

Foster S.A. (1986) On the adaptive value of large seeds for tropical moist forests. *Botanical Review* **52**, 260–299.

Francis C.M. (1994) Vertical stratification of fruit bats (Pteropodidae) in lowland dipterocarp rainforest in Malaysia. *Journal of Tropical Ecology* **10**, 523–530.

Frank J.H. (1983) Bromeliad phytotelmata and their biota, especially mosquitos. In *Phytotelmata: Terrestrial Plants as Hosts of Aquatic Insect Communities* (eds, Frank J.H. & Lounibus L.P.). Plexus Publishing, Medford, NJ, pp. 101–128.

Frank J.H. & Curtis G.A. (1981) On the bionomics of bromeliad-inhabiting mosquitos. VI. A review of the bromeliad-inhabiting species. *Journal of the Florida Anti-Mosquito Association* **52**, 4–23.

Franks N.R., Sendova-Franks A.B. & Anderson C. (2001) Division of labour within teams of New World and Old World army ants. *Animal Behaviour* **62**, 635–642.

Freeman P.W. (1998) Form, function, and evolution in the skulls and teeth of bats. In *Bat Biology and Conservation* (eds, Kunz T.H. & Racey P.A.). Smithsonian Institution Press, Washington, DC, pp. 140–156.

Freeman P.W. (2000) Macroevolution in Microchiroptera: recoupling morphology and ecology with phylogeny. *Evolutionary Ecology Research* **2**, 317–335.

Frith C.B. (1998) Bowerbirds and birds of paradise. In *Encyclopedia of Birds* (ed, Forshaw J.). Academic Press, San Diego, CA, pp. 228–231.

Frith C.B. & Beehler B.M. (1998) *The Birds of Paradise*. Oxford University Press, Oxford.

Fujita M.S. & Tuttle M.D. (1991) Flying foxes (Chiroptera: Pteropodidae): threatened animals of key ecological and economic importance. *Conservation Biology* **5**, 455–463.

Gadgil M. & Guha R. (1992) *This Fissured Land: an Ecological History of India*. Oxford University Press, Delhi, India.

Galetti M. (2001) Indians within conservation units: lessons from the Atlantic forest. *Conservation Biology* **15**, 798–799.

Galindo-Leal C. & Camara I.d.G. (2003) *The Atlantic Forest of South America*. Island Press, Washington, DC.

Gamble C. (1994) *Timewalkers: the Prehistory of Global Colonization*. Harvard University Press, Cambridge, MA.

Ganzhorn J.U. & Sorg J.P (eds) (1997) *Ecology and Economy of a Tropical Dry Forest in Madagascar, Primate Report 46–1*. German Primate Center, Gottingen, Germany.

Garbutt N. (1999) *Mammals of Madagascar*. Pica Press, East Sussex, UK.

Garnett S. (1998) Logrunners. In *Encylopedia of Birds* (ed, Forshaw J.). Academic Press, San Diego, CA, p. 198.

Gartrell B.D. & Jones S.M. (2001) Eucalyptus pollen grain emptying by two Australian nectarivorous psittacines. *Journal of Avian Biology* **32**, 224–230.

Gates D.M. (1993) *Climate Change and its Biological Consequences*. Sinauer Associates, New York.

Gathorne-Hardy F.J., Jones D.T. & Syaukani (2002) A regional perspective on the effects of human disturbance on the termites of Sundaland. *Biodiversity and Conservation* **11**, 1991–2006.

Gaubert P. & Veron G. (2003) Exhaustive sample set among Viverridae reveals the sister-group of felids: the linsangs as a case of extreme morphological convergence within Feliformia. *Proceedings of the Royal Society of London, Series B* **270**, 2523–2530.

Gayot M., Henry O., Dubost G. & Sabatier D. (2004) Comparative diet of two forest cervids of the genus *Mazama* in French Guiana. *Journal of Tropical Ecology* **20**, 31–43.

Gentry A.H. (1982) Patterns of neotropical plant species diversity. *Evolutionary Biology* **15**, 1–84.

Gentry A.H. (1986) Endemism in tropical versus temperate plant communities. In *Conservation Biology: the Science of Scarcity and Diversity* (ed, Soulé M.E.). Sinauer Associates, Sunderland, MA, pp. 153–181.

Gentry A.H. (1990) *Four Neotropical Rainforests*. Yale University Press, New Haven, CT.

Gentry A.H. (1993) Diversity and floristic composition of lowland tropical forest in Africa and South America. In *Biological Relationships between Africa and South America* (ed, Goldblatt P.). Yale University Press, New Haven, CT, pp. 500–548.

Gentry A.H. & Dodson C. (1987) Contribution of nontrees to species richness of a tropical rain forest. *Biotropica* **19**, 149–156.

George W. & Lavocat R. (eds) (1993) *The Africa–South America Connection*. Clarendon Press, Oxford.

Giannini N.P. & Kalko E.K.V. (2004) Trophic structure in a large assemblage of phyllostomid bats in Panama. *Oikos* **105**, 209–220.

Gibbs D., Barnes E. & Cox J. (2001) *Pigeons and Doves*. Yale University Press, New Haven, CT.

Gilbert L.E. (1972) Pollen feeding and reproductive biology of *Heliconius* butterflies. *Proceedings of the National Academy of Sciences of the United States of America* **69**, 1403–1407.

Givnish T.J. & Renner S.S. (2004) Tropical intercontinental disjunctions: Gondwana breakup, immigration from the boreotropics, and transoceanic dispersal. *International Journal of Plant Sciences* **165** (suppl.), 1–6.

Givnish T.J., Millam K.C., Evans T.M., Hall J.C., Pires J.C., Berry P.E. & Sytsma K.K. (2004) Ancient vicariance or recent long-distance dispersal? Inferences about phylogeny and South American–African disjunctions in Rapataceae and Bromeliaceae based on *ndhF* sequence data. *International Journal of Plant Sciences* **165** (suppl.), 35–54.

Glastra R., Wakker E. & Richert W. (2002) *Oil Palm Plantations and Deforestation in Indonesia: What Role do Europe and Germany Play?* WWF Schweiz, Zurich, Switzerland.

Glenn M.E. & Cords M. (eds) (2002) *The Guenons: Diversity and Adaptation in African Monkeys*. Kluwer Academic, New York.

Godfrey L.R. & Jungers W.L. (2002) Quaternary fossil lemurs. In *The Primate Fossil Record* (ed, Hartwig W.C.). Cambridge University Press, Cambridge, UK, pp. 97–121.

Godfrey L.R., Jungers W.L., Reed K.E., Simons E.L. & Chatrath P.S. (1997) Inferences about past and present primate communities in Madagascar. In *Natural Change and Human Impact in Madagascar* (eds, Goodman S.M. & Patterson B.D.). Smithsonian Institution Press, Washington, DC, pp. 218–256.

Goldblatt P. (1993) Biological relationships between Africa and South America: an overview. In *Biological Relationships between Africa and South America* (ed, Goldblatt P.). Yale University Press, New Haven, CT, pp. 3–14.

Gómez-Pompa A. & Kaus A. (1992) Taming the wilderness myth. *Bioscience* **42**, 271–279.

Goodman S.M. & Ganzhorn J.U. (1997) Rarity of figs (*Ficus*) on Madagascar and its relationship to a depauperate frugivore community. *Revue d'Ecologie* **52**, 321–329.

Goodman S.M., Kerridge F.J. & Ralisoamalala R.C. (2003) A note on the diet of *Fossa fossana* (Carnivora) in the central eastern humid forests of Madagascar. *Mammalia* **67**, 595–598.

Goodman S.M. & Patterson B.D. (eds) (1997) *Natural Change and Human Impact in Madagascar*. Smithsonian Institution Press, Washington, DC.

Gorog A.J., Sinaga M.H. & Engstrom M.D. (2004) Vicariance or dispersal? Historical biogeography of three Sunda shelf murine rodents (*Maxomys surifer, Leopoldamys sabanus* and *Maxomys whiteheadi*). *Biological Journal of the Linnean Society* **81**, 91–109.

Gotwald W.H. (1995) *Army Ants: the Biology of Social Predation*. Cornell University Press, Ithaca, NY.

Goulding M. (1989) *Amazon: the Flooded Forest*. BBC Books, London.

Graham A. (1999) The Tertiary history of the northern temperate element in the northern Latin American biota. *American Journal of Botany* **86**, 32–38.

Graham E.A., Mulkey S.S., Kitajima K., Phillips N.G. & Wright S.J. (2003) Cloud cover limits net $CO_2$ uptake and growth of a rainforest tree during tropical rainy seasons. *Proceedings of the National Academy of Sciences of the United States of America* **100**, 572–576.

Green J.J. & Newbery D.M. (2002) Reproductive investment and seedling survival of the mast-fruiting rain forest tree, *Microberlinia bisulcata* A. Chev. *Plant Ecology* **162**, 169–183.

Greene H.W. (1988) Species richnesss in tropical predators. In *Tropical Rainforest: Diversity and Conservation* (eds, Almeda F. & Pringle C.M.). California Academy of Sciences, San Francisco, CA, pp. 259–280.

Groves C.P. (2001) *Primate Taxonomy*. Smithsonian Series in Comparative Evolutionary Biology. Smithsonian Institution Press, Washington, DC.

Groves C.P. & Schaller G.B. (2000) The phylogeny and biogeography of the newly discovered Annamite artiodactyls. In *Antelopes, Deer, and Relatives* (eds, Vrba E.S. & Schaller G.B.). Yale University Press, New Haven, CT, pp. 261–282.

Grubb P.J. (2003) Interpreting some outstanding features of the flora and vegetation of Madagascar. *Perspectives in Plant Ecology Evolution and Systematics* **6**, 125–146.

Gullison R.E. (2003) Does forest certification conserve biodiversity? *Oryx* **37**, 153–165.

Haffer J. (1997) Alternative models of vertebrate speciation in Amazonia: an overview. *Biodiversity and Conservation* **6**, 451–476.

Hamilton L.S., Juvik J.O. & Scatena F.N. (eds) (1995) *Tropical Montane Cloud Forests*. Springer-Verlag, New York.

Handley C.O., Wilson D.E. & Gardner A.L. (1991) *Demography and Natural History of the Common Fruit Bat, Artibeus jamaicensis, on Barro Colorado Island, Panama*. Smithsonian Institution Press, Washington, DC.

Happold D.C. (1996) Mammals of the Guinea–Congo rain forest, West Africa. In *Essays on the Ecology of the Guinea–Congo Rain Forest* (eds, Alexander I.J., Swaine M.D. & Watling R.). The Royal Society of Edinburgh, Edinburgh, UK, pp. 243–284.

Hardin G. (1993) *Living Within Limits: Ecology, Economics, and Population Taboos*. Oxford University Press, New York.

Harrison R.D. (2001) Drought and the consequences of El Niño in Borneo: a case study of figs. *Population Ecology* **43**, 63–75.

Harrison R.D., Hamid A.A., Kenta T., LaFrankie J., Lee H.S., Nagamasu H., Nakashizuka T. & Palmiotto P. (2003) The diversity of hemi-epiphytic figs (*Ficus*; Moraceae) in a Bornean lowland rain forest. *Biological Journal of the Linnean Society* **78**, 439–455.

Hart J.A., Katembo M. & Punga K. (1996) Diet, prey selection and ecological relations of leopard and golden cat in the Ituri forest, Zaire. *African Journal of Ecology* **34**, 364–379.

Hart T.B. (1990) Monospecific dominance in tropical rain forests. *Trends in Ecology and Evolution* **5**, 6–11.

Hart T.B. (1995) Seed, seedling and sub-canopy survival in monodominant and mixed forests of the Ituri Forest, Africa. *Journal of Tropical Ecology* **11**, 443–459.

Hart T.B., Hart J.A. & Murphy P.G. (1989) Monodominant and species-rich forests of the humid tropics: causes for their co-occurrence. *American Naturalist* **133**, 613–633.

Hartwig W.C. (ed.) (2002) *The Primate Fossil Record*. Cambridge University Press, Cambridge, UK.

Hassanin A. & Douzery E.J.P. (2003) Molecular and morphological phylogenies of Ruminantia and the alternative position of the Moschidae. *Systematic Biology* **52**, 206–228.

Hay-Roe M.M. & Mankin R.W. (2004) Wing-click sounds of *Heliconius cydno alithea* (Nymphalidae: Heliconiinae) butterflies. *Journal of Insect Behavior* **17**, 329–335.

Heaney L.E. & Regalado J.C. (1998) *Vanishing Treasures of the Philippine Rain Forest*. The Field Museum, Chicago, IL.

Heaney L.R. & Heideman P.D. (1987) Philippine fruit bats: endangered and extinct. *Bats* **5**, 3–5.

Hedin L.O., Vitousek P.M. & Matson P.A. (2003) Nutrient losses over four million years of tropical forest development. *Ecology* **84**, 2231–2255.

Heideman P.D. & Heaney L.R. (1989) Population biology and estimates of abundance of fruit bats (Pteropodidae) in Philippine submontane rain-forest. *Journal of Zoology* **218**, 565–586.

Heil M., Fiala B., Linsenmair K.E., Zotz G., Menke P. & Maschwitz U. (1997) Food body production in *Macaranga triloba* (Euphorbiaceae): a plant investment in anti-herbivore defence via symbiotic ant partners. *Journal of Ecology* **85**, 847–861.

Hemphill A.H., Murdoch J.D., Mittermeier R.A., Konstant W.R., Ottenwalder J.A., Akre T.S.B., Mittermeier C.G. & Mast R.B. (1999) Caribbean. In *Hotspots: Earth's Biologically Richest and Most Endangered Terrestrial Ecoregions* (eds, Mittermeier R.A., Myers N., Gil P.R. & Mittermeier C.G.). CEMEX, Mexico City, Mexico, pp. 108–121.

Henkel T.W. (2003) Monodominance in the ectomycorrhizal *Dicymbe corymbosa* (Caesalpiniaceae) from Guyana. *Journal of Tropical Ecology* **19**, 417–437.

Herrera L.G., Gutiérrez E., Hobson K.A., Altube B., Díaz W.G. & Sánchez-Cordero V. (2002) Sources of assimilated protein in five species of New World frugivorous bats. *Oecologia* **133**, 280–287.

Heymann E.W. (1998) Giant fossil New World primates: arboreal or terrestrial? *Journal of Human Evolution* **34**, 99–101.

Heywood V.H. (ed.) (1993) *Flowering Plants of the World*. Mayflower Books, New York.

Hilbert D.W., Ostendorf B. & Hopkins M.S. (2001) Sensitivity of tropical forests to climate change in the humid tropics of north Queensland. *Austral Ecology* **26**, 590–603.

Hnatiuk R.J., Smith J.M.B. & McVean D.N. (1976) *The Climate of Mount Wilhelm*. Australian National University, Canberra.

Hodgkison R., Balding S.T., Akbar Z. & Kunz T.H. (2003) Roosting ecology and social organization of the spotted-winged fruit bat, *Balionycteris maculata* (Chiroptera: Pteropodidae), in a Malaysian lowland dipterocarp forest. *Journal of Tropical Ecology* **19**, 667–676.

Hölldobler B. & Wilson E.O. (1990) *The Ants*. Springer-Verlag, Berlin.

Hölldobler B. & Wilson E.O. (1994) *Journey to the Ants: a Story of Scientific Exploration*. Harvard University Press, Cambridge, MA.

Holyoak D.T. (2001) *Nightjars and their Allies: Caprimulgiformes*. Oxford University Press, Oxford.

Homer-Dixon T.F., Boutwell H.H. & Rathjens G.W. (1993) Environmental change and violent conflict. *Scientific American* **268**, 38–45.

Horwich R.H. & Lyon J. (1998) *A Belizean Rainforest: the Community Baboon Sanctuary*. Orangutan Press, Gay Mills, WI.

Houghton J.T., Meira Filho L.G., Callander B.A., Harris N., Kattenberg A. & Maskell K. (eds) (1996) *Climate Change 1995: the Science of Climate Change*. Cambridge University Press, Cambridge, UK.

Houghton R.A. (2003) Why are estimates of the terrestrial carbon balance so different? *Global Change Biology* **9**, 500–509.

Houston D.C. (1994) Family Cathartidae (New World vultures). In *Handbook of the Birds of the World*, Vol. 2. *New World Vultures to Guineafowl* (eds, del Hoyo J., Elliott A. & Sargatal J.). Lynx Edicions, Barcelona, Spain, pp. 24–41.

Howe H.F. (1985) Gomphothere fruits: a critique. *American Naturalist* **125**, 853–865.

Huchon D. & Douzery E.J.P. (2001) From the old world to the new world: a molecular chronicle of the phylogeny and biogeography of Hystricognath rodents. *Molecular Phylogenetics and Evolution* **20**, 238–251.

Hume I.D. (1999) *Marsupial Nutrition*. Cambridge University Press, Cambridge, UK.

Hunt J.H. (2003) Cryptic herbivores of the rainforest canopy. *Science* **300**, 916–917.

Hunt R.M. (1996) Biogeography of the Order Carnivora. In *Carnivore Behavior, Ecology and Evolution*, Vol. 2 (ed, Gittleman J.L.). Cornell University Press, Ithaca, NY, pp. 485–541.

Huxley J. (1976) The colouration of *Papilio zalmoxis* and *P. antimachus*, and the discovery of Tyndall blue in butterflies. *Proceedings of the Royal Society of London, UK, Series B* **193**, 441–453.

Hyodo F., Tayasu I., Inoue T., Azuma J.I., Kudo T. & Abe T. (2003) Differential role of symbiotic fungi in lignin degradation and food provision for fungus-growing termites (Macrotermitinae: Isoptera). *Functional Ecology* **17**, 186–193.

Inger R.F. (1980) Densities of floor-dwelling frogs and lizards in lowland forests of Southeast Asia and Central America. *American Naturalist* **115**, 761–770.

IPCC (Intergovernmental Panel on Climate Change) (2001) *Climate Change 2001: Synthesis Report*. Cambridge University Press, Cambridge, UK.

Itioka T., Inoue T., Kaliang H., Kato M., Nagamitsu T., Momose K., Sakai S., Yumoto T., Mohamad S.U., Hamid A.A. & Yamane S. (2001) Six-year population fluctuation of the giant honey bee *Apis dorsata* (Hymenoptera: Apidae) in a tropical lowland dipterocarp forest in Sarawak. *Annals of the Entomological Society of America* **94**, 545–549.

Jablonski N.G. (2002) Fossil Old World monkeys: the late Neogene radiation. In *The Primate Fossil Record* (ed, Hartwig W.C.). Cambridge University Press, Cambridge, UK, pp. 255–299.

Jacobs G.H. (1995) Variations in primate color vision: mechanisms and utility. *Evolutionary Anthropology* **3**, 196–205.

Janzen D.H. (1974) Tropical blackwater rivers, animals, and mast fruiting by the Dipterocarpaceae. *Biotropica* **6**, 69–103.

Janzen D.H. & Martin P.S. (1982) Neotropical anachronisms: the fruits the gomphotheres ate. *Science* **215**, 19–27.

Jensen M.N. (2004) Climate warming shakes up species. *BioScience* **54**, 722–729.

Jepson P., Jarvie J.K., MacKinnon K. & Monk K.A. (2001) The end for Indonesia's lowland forests? *Science* **292**, 859–861.

Johns A.G. (1997) *Timber Production and Biodiversity Conservation in Tropical Rain Forests.* Cambridge University Press, New York.

Johnsgard P.A. (1994) *Arena Birds: Sexual Selection and Behavior.* Smithsonian Institution Press, Washington, DC.

Johnson S. (2002) Documenting the undocumented. *Tropical Forest Update* **12**, 6–9.

Jones D.N., Dekker R.W.R.J. & Roselaar C. (1995) *The Megapodes: Megapodiidae.* Oxford University Press, Oxford.

Jones M.E. & Stoddart D.M. (1998) Reconstruction of the predatory behaviour of the extinct marsupial thylacine (*Thylacinus cynocephalus*). *Journal of Zoology* **246**, 239–246.

Jousselin E., Rasplus J.Y. & Kjellberg F. (2003) Convergence and coevolution in a mutualism: evidence from a molecular phylogeny of *Ficus*. *Evolution* **57**, 1255–1269.

Jullien M. & Thiollay J.M. (1998) Multi-species territoriality and dynamics of Neotropical understorey birds flocks. *Journal of Animal Ecology* **67**, 227–252.

Juniper T. & Parr M. (1998) *Parrots. A Guide to Parrots of the World.* Yale University Press, New Haven, CT.

Kapan D.D. (2001) Three-butterfly system provides a field test of mullerian mimicry. *Nature* **409**, 338–340.

Kappeler P.M. (2000) Lemur origins: rafting by groups of hibernators? *Folia Primatologica* **71**, 422–425.

Kappeler P.M. & Heymann E.W. (1996) Non-convergence in the evolution of primate life history and socio-ecology. *Biological Journal of the Linnean Society* **59**, 297–326.

Karl T.R., Nicholls N. & Gregory J. (1997) The coming climate. *Scientific American* **276**, 78–83.

Karr J.R. (1976a) Seasonality, resource availability and community diversity in tropical bird communities. *American Naturalist* **110**, 973–994.

Karr J.R. (1976b) Within- and between-habitat avian diversity in African and neotropical lowland habitats. *Ecological Monographs* **46**, 457–481.

Karr J.R. (1980) Patterns in the migration systems between the north temperate zone and the tropics. In *Migrant Birds in the Neotropics: Ecology, Behavior, Distribution and Conservation* (eds, Keast A. & Morton E.S.). Smithsonian Institution Press, Washington, DC, pp. 519–543.

Karr J.R. (1989) Birds. In *Tropical Rain Forest Ecosystems* (eds, Lieth H. & Werger M.J.A.). Elsevier Scientific Publishing, Amsterdam, Netherlands, pp. 401–416.

Karr J.R. (1990) Birds of tropical rainforest: comparative biogeography and ecology. In *Biogeography and Ecology of Forest Bird Communities* (ed, Keast A.). SPB Academic Publishing, The Hague, Netherlands, pp. 215–228.

Kay R.F. & Madden R.H. (1997) Paleogeography and paleoecology. In *Vertebrate Paleontology in the Neotropics: the Miocene Fauna of La Venta, Colombia* (eds, Kay R.F., Madden R.H., Cifelli R.L. & Flynn J.J.). Smithsonian Institution Press, Washington, DC, pp. 520–550.

Kay R.F., Madden R.H., van Schaik C.P. & Higdon D. (1997) Primate species richness is determined by plant productivity: implications for conservation. *Proceedings of the National Academy of Sciences of the United States of America* **94**, 13023–13027.

Kelly D. (1994) The evolutionary ecology of mast seeding. *Trends in Ecology and Evolution* **9**, 465–470.

Kiltie R.A. & Terborgh J. (1983) Observations on the behavior of rain forest peccaries in Peru: why do white-lipped peccaries form herds? *Zeitschrift für Tierpsychologie* **62**, 241–255.

Kingdon J. (1989) *Island Africa: the Evolution of Africa's Rare Plants and Animals.* Princeton University Press, Princeton, NJ.

Kingdon J. (1997) *The Kingdon Field Guide to African Mammals.* Princeton University Press, Princeton, NJ.

Kinoshita S., Yoshioka S. & Kawagoe K. (2002) Mechanisms of structural colour in the morpho butterfly: cooperation of regularity and irregularity in an iridescent scale. *Proceedings of the Royal Society of London, Series B, Biological Sciences* **269**, 1417–1421.

Knogge C. & Heymann E.W. (2003) Seed dispersal by sympatric tamarins *Saguinus mystax* and *Saguinus fuscicollis*: diversity and characteristics of plant species. *Folia Primatologica* **74**, 33–47.

Knogge C., Heymann E.W. & Herrera E.R.T. (1998) Seed dispersal of *Asplundia peruviana* (Cyclanthaceae) by the primate *Saguinus fuscicollis*. *Journal of Tropical Ecology* **14**, 99–102.

Knott C.D. (1998) Changes in orangutan caloric intake, energy balance, and ketones in response to fluctuating fruit availability. *International Journal of Primatology* **19**, 1061–1079.

Konig C., Weick F. & Becking J.H. (1999) *Owls: a Guide to the Owls of the World*. Yale University Press, New Haven, CT.

Kress W.J. (1985) Bat pollination of an Old World *Heliconia*. *Biotropica* **17**, 302–308.

Kress W.J. & Beach J.H. (1994) Flowering plant reproductive systems. In *La Selva: Ecology and Natural History of a Neotropical Rainforest* (eds, McDade L.A., Bawa K.S., Hespenheide H.A. & Hartshorn G.S.). University of Chicago Press, Chicago, IL, pp. 161–182.

Kress W.J., Schatz G.E., Andrianifahanana M. & Morland H.S. (1994) Pollination in *Ravenala madagascariensis* (Strelitziaceae) by lemurs in Madagascar: evidence for an archaic coevolutionary system? *American Journal of Botany* **81**, 542–551.

Kricher J. (1997) *A Neotropical Companion*, 2nd edn. Princeton University Press, Princeton, NJ.

Kumar A., Konstant W.R. & Mittermeier R.A. (1999) Western Ghats and Sri Lanka. In *Hotspots: Earth's Biologically Richest and Most Endangered Terrestrial Ecoregions* (eds, Mittermeier R.A., Myers N., Gil P.R. & Mittermeier C.G.). CEMEX, Mexico City, Mexico, pp. 352–365.

Kummer D.M. & Turner B.L. (1994) The human causes of deforestation in Southeast Asia: the recurrent pattern is that of large-scale logging for exports, followed by agricultural expansion. *BioScience* **44**, 323–328.

Kunz T.H. & Fenton M.B. (eds) (2003) *Bat Ecology*. University of Chicago Press, Chicago, IL.

Kunz T.H. & Jones D. (2000) *Pteropus vampyrus*. *Mammalian Species* **642**, 1–6.

Kunz T.H. & McCracken G.F. (1996) Tents and harems: apparent defence of foliage roosts by tent-making bats. *Journal of Tropical Ecology* **12**, 121–137.

Kunz T.H. & Racey P.A. (eds) (1998) *Bat Biology and Conservation*. Smithsonian Institution Press, Washington, DC.

Laman T. (2000) Wild gliders: the creatures of Borneo's rain forest go airborne. *National Geographic* **Oct**, 68–85.

Lambert F. & Woodcock M. (1996) *Pittas, Broadbills and Asities*. Pica Press, East Sussex, UK.

Lambert J.E. (1999) Seed handling in chimpanzees (*Pan troglodytes*) and redtail monkeys (*Cercopithecus ascanius*): implications for understanding hominoid and cercopithecine fruit-processing strategies and seed dispersal. *American Journal of Physical Anthropology* **109**, 365–386.

Laurance S.G.W., Stouffer P.C. & Laurance W.F. (2004) Effects of road clearings on movement patterns of understorey rainforest birds in Central Amazonia. *Conservation Biology* **18**, 1099–1109.

Laurance W.F. (2004) The perils of payoff: corruption as a threat to global biodiversity. *Trends in Ecology and Evolution* **19**, 399–401.

Laurance W.F., Albernaz A.K.M., Schroth G., Fearnside P.M., Bergen S., Venticinque E.M. & Da Costa C. (2002) Predictors of deforestation in the Brazilian Amazon. *Journal of Biogeography* **29**, 737–748.

Laurance W.F. & Bierregaard R.O. (1997) *Tropical Forest Remnants: Ecology, Management, and Conservation of Fragmented Communities*. University of Chicago Press, Chicago, IL.

Laurance W.F., Cochrane M.A., Bergen S., Fearnside P.M., Delamonica P., Barber C.V., D'Angelo S. & Tito F. (2001) The future of the Brazilian Amazon. *Science* **291**, 438–439.

Laurance W.F. & Williamson G.B. (2001) Positive feedbacks among forest fragmentation, drought, and climate change in the Amazon. *Conservation Biology* **15**, 1529–1535.

Lee D.W. (2001) Leaf colour in tropical plants. *Malayan Nature Journal* **55**, 117–131.

Lee H.S., Davies S.J., LaFrankie J.V., Tan S., Yamakura T., Itoh A., Ohkubo T. & Ahston P.S. (2002) Floristic and structural diversity of mixed dipterocarp forest in Lambir Hills National Park, Sarawak, Malaysia. *Journal of Tropical Forest Science* **14**, 379–400.

Leslie A., Sarre A., Filho M.S. & Buang A.B. (2002) Forest certification and biodiversity. *Tropical Forest Update* **12**, 13–15.

Levey D.J., Moermond T.C. & Denslow J.S. (1994) Frugivory: an overview. In *La Selva: Ecology and Natural History of a Neotropical Rain Forest* (eds, McDade L.A., Bawa K.S., Hespenheide H.A. & Hartshorn G.S.). University of Chicago Press, Chicago, IL, pp. 287–294.

Lim B.L. (1966) Abundance and distribution of Malaysian bats in different ecological conditions. *Federated Museums Journal* **11**, 61–76.

Lin Y.H., McLenachan P.A., Gore A.R., Phillips M.J., Ota R., Hendy M.D. & Penny D. (2002) Four new mitochondrial genomes and the increased stability of evolutionary trees of mammals from improved taxon sampling. *Molecular Biology and Evolution* **19**, 2060–2070.

Lindsey T. (1998) Honeyeaters and their allies. In *Encyclopedia of Birds* (ed, Forshaw J.). Academic Press, San Diego, CA, pp. 205–208.

Liu W., Meng F.R., Zhang Y., Liu Y. & Li H. (2004) Water input from fog drip in the tropical seasonal rain forest of Xishuangbanna, South-West China. *Journal of Tropical Ecology* **20**, 517–524.

Long J., Archer M., Flannery T. & Hand S. (2002) *Prehistoric Mammals of Australia and New Guinea: One Hundred Million Years of Evolution*. Johns Hopkins University Press, Baltimore, MD.

Long J.L. (2003) *Introduced Mammals of the World*. CSIRO Publishing, Collingwood, Australia.

Loope L.L., Howarth F.G., Kraus F. & Pratt T.K. (2001) Newly emergent and future threats of alien species to Pacific birds and ecosystems. *Studies in Avian Biology* **22**, 291–304.

Lopez O.R. & Kursar T.A. (2003) Does flood tolerance explain tree species distribution in tropical seasonally flooded habitats? *Oecologia* **136**, 193–204.

Losos E.C. & Leigh E.G. Jr. (2004) *Tropical Forest Diversity and Dynamism: Findings from a Large-Scale Plot Network*. Univeristy of Chicago Press, Chicago, IL.

Lovett J.C. & Wasser S.K. (eds) (1993) *Biogeography and Ecology of the Rain Forests of Eastern Africa*. Cambridge University Press, Cambridge, UK.

Lowman M.D. & Nadkarni N.M. (eds) (1995) *Forest Canopies*. Academic Press, San Diego, CA.

MacArthur R.H. & Wilson E.O. (1967) *The Theory of Island Biogeography*. Princeton University Press, Princeton, NJ.

Macdonald D.W. & Sillero-Zubiri C. (2004) Wild canids—an introduction and *dramatis personae*. In *The Biology and Conservation of Wild Canids* (eds, Macdonald D.W. & Sillero-Zubiri C.). Oxford University Press, Oxford, pp. 3–36.

Mack A.L. (1993) The sizes of vertebrate-dispersed fruits: a neotropical–paleotropical comparison. *American Naturalist* **142**, 840–856.

Mack A.L. & Jones J. (2003) Low-frequency vocalizations by cassowaries (*Casuarius* spp.). *Auk* **120**, 1062–1068.

Mack A.L. & Wright D.D. (1998) The vulturine parrot, *Psittrichas fulgidus*, a threatened New Guinea endemic: notes on its biology and conservation. *Bird Conservation International* **8**, 185–194.

MacKinnon J. (2000) New mammals in the 21st century? *Annals of the Missouri Botanical Garden* **87**, 63–66.

MacPhee R.D.E. & Horovitz I. (2002) Extinct Quaternary platyrrhines of the Greater Antilles and Brazil. In *The Primate Fossil Record* (ed, Hartwig W.C.). Cambridge University Press, Cambridge, UK, pp. 189–200.

Madge S. & McGowan P. (2002) *Pheasants, Partridges and Grouse*. Christopher Helm, London.

Madsen O., Scally M., Douady C.J., Kao D.J., DeBry R.W., Adkins R., Amrine H.M., Stanhope M.J., de Jong W.W. & Springer M.S. (2001) Parallel adaptive radiations in two major clades of placental mammals. *Nature* **409**, 610–614.

Magallón S.A. (2004) Dating lineages: molecular and paleontological approaches to the temporal framework of clades. *International Journal of Plant Sciences* **165** (suppl.), 7–21.

Magnusson W.E. & Lima A.P. (1991) The ecology of a cryptic predator, *Paleosuchus trigonatus*, in a tropical rainforest. *Journal of Herpetology* **25**, 41–48.

Maley J. (2001) The impact of arid phases on the African rain forest through geological history. In *African Rain Forest Ecology and Conservation: an Interdisciplinary Perspective* (eds, Weber W., White L.J.T., Vedder A. & Naughton-Treves L.). Yale University Press, New Haven, CT, pp. 68–87.

Mallet J. & Gilbert L.E. (1995) Why are there so many mimicry rings? Correlations between habitat, behaviour and mimicry in *Heliconius* butterflies. *Biological Journal of the Linnean Society* **55**, 159–80.

Mann C.C. (2002) Agriculture—the real dirt on rainforest fertility. *Science* **297**, 920–923.

Manokaran N., LaFrankie J.V., Kochummen K.M., Quah E.S., Klahn J.E., Ashton P.S. & Hubbell S.P. (1992) Stand table and distribution of species in the fifty hectare research plot at Pasoh Forest Reserve. *Forest Research Institute of Malaysia, Research Data* **1**, 1–454.

Marshall A.G. (1983) Bats, flowers and fruit: evolutionary relationships in the Old World. *Biological Journal of the Linnean Society* **20**, 115–135.

Marshall A.G. (1985) Old World phytophagous bats (Megachiroptera) and their food plants: a survey. *Zoological Journal of the Linnean Society* **29**, 115–135.

Marshall L.G. (1988) Land mammals and the Great American Interchange. *American Scientist* **76**, 380–388.

Martin C. (1991) *The Rainforests of West Africa: Ecology, Threats, Conservation*. Birkhauser Verlag, Basel.

Martin P.S. (2001) Mammals (late Quaternary), extinctions of. In *Encyclopedia of Biodiversity*, Vol. 3 (ed, Levin S.A.). Academic Press, San Diego, CA, pp. 825–840.

Martin P.S. & Steadman D.W. (1999) Prehistoric extinctions on islands and continents. In *Extinctions in Near Time: Causes, Contexts, and Consequences* (ed, MacPhee R.D.E.). Kluwer Academic, New York, pp. 17–55.

Martius C. (1994) Diversity and ecology of termites in Amazonian forests. *Pedobiologia* **38**, 407–428.

Maser C. (1997) *Sustainable Community Development: Principles and Concepts*. St Lucie Press, Boca Rotan, FL.

Mast R.B., Mahecha J.V.R., Mittermeier R.A., Hemphill A.H. & Mittermeier C.G. (1999) Choco-Darien-Western Ecuador. In *Hotspots: Earth's Biologically Richest and Most Endangered Terrestrial Ecoregions* (eds, Mittermeier R.A., Myers N., Gil P.R. & Mittermeier C.G.). CEMEX, Mexico City, Mexico, pp. 122–131.

Matthews E. (2001) *Understanding the FRA 2000*. World Resources Institute, Washington, DC.

Mayle F.E., Beerling D.J., Gosling W.D. & Bush M.B. (2004) Responses of Amazonian ecosystems to climatic and atmospheric carbon dioxide changes since the last glacial maximum. *Philosophical Transactions of the Royal Society of London, Series B* **359**, 499–514.

McClure H.E. (1974) *Migration and Survival of the Birds of Asia*. Applied Scientific Research Corporation of Thailand, Bangkok.

McDonald G. & Lane M. (2002) *Securing the Wet Tropics?* Federation Press, Leichhardt, Australia.

McGowan P.J.K. & Garson P.J. (2002) The Galliformes are highly threatened: should we care? *Oryx* **36**, 311–312.

McGregor G.R. & Nieuwolt S. (1998) *Tropical Climatology: an Introduction to the Climates of Low Latitudes*, 2nd end. John Wiley & Sons, Chichester, UK.

McKenzie N.L., Gunnell A.C., Yani M. & Williams M.R. (1995) Correspondence between flight morphology and foraging ecology in some palaeotropical bats. *Australian Journal of Zoology* **43**, 241–257.

McLoughlin S. (2001) The breakup history of Gondwana and its impact on pre-Cenozoic floristic provincialism. *Australian Journal of Botany* **49**, 271–300.

Medway L. (1983) *The Wild Mammals of Malaya (Peninsular Malaysia) and Singapore*. Oxford University Press, Kuala Lumpur, Malaysia.

Meijaard E. & Nijman V. (2003) Primate hotspots on Borneo: predictive value for general biodiversity and the effects of taxonomy. *Conservation Biology* **17**, 725–732.

Mercader J. (2003) Introduction: the Paleolithic settlement of rain forests. In *Under the Canopy: the Archaeology of Tropical Rain Forests* (ed, Mercader J.). Rutgers University Press, New Brunswick, NJ, pp. 1–31.

Mercader J., Panger M. & Boesch C. (2002) Excavation of a chimpanzee stone tool site in the African rainforest. *Science* **296**, 1452–1455.

Metcalfe I. (1998) Palaeozoic and Mesozoic geological evolution of the SE Asian region: multidisciplinary constraints and implications for biogeography. In *Biogeography and Geological Evolution of SE Asia* (eds, Hall R. & Holloway J.D.). Backhuys Publishers, Leiden, Netherlands, pp. 25–41.

Meynecke J.-O. (2004) Effects of global climate change on geographic distributions of vertebrates in North Queensland. *Ecological Modelling* **174**, 347–357.

Michener C.D. (2000) *The Bees of the World*. Johns Hopkins University Press, Baltimore, MD.

Mickleburgh S.P., Hutson A.M. & Racey P.A. (1992) *Old World Fruit Bats: an Action Plan for their Conservation*. IUCN, Gland, Switzerland.

Milner-Gulland E.J. & Bennett E.L. (2003) Wild meat: the bigger picture. *Trends in Ecology and Evolution* **18**, 351–357.

Minnemeyer S. (2002) *An Analysis of Access into Central Africa's Rainforests*. World Resources Institute, Washington, DC.

Mittermeier R.A., Myers N., Gil P.R. & Mittermeier C.G. (1999) *Hotspots: Earth's Biologically Richest and Most Endangered Terrestrial Ecoregions*. CEMEX, Mexico City, Mexico.

Miura T. & Matsumoto T. (1998) Open-air litter foraging in the nasute termite *Longipeditermes longipes* (Isoptera: Termitidae). *Journal of Insect Behavior* **11**, 179–189.

Moffett M.W. (1987) Division of labor and diet in the extremely polymorphic ant *Pheidologeton diversus*. *National Geographic Research* **3**, 282–304.

Momose K., Yumoto T., Nagamitsu T., Kato M., Nagamasu H., Sakai S., Harrison R.D., Itioka T., Hamid A.A. & Inoue T. (1998) Pollination biology in a lowland dipterocarp forest in Sarawak, Malaysia. I. Characteristics of the plant-pollinator community in a lowland dipterocarp forest. *American Journal of Botany* **85**, 1477–1501.

Moritz C., Patton J.L., Schneider C.J. & Smith T.B. (2000) Diversification of rainforest faunas: an integrated molecular approach. *Annual Review of Ecology and Systematics* **31**, 533–563.

Morley R.J. (2000) *Origin and Evolution of Tropical Rain Forests*. Wiley, Chichester, UK.

Morley R.J. (2003) Interplate dispersal paths for megathermal angiosperms. *Perspectives in Plant Ecology Evolution and Systematics* **6**, 5–20.

Morris P. & Hawkins F. (1998) *Birds of Madagascar*. Yale University Press, New Haven, CT.

Morrison D.W. (1978) Foraging ecology and energetics of the frugivorous bat *Artibeus jamaicensis*. *Ecology* **59**, 716–723.

Moy C.M., Seltzer G.O., Rodbell D.T. & Anderson D.M. (2002) Variability of El Niño/Southern Oscillation activity at millennial timescales during the Holocene epoch. *Nature* **420**, 162–165.

Munn C.A. (1985) Permanent canopy and understory flocks in Amazonia: species composition and population density. *Ornithological Monographs* **36**, 683–712.

Myers N. (1984) *The Primary Source: Tropical Forests and Our Future*. Norton, New York.

Myers N. (1986) Tropical forests: patterns of depletion. In *Tropical Rain Forests and the World Atmosphere* (ed, Prance G.T.). Westview Press, Boulder, CO, pp. 9–22.

Myers N. (1997) Consumption in relation to population, environment and development. *The Environmentalist* **17**, 33–44.

Myers N. (1998) Global population and emergent pressures. In *Population and Global Security* (ed, Polunin N.). Cambridge University Press, Cambridge, UK, pp. 17–46.

Myers N. (2003) Biodiversity hotspots revisited. *BioScience* **53**, 916–917.

Nadkarni N.M. & Wheelwright N.T. (eds) (2000) *Monteverde: Ecology and Conservation of a Tropical Cloud Forest*. Oxford University Press, New York.

Nash J.M. (2002) *El Niño: Unlocking the Secrets of the Master Weather-maker*. Warner Books, New York.

Nchanji A.C. & Plumptre A.J. (2003) Seed germination and early seedling establishment of some elephant-dispersed species in Banyang-Mbo Wildlife Sanctuary, south-western Cameroon. *Journal of Tropical Ecology* **19**, 229–237.

Nelson B.W., Kapos V., Adams J.B., Oliveira W.J., Braun O.P.G. & do Amaral I.L. (1994) Forest disturbance by large blowdowns in the Brazilian Amazon. *Ecology* **75**, 853–858.

Newbery D.M., Alexander I.J. & Rother J.A. (2000) Does proximity to conspecific adults influence the establishment of ectomycorrhizal trees in rain forest? *New Phytologist* **147**, 401–409.

Newman A. (1990) *Tropical Rainforest*. Facts on File, New York.

Nieder J., Prosperi J. & Michaloud G. (2001) Epiphytes and their contribution to canopy diversity. *Plant Ecology* **153**, 51–63.

Nieh J.C. (2004) Recruitment communication in stingless bees (Hymenoptera, Apidae, Meliponini). *Apidologie* **35**, 159–182.

Nilsson L.A. (1998) Deep flowers for long tongues. *Trends in Ecology and Evolution* **13**, 259–260.

Nogueira M.R. & Peracchi A.L. (2003) Fig-seed predation by 2 species of *Chiroderma*: discovery of a new feeding strategy in bats. *Journal of Mammalogy* **84**, 225–233.

Norberg U.M. & Rayner J.M.V. (1987) Ecological morphology and flight in bats (Mammalia; Chiroptera): wing adaptations, flight performance, foraging strategy and echolocation. *Philosophical Transactions of the Royal Society of London, Series B, Biological Sciences* **316**, 335–427.

Nott J. & Hayne M. (2001) High frequency of 'super-cyclones' along the Great Barrier reef over the past 5000 years. *Nature* **413**, 508–512.

Novotny V., Tonner M. & Spitzer K. (1991) Distribution and flight behaviour of the junglequeen butterfly, *Stichophthalma louisa* (Lepidoptera: Nymphalidae), in an Indochinese montane rainforest. *Journal of Research on the Lepidoptera* **30**, 279–288.

Nowak R.M. (1994) *Walker's Bats of the World*. Johns Hopkins University Press, Baltimore, MD.

Nowak R.M. (1999) *Walker's Mammals of the World*, 6th edn. Johns Hopkins University Press, Baltimore, MD.

Numata S., Yasuda M., Okuda T., Kachi N. & Noor N.S.M. (2003) Temporal and spatial patterns of mass flowerings on the Malay Peninsula. *American Journal of Botany* **90**, 1025–1031.

Oates J.F. (1999) *Myth and Reality in the Rain Forest: How Conservation Strategies are Failing in West Africa*. University of California Press, Berkeley, CA.

Ochoa J. (2000) Effects of logging on small-mammal diversity in the lowland forests of the Venezuelan Guyana region. *Biotropica* **32**, 146–164.

Ødegaard F. (2000) How many species of arthropods? Erwin's estimate revised. *Biological Journal of the Linnean Society* **71**, 583–597.

Osada N., Takeda H., Furukawa A. & Awang M. (2001) Fruit dispersal of two dipterocarp species in a Malaysian rain forest. *Journal of Tropical Ecology* **17**, 911–917.

Osorio D. & Vorobyev M. (1996) Colour vision as an adaptation to frugivory in primates. *Proceedings of the Royal Society of London, Series B, Biological Sciences* **263**, 593–599.

Owen-Smith R.N. (1992) *Megaherbivores: the Influence of Very Large Body Size on Ecology*. Cambridge University Press, Cambridge, UK.

Pace M.L., Cole J.J., Carpenter S.R. & Kitchell J.F. (1999) Trophic cascades revealed in diverse ecosystems. *Trends in Ecology and Evolution* **14**, 483–488.

Pacheco L.F. & Simonetti J.A. (2000) Genetic structure of a mimosoid tree deprived of its seed disperser, the spider monkey. *Conservation Biology* **14**, 1766–1775.

Page S.E., Siegert F., Rieley J.O., Böhm H.-D.V., Jaya A. & Limin S. (2002) The amount of carbon released from peat and forest fires in Indonesia during 1997. *Nature* **420**, 61–65.

Paijmans K. (1976) Vegetation. In *New Guinea Vegetation* (ed, Paijmans K.). Australian National University Press, Canberra, pp. 23–105.

Paini D.R. (2004) Impact of the introduced honey bee (*Apis mellifera*) (Hymenoptera: Apidae) on native bees: a review. *Austral Ecology* **29**, 399–407.

Palmiotto P.A., Davies S.J., Vogt K.A., Ashton M.S., Vogt D.J. & Ashton P.S. (2004) Soil-related habitat specialization in dipterocarp rain forest tree species in Borneo. *Journal of Ecology* **92**, 609–623.

Parsons M. (1999) *The Butterflies of Papua New Guinea: Systematics and Biology*. Academic Press, San Diego, CA.

Paton D.C. & Collins B.G. (1989) Bills and tongues of nectar-feeding birds: a review of morphology, function and performance, with intercontinental comparisons. *Australian Journal of Ecology* **14**, 473–506.

Payne J. (1995) Links between vertebrates and the conservation of Southeast Asian rainforests. In *Ecology, Conservation, and Management of Southeast Asian Rainforests* (eds, Primack R.B. & Lovejoy T.E.). Yale University Press, New Haven, CT, pp. 54–65.

Pearce D. & Brown K. (1994) Saving the world's tropical forests. In *The Causes of Tropical Deforestation: the Economic and Statistical Analysis of Factors Giving Rise to the Loss of Tropical Forests* (eds, Brown K. & Pearce D.). University College Press, London, pp. 2–26.

Pearce D., Putz F.E. & Vanclay J.K. (2003) Sustainable forestry in the tropics: panacea or folly? *Forest Ecology and Management* **172**, 229–247.

Pearson D.L. (1977) A pantropical comparison of bird community structure in six lowland forest sites. *Condor* **79**, 232–244.

Pennington R.T. & Dick C.W. (2004) The role of immigrants in the assembly of the South American rainforest tree flora. *Philosophical Transactions of the Royal Society of London, Series B* **359**, 1611–1622.

Peres C.A. (2000) Effects of subsistence hunting on vertebrate community structure in Amazonian forests. *Conservation Biology* **14**, 240–253.

Peres C.A., Schiesari L.C. & Dias L.C.L. (1997) Vertebrate predation of Brazil-nuts (*Bertholletia excelsa*, Lecythidaceae), an agouti-dispersed Amazonian seed crop: a test of the escape hypothesis. *Journal of Tropical Ecology* **13**, 69–79.

Peres C.A. & Terborgh J. (1995) Amazonian nature reserves: an analysis of the defensibility status of existing conservation units and design criteria for the future. *Conservation Biology* **9**, 34–46.

Perreijn K. (2002) *Symbiotic Nitrogen Fixation by Leguminous Trees in Tropical Rain Forest in Guyana*. Tropenbos-Guyana Programme, Georgetown, Guyana.

Peters H.A. (2001) *Clidemia hirta* invasion at the Pasoh Forest Reserve: an unexpected invasion in an undisturbed tropical forest. *Biotropica* **33**, 60–68.

Pfeiffer M. & Linsenmair K.E. (2001) Territoriality in the Malaysian giant ants *Camponotus gigas* (Hymenoptera/Formicidae). *Journal of Ethology* **19**, 75–85.

Phillips O.L. & Gentry A.H. (1994) Increasing turnover through time in tropical forests. *Science* **263**, 954–958.

Phillips O.L. and 33 other authors (2004) Pattern and process in Amazon tree turnover 1976–2001. *Philosophical Transactions of the Royal Society of London, Series B* **359**, 381–407.

Pierson E.D. & Racey P.A. (1998) Introduction. In *Bat Biology and Conservation* (eds, Kunz T.H. & Racey, P.A.). Smithsonian Institution Press, Washington, DC, pp. 247–248.

Poffenberger M. (ed.) (1990) *Keepers of the Forest: Land Management Alternatives in Southeast Asia*. Kumarian Press, West Hartford, CT.

Potts M.D. (2003) Drought in a Bornean everwet rain forest. *Journal of Ecology* **91**, 467–474.

Powell G.V.N. (1985) Sociobiology and adaptive significance of interspecific foraging flocks in the Neotropics. *Ornithological Monographs* **36**, 713–732.

Powell G.V.N. & Bjork R. (1995) Implications of intratropical migration on reserve design: a case study using *Pharomachrus mocinno*. *Conservation Biology* **9**, 354–362.

Powell G.V.N. & Bjork R.D. (2004) Habitat linkages and the conservation of tropical biodiversity as indicated by seasonal migrations of three-wattled bellbirds. *Conservation Biology* **18**, 500–509.

Prance G. (2001) Amazon ecosystems. In *Encyclopedia of Biodiversity*, Vol. 1 (ed, Levin S.A.). Academic Press, San Diego, CA, pp. 145–157.

Pridgeon A.M. (1994) The realm of wonder. In *Proceedings of the 14th World Orchid Conference*. HMSO, Edinburgh, UK, pp. 5–12.

Primack R.B. (1987) Relationships among flowers, fruits and seeds. *Annual Review of Ecology and Systematics* **18**, 409–430.

Primack R.B. (2002) *Essentials of Conservation Biology*. Sinauer Associates, Sunderland, MA.

Primack R.B. (2004) *A Primer of Conservation Biology*. Sinauer Associates, Sunderland, MA.

Primack R.B., Bray D., Galletti H.A. & Ponciano J. (eds) (1998) *Timber, Tourists and Temples: Conservation and Development in the Maya Forest of Belize, Guatemala and Mexico*. Island Press, Washington, DC.

Primack R.B., Chai E.O.K. & Lee H.S. (1989) Relative performance of dipterocarp trees in natural forest, managed forest, logged forest and plantations throughout Sarawak, East Malaysia. In *Growth and Yield in Tropical Mixed/Moist Forests* (eds, Wan Razali W.M., Chan H.T. & Appanah S.). Forest Research Institute, Kuala Lumpur, Malaysia, pp. 161–175.

Primack R.B. & Lovejoy T.E. (eds) (1995) *Ecology, Conservation and Management of Southeast Asian Rainforests*. Yale University Press, New Haven, CT.

Proctor J. (2003) Vegetation and soil and plant chemistry on ultramafic rocks in the tropical Far East. *Perspectives in Plant Ecology Evolution and Systematics* **6**, 105–124.

Proctor J., Brearley F.Q., Dunlop H., Proctor K., Supramono & Taylor D. (2001) Local wind damage in Barito Ulu, Central Kalimantan: a rare but essential event in a lowland dipterocarp forest? *Journal of Tropical Ecology* **17**, 473–475.

Proctor J., Haridasan K. & Smith G.W. (1998) How far north does lowland evergreen tropical rain forest go? *Global Ecology and Biogeography Letters* **7**, 141–146.

Prum R.O. (1994) Phylogenetic analysis of the evolution of alternative social behavior in the manakins (Aves: Pipridae). *Evolution* **48**, 1657–1675.

Putz F.E., Dykstra D.P. & Heinrich R. (2000) Why poor logging practices persist in the tropics. *Conservation Biology* **14**, 951–956.

Putz F.E. & Mooney H.E. (eds) (1991) *Biology of Vines*. Cambridge University Press, Cambridge, UK.

Rabinowitz A. (1991) *Chasing the Dragon's Tail: the Struggle to Save Thailand's Wild Cats*. Doubleday, New York.

Rabinowitz A. (2000) *Jaguar: One Man's Struggle to Establish the World's First Jaguar Preserve*. Island Press, Washington, DC.

Rabinowitz A., Myint T., Khaing S.T. & Rabinowitz S. (1999) Description of the leaf deer (*Muntiacus putaoensis*), a new species of muntjac from northern Myanmar. *Journal of Zoology* **249**, 427–435.

Racey P.A. & Entwistle A.C. (2003) Conservation ecology of bats. In *Bat Ecology* (eds, Kunz T.H. & Fenton M.B.). University of Chicago Press, Chicago, IL, pp. 680–743.

Radek R. (1999) Flagellates, bacteria, and fungi associated with termites: diversity and function in nutrition—a review. *Ecotropica* **5**, 183–196.

Rainey W.E. (1998) Conservation of bats on remote Indo-Pacific Islands. In *Bat Biology and Conservation* (eds, Kunz T.H. & Racey P.A.). Smithsonian Institution Press, Washington, DC, pp. 326–341.

Rakotomanana H., Hino T., Kanzaki M. & Morioka H. (2003) The role of the velvet asity *Philepitta castanea* in regeneration of understorey shrubs in Madagascan rainforest. *Ornithological Science* **2**, 49–58.

Rametsteiner E. & Simula M. (2003) Forest certification—an instrument to promote sustainable forest management? *Journal of Environmental Management* **67**, 87–98.

Rawlins D.R. & Handasyde K.A. (2002) The feeding ecology of the striped possum *Dactylopsila trivirgata* (Marsupialia: Petauridae) in far north Queensland, Australia. *Journal of Zoology* **257**, 195–206.

Redford K.H. (1992) The empty forest. *BioScience* **42**, 412–422.

Redford K.H. & Sanderson S.E. (2000) Extracting humans from nature. *Conservation Biology* **14**, 1362–1364.

Reed K.E. & Fleagle J.G. (1995) Geographical and climatic control of primate diversity. *Proceedings of the National Academy of Sciences of the United States of America* **92**, 7874–7876.

Reid N. (1991) Coevolution of mistletoes and frugivorous birds? *Australian Journal of Ecology* **16**, 457–469.

Richards P.W. (1973) Africa: the "odd man out". In *Tropical Forest Ecosystems in Africa and South America* (eds, Meggers B.J., Ayensu E.S. & Duckworth W.D.). Smithsonian Institution Press, Washington, DC, pp. 21–26.

Richards P.W. (1996a) *The Tropical Rainforest: an Ecological Study*, 2nd edn. Cambridge University Press, Cambridge, UK.

Richards P.W. (1996b) Forest indigenous peoples: concepts, critique and cases. In *Essays on the Ecology of the Guinea–Congo Rain Forest* (eds, Alexander I.J., Swaine M.D. & Watling R.). The Royal Society of Edinburgh, Edinburgh, UK, pp. 349–365.

Ricklefs R.E. (2002) Splendid isolation: historical ecology of the South American passerine fauna. *Journal of Avian Biology* **33**, 207–211.

Ridgely R.S. & Tudor G. (1994) *Birds of South America, Vol. II. The Suboscine Passerines*. Texas University Press, Austin, TX.

Ripley S.D. (1964) *The Land and Wildlife of Tropical Asia*. Time Incorporated, New York.

Ripple W.J. & Beschta R.L. (2004) Wolves and the ecology of fear: can predation risk structure ecosystem? *BioScience* **54**, 755–766.

Robbins R.K. (1981) The "false head" hypothesis: predation and wing pattern variation of lycaenid butterflies. *American Naturalist* **118**, 770–775.

Robinson J.G., Redford K.H. & Bennett E.L. (1999) Conservation—wildlife harvest in logged tropical forests. *Science* **284**, 595–596.

Roca A.L., Georgiadis N., Pecon-Slattery J. & O'Brien S.J. (2001) Genetic evidence for two species of elephant in Africa. *Science* **293**, 1473–1477.

Rodda G.H., Fritts T.H. & Conry P.J. (1992) Origin and population growth of the brown tree snake, *Boiga irregularis*, on Guam. *Pacific Science* **46**, 46–57.

Rogers H.M. & Hartemink A.E. (2001) Soil seed bank and growth rates on an invasive species, *Piper aduncum*, in the lowlands of Papua New Guinea. *Journal of Ecology* **88**, 622–633.

Roubik D.W. (1989) *Ecology and Natural History of Tropical Bees*. Cambridge University Press, Cambridge, UK.

Roubik D.W. (2000) Pollination system stability in tropical America. *Conservation Biology* **14**, 1235–1236.

Roubik D.W., Sakai S. & Gattesco F. (2003) Canopy flowers and certainty: loose niches revisited. In *Arthropods of Tropical Forests* (eds, Basset Y., Novotny V., Miller S.E. & Kitching R.L.). Cambridge University Press, Cambridge, UK, pp. 360–368.

Roubik D.W. & Wolda H. (2001) Do competing honeybees matter? Dynamics and abundance of native bees before and after honey bee invasion. *Population Ecology* **43**, 53–62.

Rudel T. & Roper J. (1996) Regional patterns and historical trends in tropical deforestation, 1976–1990: a qualitative comparative approach. *Ambio* **25**, 160–166.

Sagers C.L., Ginger S.M. & Evans R.D. (2000) Carbon and nitrogen isotopes trace nutrient exchange in an ant–plant mutualism. *Oecologia* **123**, 582–586.

Sakai S. (2002) General flowering in lowland mixed dipterocarp forests of South-east Asia. *Biological Journal of the Linnean Society* **75**, 233–247.

Sakai S., Momose K., Yumoto T., Nagamitsu T., Nagamasu H., Hamid A.A. & Nakashizuka T. (1999) Plant reproductive phenology over four years including an episode of general flowering in a lowland dipterocarp forest, Sarawak, Malaysia. *American Journal of Botany* **86**, 1414–1436.

Salafsky N., Cauley H., Balachander G., Cordes B., Parks J., Margoluis C., Bhatt S., Encarnación C., Russell D. & Margoluis R. (2001) A systematic test of an enterprise strategy for community-based biodiversity conservation. *Conservation Biology* **15**, 1585–1595.

Sánchez-Villagra M.R., Aguilera O. & Horovitz I. (2003) The anatomy of the world's largest extinct rodent. *Science* **301**, 1708–1710.

Sauquet H., Doyle J.A., Scharaschkin T., Borsch T., Hilu K.W., Chatrou L.W. & le Thomas A. (2003) Phylogenetic analysis of Magnoliales and Myristicaceae based on multiple data sets: implications for character evolution. *Botanical Journal of the Linnean Society* **142**, 125–186.

Savolainen P., Leitner T., Wilton A.N., Matisoo-Smith E. & Lundeberg J. (2004) A detailed picture of the origin of the Australian dingo, obtained from the study of mitochondrial DNA. *Proceedings of the National Academy of Sciences of the United States of America* **101**, 12387–12390.

Sayer J.A. & Whitmore T.C. (1991) Tropical moist forest: destruction and species extinction. *Biological Conservation* **55**, 199–213.

Schatz G.E. (2001) *Generic Tree Flora of Madagascar*. Royal Botanic Gardens, Kew, UK.

Schelhas J. & Greenberg R. (1996) *Forest Patches in Tropical Landscapes*. Island Press, Washington, DC.

Schnitzer S.A. & Bongers F. (2002) The ecology of lianas and their role in forests. *Trends in Ecology and Evolution* **17**, 223–230.

Schuchmann K.L. (1999) Family Trochilidae (Hummingbirds). In *Handbook of the Birds of the World*, Vol. 5. *Barn-owls to Hummingbirds* (eds, del Hoyo J., Elliott A. & Sargatal J.). Lynx Edicions, Barcelona, Spain, pp. 468–680.

Schulenberg T.S. (2003) The radiations of passerine birds on Madagascar. In *The Natural History of Madagascar* (eds, Goodman S.M. & Benstead J.P.). University of Chicago Press, Chicago, IL, pp. 1130–1134.

Schulze M.D., Seavy N.E. & Whitacre D.F. (2000) A comparison of the phyllostomid bat assemblages in undisturbed neotropical forest and in forest fragments of a slash-and-burn farming mosaic in Peten, Guatemala. *Biotropica* **32**, 174–184.

Schwartzman S., Moreira A. & Nepstad D. (2000) Rethinking tropical forest conservation: perils in parks. *Conservation Biology* **14**, 1351–1357.

Semah F., Semah A.-M. & Simanjuntak T. (2003) More than a million years of human occupation in insular Southeast Asia. In *Under the Canopy: the Archaeology of Tropical Rain Forests* (ed, Mercader J.). Rutgers University Press, New Brunswick, NJ, pp. 161–190.

Serrão J.E., Cruz-Landim C.D. & Silva-de-Moraes R.L.M. (1997) Morphological and biochemical analyses of the stored and larval food of an obligate necrophagous bee, *Trigona hypogea*. *Insectes Sociaux* **44**, 337–344.

Shanahan M., So S., Compton S.G. & Corlett R.T. (2001) Fig-eating by vertebrate frugivores: a global review. *Biological Reviews* **76**, 529–572.

Shapiro B., Sibthorpe D., Rambaut A., Austin J., Wragg G.M., Bininda-Emonds O.R.P., Lee P.L.M. & Cooper A. (2002) Flight of the dodo. *Science* **295**, 1683–1683.

Sibley C.G. & Ahlquist J.A. (1990) *Phylogeny and Classification of Birds: a Study in Molecular Evolution*. Yale University Press, New Haven, CT.

Siegert F., Rucker G., Hinrichs A. & Hoffmann A. (2001) Increased damage from fires in logged forests during droughts caused by El Niño. *Nature* **414**, 437–440.

Simmons N.B. & Conway T.M. (2003) Evolution of ecological diversity in bats. In *Bat Ecology* (eds, Kunz T.H. & Fenton M.B.). University of Chicago Press, Chicago, IL, pp. 493–535.

Simmons N.B. & Voss R.S. (1998) The mammals of Paracou, French Guiana: a neotropical lowland rainforest fauna. Part 1. Bats. *Bulletin of the American Museum of Natural History* **237**, 1–219.

Sizer N. & Plouvier D. (2000) *Increased Investment and Trade by Transnational Logging Companies in Africa, the Caribbean and the Pacific*. WWF Belgium, Brussels, Belgium.

Skole D.L., Chomentowski W.H., Salas W.A. & Nobre A.D. (1994) Physical and human dimensions of deforestation in Amazonia: in the Brazilian Amazon, regional trends are influenced by large-scale external forces mediated by local conditions. *BioScience* **44**, 314–322.

Sledge M.F., Dani F.R., Fortunato A., Maschwitz U., Clarke S.R., Francescato E., Hashim R., Morgan E.D., Jones G.R. & Turillazzi S. (1999) Venom induces alarm behaviour in the social wasp *Polybioides raphigastra* (Hymenoptera: Vespidae): an investigation of alarm behaviour, venom volatiles and sting autonomy. *Physiological Entomology* **24**, 234–239.

Smith A.P. & Ganzhorn J.U. (1996) Convergence in community structure and dietary adaptation in Australian possums and gliders and Malagasy lemurs. *Australian Journal of Ecology* **21**, 31–46.

Smith R.J., Muir R.D.J., Walpole M.J., Balmford A. & Leader-Williams N. (2003) Governance and the loss of biodiversity. *Nature* **426**, 67–70.

Smith W. (2002) The global problem of illegal logging. *Tropical Forest Update* **12**, 3–5.

Snow D.W. (1981) Tropical frugivorous birds and their food plants: a world survey. *Biotropica* **13**, 1–14.

Socha J.J. (2002) Kinematics—gliding flight in the paradise tree snake. *Nature* **418**, 603–604.

Sollins P. (1998) Factors influencing species composition in tropical lowland rain forest: does soil matter? *Ecology* **79**, 23–30.

Sombroek W. (2001) Spatial and temporal patterns of Amazon rainfall: consequences for the planning of agricultural occupation and the protection of primary forests. *Ambio* **30**, 388–396.

Springer M.S., Stanhope M.J., Madsen O. & de Jong W.W. (2004) Molecules consolidate the placental mammal tree. *Trends in Ecology and Evolution* **19**, 430–438.

Srygley R.B. & Penz C.M. (1999) Lekking in neotropical owl butterflies, *Caligo illioneus* and *C. oileus* (Lepidoptera: Brassolinae). *Journal of Insect Behavior* **12**, 81–103.

Start A.N. & Marshall A.G. (1976) Nectarivorous bats as pollinators of trees in West Malaysia. *Linnean Society Symposium Series* **2**, 141–150.

Stein B.A. (1992) Sicklebill hummingbirds, ants and flowers. *BioScience* **42**, 27–33.

Stiles F.G. (1978) Possible specialization for hummingbird-hunting in the tiny hawk. *Auk* **95**, 550–553.

Stiles F.G. (1981) Geographical aspects of bird–flower coevolution, with particular reference to Central America. *Annals of the Missouri Botanical Garden* **68**, 323–351.

Stocker G.C. & Irvine A.K. (1983) Seed dispersal by cassowaries (*Casuarius casuarius*) in North Queensland's rainforests. *Biotropica* **15**, 170–176.

Stotz D.F., Fitzpatrick J.W., Parker T.A. & Moskovits D.K. (eds) (1996) *Neotropical Birds: Ecology and Conservation*. University of Chicago Press, Chicago, IL.

Struhsaker T.T. & Leakey M. (1990) Prey selectivity by crowned hawk-eagles on monkeys in the Kibale Forest, Uganda. *Behavioral Ecology and Sociobiology* **26**, 435–444.

Styring A.P. & Ickes K. (2003) Woodpeckers (Picidae) at Pasoh: foraging ecology, flocking and the impacts of logging on abundance and diversity. In *Pasoh: Ecology of a Lowland Rain Forest in Southeast Asia* (eds, Okuda T., Manokaran N., Matsumoto Y., Niiyama K., Thomas S.C. & Ashton P.S.). Springer-Verlag, Tokyo, Japan, pp. 547–557.

Sunquist M. & Sunquist F. (2002) *Wild Cats of the World*. University of Chicago Press, Chicago, IL.

Swartz M.B. (2001) Bivouac checking, a novel behavior distinguishing obligate from opportunistic species of army-ant-following birds. *Condor* **103**, 629–633.

Swingland I.R. (2003) *Capturing Carbon and Conserving Biodiversity*. Earthscan Publications, London.

Tello J.G. (2003) Frugivores at a fruiting *Ficus* in south-eastern Peru. *Journal of Tropical Ecology* **19**, 717–721.

Temeles E.J., Pan I.L., Brennan J.L. & Horwitt J.N. (2000) Evidence for ecological causation of sexual dimorphism in a hummingbird. *Science* **289**, 441–443.

Teo D.H.L., Tan H.T.W., Corlett R.T., Wong C.M. & Lum S.K.Y. (2003) Continental rain forest fragments in Singapore resist invasion by exotic plants. *Journal of Biogeography* **30**, 305–310.

ter Steege H., Pitman N., Sabatier D., Castellanos H., Van der Hout P., Daly D.C., Silveira M., Phillips O., Vásquez R., Van Andel T., Duivenvoorden J., De Oliveira A.A., Ek R., Lilwah R., Thomas R., Van Essen J., Baider C., Maas P., Mori S., Terborgh J., Vargas P.N., Mogollón H. & Morawetz W. (2003) A spatial model of tree alpha-diversity and tree density for the Amazon. *Biodiversity and Conservation* **12**, 2255–2277.

ter Steege H., Sabatier D., Castellanos H., Van Andel T., Duivenvoorden J., De Oliveira A.A., Ek R., Lilwah R., Maas P. & Mori S. (2000) An analysis of the floristic composition and diversity of Amazonian forests including those of the Guiana Shield. *Journal of Tropical Ecology* **16**, 801–828.

Terborgh J. (1983) *Five New World Primates: a Study in Comparative Ecology*. Princeton University Press, Princeton, NJ.

Terborgh J. (1992) *Diversity and the Tropical Rain Forest*. Scientific American Library, New York.

Terborgh J. (1999) *Requiem for Nature*. Island Press, Washington, DC.

Terborgh J., Lopez L., Nunez P., Rao M., Shahabuddin G., Orihuela G., Riveros M., Ascanio R., Adler G.H., Lambert T.D. & Balbas L. (2001) Ecological meltdown in predator-free forest fragments. *Science* **294**, 1923–1926.

Terborgh J. & Peres C.A. (2002) The problem of people in parks. In *Making Parks Work: Strategies for Preserving Tropical Nature* (eds, Terborgh J., van Schaik C.P., Davenport L. & Rao M.). Island Press, Washington, DC, pp. 307–319.

Terborgh J., van Schaik C.P., Davenport L. & Rao M. (eds) (2002) *Making Parks Work: Strategies for Preserving Tropical Nature*. Island Press, Washington, DC.

Thiollay J.M. (1985) Species diversity and comparative ecology of rainforest falconiforms on three continents. In *Conservation Studies on Raptors* (eds, Newton I. & Chancellor R.D.). International Council for Bird Preservation, Cambridge, UK, pp. 167–189.

Thiollay J.M. (1991) Foraging, home range use and social behaviour of a group-living rainforest raptor, the red-throated caracara, *Daptrius americanus*. *Ibis* **133**, 382–393.

Thiollay J.M. (1999) Frequency of mixed species flocking in tropical forest birds and correlates of predation risk: an intertropical comparison. *Journal of Avian Biology* **30**, 282–294.

Thiollay J.M. (2003) Comparative foraging behavior between solitary and flocking insectivores in a Neotropical forest: does vulnerability matter? *Ornitologia Neotropical* **14**, 47–65.

Thomas B.T. (1999) Family Steatornithidae (oilbird). In *Handbook of the Birds of the World*, Vol. 5. *Barn-owls to Hummingbirds* (eds, del Hoyo J., Elliott A. & Sargatal J.). Lynx Edicions, Barcelona, Spain, pp. 244–251.

Thomas S.C. (2004) Ecological correlates of tree species persistence in tropical forest fragments. In *Tropical Forest Diversity and Dynamism: Findings from a Large-scale Plot Network* (eds, Losos E.C. & Leigh E.G. Jr.). University of Chicago Press, Chicago, IL, pp. 279–313.

Times Books (1994) *The Times Atlas of the World, Comprehensive Edition*. Times Books Ltd, London.

Torti S.D., Coley P.D. & Kursar T.A. (2001) Causes and consequences of monodominance in tropical lowland forests. *American Naturalist* **157**, 141–153.

Trail P.W. (1987) Predation and antipredator behavior at Guianan cock-of-the-rock leks. *Auk* **104**, 496–507.

Tschapka M. & Dressler S. (2002) Chiropterophily: on bat-flowers and flower-bats. *Curtis's Botanical Magazine* **19**, 114–125.

Tuomisto H., Ruokolainen K. & Yli-Halla M. (2003) Dispersal, environment, and floristic variation of western Amazonian forests. *Science* **299**, 241–244.

Turner D.A. (1997a) Family Musophagidae (Turacos). In *Handbook of the Birds of the World*, Vol. 4. *Sandgrouse to Cuckoos* (eds, del Hoyo J., Elliott A. & Sargatal J.). Lynx Edicions, Barcelona, Spain, pp. 480–506.

Turner I.M. (1997b) A tropical flora summarized: a statistical analysis of the vascular plant diversity of Malaya. *Flora* **192**, 157–163.

Turner I.M. (2001a) Rainforest ecosystems, plant diversity. In *Encyclopedia of Biodiversity*, Vol. 5 (ed, Levin S.A.). Academic Press, San Diego, CA, pp. 13–23.

Turner I.M. (2001b) *The Ecology of Trees in the Tropical Rain Forest*. Cambridge University Press, Cambridge, UK.

Turner I.M. & Corlett R.T. (1996) The conservation value of small, isolated fragments of lowland tropical rainforest. *Trends in Ecology and Evolution* **11**, 330–333.

Uhl C.J., Kauffman J.B. & Cummings D.L. (1988) Fire in the Venezuelan Amazon: 2. Environmental conditions necessary for forest fires in the evergreen rainforest of Venezuela. *Oikos* **53**, 176–184.

UNDP (United Nations Development Programme) (2002) *Human Development Report 2002*. Oxford University Press, New York.

United Nations Population Division (2001) *World Population Prospects: the 2000 Revision: Highlights*. United Nations, New York.

Van Bael S.A., Brawn J.D. & Robinson S.K. (2003) Birds defend trees from herbivores in a Neotropical forest canopy. *Proceedings of the National Academy of Sciences of the United States of America* **100**, 8304–8307.

Van Den Bussche R.A. & Hoofer S.R. (2004) Phylogenetic relationships among recent chiropteran families and the importance of choosing appropriate out-group taxa. *Journal of Mammalogy* **85**, 321–330.

van Riper C., van Riper S.G. & Hansen W.R. (2002) Epizootiology and effect of avian pox on Hawaiian forest birds. *Auk* **119**, 929–942.

van Schaik C.P., Ancrenaz M., Borgen G., Galdikas B., Knott C.D., Singleton I., Suzuki A., Utami S.S. & Merrill M. (2003) Orangutan cultures and the evolution of material culture. *Science* **299**, 102–105.

van Schaik C.P. & Rijksen H.D. (2002) Integrated conservation and development projects: problems and potential. In *Making Parks Work: Strategies for Preserving Tropical Nature* (eds, Terborgh J., van Schaik C.P., Davenport L. & Rao M.). Island Press, Washington, DC, pp. 15–29.

van Schaik C.P., Terborgh J. & Wright S.J. (1993) The phenology of tropical forests: adaptive significance and consequences for primary consumers. *Annual Review of Ecology and Systematics* **24**, 353–377.

Velthuis H.H.W. (1997) *Biology of the Stingless Bees*. University of Utrecht, Utrecht, Netherlands.

Wagner W.L., Herbst D.R. & Sohmer S.H. (1990) *Manual of the Flowering Plants of Hawai'i*. University of Hawaii Press, Honolulu, Hawaii.

Wallace A.R. (1859) *On the Zoological Geography of the Malay Archipelago*. Linnean Society, London.

Wallace R.B., Painter R.L.E. & Saldania A. (2002) An observation of the bush dog (*Speothos venaticus*) hunting behaviour. *Mammalia* **66**, 309–311.

Walsh P.D., Abernethy K.A., Bermejo M., Beyersk R., De Wachter P., Akou M.E., Huljbregis B., Mambounga D.I., Toham A.K., Kilbourn A.M., Lahm S.A., Latour S., Maisels F., Mbina C., Mihindou Y., Obiang S.N., Effa E.N., Starkey M.P., Telfer P., Thibault M., Tutin C.E.G., White L.J.T. & Wilkie D.S. (2003) Catastrophic ape decline in western equatorial Africa. *Nature* **422**, 611–614.

Walsh R.P.D. (1996) Climate. In *The Tropical Rain Forest: an Ecological Study* (eds, Richards P.W., Walsh R.P.D., Baillie I.C. & Greig-Smith P.). Cambridge University Press, Cambridge, UK, pp. 159–205.

Walsh R.P.D. & Newbery D.M. (1999) The ecoclimatology of Danum, Sabah, in the context of the world's rainforest regions, with particular reference to dry periods and their impact. *Philosophical Transactions of the Royal Society of London, Series B, Biological Sciences* **354**, 1869–1883.

Ward P. & Kynaston S. (1995) *Bears of the World*. Blandford, London.

Webb S.D. (1997) The great American faunal interchange. In *Central America: a Natural and Cultural History* (ed, Coates A.G.). Yale University Press, New Haven, CT, pp. 97–122.

Weber W., White L.J.T., Vedder A. & Naughton-Treves L. (eds) (2001) *African Rain Forest Ecology and Conservation: an Interdisciplinary Perspective*. Yale University Press, New Haven, CT.

Webster D. (1997) The looting and smuggling and fencing and hoarding of impossibly precious feathered and scaly wild things. *New York Times Magazine* **Feb 16**, 26–33, 48–49, 53, 61.

Wells D.R. (1990) Migratory birds and tropical forest in the Sunda region. In *Biogeography and Ecology of Forest Bird Communities* (ed, Keast A.). SPB Academic Publishing, The Hague, Netherlands, pp. 357–369.

Westerkamp C. (1990) Bird-flowers: hovering versus perching exploitation. *Botanica Acta* **103**, 366–371.

Western D. (1997) *In the dust of Kilimanjaro*. Island Press, Washington, DC.

Western D., Wright M. & Strum S. (eds) (1994) *Natural Connections: Perspectives in Community-Based Conservation*. Island Press, Washington, DC.

White L.J.T. (2001) The African rain forest: climate and vegetation. In *African Rain Forest Ecology: an Interdisciplinary Perspective* (eds, Weber W., White L.J.T., Vedder A. & Naughton-Treves L.). Yale University Press, New Haven, CT, pp. 3–29.

White L.J.T., Tutin C.E.G. & Fernandez M. (1993) Group composition and diet of forest elephants, *Loxodonta africana cyclotis* Matschie 1900, in the

Lope Reserve, Gabon. *African Journal of Ecology* **31**, 181–199.

Whiten A., Goodall J., McGrew W.C., Nishida T., Reynolds V., Sugiyama Y., Tutin C.E.G., Wrangham R.W. & Boesch C. (1999) Cultures in chimpanzees. *Nature* **399**, 682–685.

Whitmore T.C. (1984) *Tropical Rain Forests of the Far East*, 2nd edn. Clarendon Press, Oxford, UK.

Whitmore T.C. (1987) *Biogeographical Evolution of the Malay Archipelago*. Clarendon Press, Oxford, UK.

Whitmore T.C. (1998) *An Introduction to Tropical Rain Forests*, 2nd edn. Oxford University Press, Oxford, UK.

Whitmore T.C. (1999) Arguments on the forest frontier. *Biodiversity and Conservation* **8**, 865–868.

Whitmore T.C. & Prance G.T. (eds) (1987) *Biogeography and Quaternary History in Tropical America*. Oxford University Press, New York.

Whittaker R.J. (1998) *Island Biogeography: Ecology, Evolution and Conservation*. Oxford University Press, Oxford, UK.

Whitten T., Whitten J., Mittermeier R.A., Mittermeier C.G., Supriatna J. & Van Dijk P.P. (1999) Wallacea. In *Hotspots: Earth's Biologically Richest and Most Endangered Terrestrial Ecoregions* (eds, Mittermeier R.A., Myers N., Gil P.R. & Mittermeier C.G.). CEMEX, Mexico City, Mexico, pp. 296–307.

Wich S.A. & van Schaik C.P. (2000) The impact of El Niño on mast fruiting in Sumatra and elsewhere in Malesia. *Journal of Tropical Ecology* **16**, 563–577.

Wilkie D.S. & Carpenter J.F. (1999) Bushmeat hunting in the Congo Basin: an assessment of impacts and options for mitigation. *Biodiversity and Conservation* **8**, 927–955.

Williams G.A. & Adam P. (1997) The composition of the bee (Apoidea: Hymenoptera) fauna visiting flowering trees in New South Wales lowland subtropical rainforest remnants. *Proceedings of the Linnean Society of New South Wales* **118**, 69–95.

Williamson G.B. & Ickes K. (2002) Mast fruiting and ENSO cycles—does the cue betray a cause? *Oikos* **97**, 459–461.

Williamson G.B., Laurance W.F., Oliveira A.A., Delamonica P., Gascon C., Lovejoy T.E. & Pohl L. (2000) Amazonian tree mortality during the 1997 El Niño drought. *Conservation Biology* **14**, 1538–1542.

Willis E.O. & Oniki Y. (1978) Birds and army ants. *Annual Review of Ecology and Systematics* **9**, 245–264.

Wilms W. & Wiechers B. (1997) Floral resource partitioning between native *Melipona* bees and the introduced Africanized honey bee in the Brazilian Atlantic rain forest. *Apidologie* **28**, 339–355.

Wilson D.E. & Sandoval A. (eds) (1996) *Manu: the Biodiversity of Southeastern Peru*. Smithsonian Institution Press, Washington, DC.

Wilson E.O. (1972) *The Insect Societies*. Harvard University Press, Cambridge, MA.

Wilson E.O. (1996) Hawaii: a world without social insects. *Bishop Museum Occasional Papers* **45**, 3–7.

Winkler H. & Christie D.A. (2002) Family Picidae (woodpeckers). In *Handbook of the Birds of the World*, Vol. 7. *Jacamars to Woodpeckers* (eds, del Hoyo J., Elliott A. & Sargatal J.). Lynx Edicions, Barcelona, Spain, pp. 296–555.

Winston M.L. (1993) *Killer Bees: the Africanized Honey Bee in the Americas*. Harvard University Press, Cambridge, MA.

Wirth R., Herz H., Beysschlag W. & Hölldobler B. (2003) *Herbivory of Leaf-cutting Ants*. Springer-Verlag, Berlin.

Wong M. (1986) Trophic organization of understory birds in a Malaysian dipterocarp forest. *Auk* **103**, 100–116.

Woodcock M. & Kemp A.C. (1996) *The Hornbills: Bucerotiformes (Bird Familes of the World)*. Oxford University Press, Oxford, UK.

World Bank (2002) *Papua New Guinea Environmental Monitor*. World Bank, Washington, DC.

WRI (World Resources Institute) (1998) *World Resources 1998–1999*. Oxford University Press, New York.

WRI (World Resources Institute) (2000) *World Resources, 2000–2001: People and Ecosystems: the Fraying Web of Life*. Elsevier Science, Amsterdam, Netherlands.

WRI (World Resources Institute) (2003) *World Resources 2002–2004: Decisions for the Earth: Balance, Voice, and Power*. World Resources Institute, Washington, DC.

Wright D.D., Jessen J.H., Burke P. & Gomez de Silva Garza H. (1997) Tree and liana enumeration and diversity on a one-hectare plot in Papua New Guinea. *Biotropica* **29**, 250–260.

Wright P.C. (1997) Behavioral and ecological comparisons of Neotropical and Malagasy primates. In *New World Primates: Ecology, Evolution, and Behavior* (ed, Kinzey W.G.). Aldine de Gruyter, New York, pp. 127–141.

Wright P.C. (1998) Impact of predation risk on the behaviour of *Propithecus diadema edwardsi* in the rain forest of Madagascar. *Behaviour* **135**, 483–512.

Wright S.J. (2003) The myriad consequences of hunting for vertebrates and plants in tropical forests. *Perspectives in Plant Ecology Evolution and Systematics* **6**, 73–86.

Wright S.J., Carrasco C., Calderon O. & Paton S. (1999) The El Niño Southern Oscillation, variable fruit production, and famine in a tropical forest. *Ecology* **80**, 1632–1647.

Yack J.E. & Fullard J.H. (2000) Ultrasonic hearing in nocturnal butterflies. *Nature* **403**, 265–266.

Yack J.E., Otero L.D., Dawson J.W., Surlykke A. & Fullard J.H. (2000) Sound production and hearing in the blue cracker butterfly *Hamadryas feronia* (Lepidoptera, Nymphalidae) from Venezuela. *Journal of Experimental Biology* **203**, 3689–3702.

Yamagishi S. & Eguchi K. (1996) Comparative foraging ecology of the Madagascan vangids (Vangidae). *Ibis* **138**, 283–290.

Yamagishi S., Honda M., Eguchi K. & Thorstrom R. (2001) Extreme endemic radiation of the Malagasy vangas (Aves: Passeriformes). *Journal of Molecular Evolution* **53**, 39–46.

Yasuda M., Matsumoto J., Osada N., Ichikawa S., Kachi N., Tani M., Okuda T., Furukawa A., Nik A.R. & Manokaran N. (1999) The mechanism of general flowering in Dipterocarpaceae in the Malay Peninsula. *Journal of Tropical Ecology* **15**, 437–449.

Yoder A.D., Burns M.M., Zehr S., Delefosse T., Veron G., Goodman S.M. & Flynn J.J. (2003) Single origin of Malagasy Carnivora from an African ancestor. *Nature* **421**, 734–737.

Yoder A.D. & Yang Z. (2004) Divergence dates for Malagasy lemurs estimated from multiple gene loci: geological and evolutionary context. *Molecular Ecology* **13**, 757–773.

Yumoto T. (1999) Seed dispersal by Salvin's curassow, *Mitu salvini* (Cracidae), in a tropical forest of Colombia: direct measurements of dispersal distance. *Biotropica* **31**, 654–660.

Yumoto T., Maruhashi T., Yamagiwa J. & Mwanza N. (1995) Seed dispersal by elephants in a tropical rain forest in Kahuzi-Biega National Park, Zaire. *Biotropica* **27**, 257–265.

Zhu H. (1997) Ecological and biogeographical studies on the tropical rain forest of south Yunnan, SW China with a special reference to its relation with rain forests of tropical Asia. *Journal of Biogeography* **24**, 647–662.

# Index

Page numbers in *italics* refer to figures; those in **bold** to tables